Yeliz Karaca, Carlo Cattani
Computational Methods for Data Analysis

Also of interest

Machine Learning for Big Data Analysis
S. Bhattacharyya et al. (Eds.) 2018
ISBN 978-3-11-055032-0, e-ISBN (PDF) 978-3-11-055143-3,
e-ISBN (EPUB) 978-3-11-055077-1

Intelligent Multimedia Data Analysis
S. Bhattacharyya et al. (Eds.) 2018
ISBN 978-3-11-055031-3, e-ISBN (PDF) 978-3-11-055207-2,
e-ISBN (EPUB) 978-3-11-055033-7

Advanced Data Management
Lena Wiese, 2015
ISBN 978-3-11-044140-6, e-ISBN (PDF) 978-3-11-044141-3,
e-ISBN (EPUB) 978-3-11-043307-4

Mathematical Demoeconomy
Y. Popkov, 2014
ISBN 978-3-11-033902-4, e-ISBN (PDF) 978-3-11-033916-1,
e-ISBN (EPUB) 978-3-11-037084-3

Yeliz Karaca, Carlo Cattani

Computational Methods for Data Analysis

DE GRUYTER

Authors
Assist. Prof. Yeliz Karaca, PhD.
University of Massachusetts Medical School,
Worcester, MA 01655, USA
yeliz.karaca@umassmemorial.org

Prof. Dr. Carlo Cattani
Engineering School, DEIM,
University of Tuscia,
Viterbo, Italy
cattani@unitus.it

ISBN 978-3-11-049635-2
e-ISBN (PDF) 978-3-11-049636-9
e-ISBN (EPUB) 978-3-11-049360-3

Library of Congress Control Number: 2018951307

Bibliographic information published by the Deutsche Nationalbibliothek
The Deutsche Nationalbibliothek lists this publication in the Deutsche Nationalbibliografie; detailed bibliographic data are available on the Internet at http://dnb.dnb.de.

Typesetting: Integra Software Services Pvt. Ltd.
Printing and binding: CPI books GmbH, Leck
Cover image: Yeliz Karaca and Carlo Cattani

www.degruyter.com

Preface

The advent of computerization has improved our capabilities in terms of generating and collecting data from myriad of sources to a large extent. A huge amount of data has inundated nearly in all walks of lives. Such growth in data has led to an immediate need for the development of new tools, which can be of help to us in an intelligent manner. In the light of all these developments, this book dwells on neural learning methods and it aims at shedding light on those applications where sample data are available but algorithms for analysis are missing.

Successful applications of artificial intelligence have already been extensively introduced into our lives. There already exist some commercial software developed for the recognition of handwriting and speech. In business life, retailer companies can examine the past data and use such data for the enhancement of their customer relationship management. Financial institutions, on the other hand, are able to conduct risk analyses for loans from what they have learned from their customers' past behavior. In particular, data are increasing every day and we can convert such data into knowledge only after examining them through a computer.

In this book, we will discuss some of the methods related to mathematics, statistics, engineering, medical applications, economics as well as finance. Our aim is to come up with a common ground and interdisciplinary approach for both the problems and suggest solutions in an integrative way.

In addition, we provide simple and to-the-point explanations so that it will be possible for our readers to make conversions from mathematical operations into programme dimension. The applications in the book cover three different datasets relating to the algorithms in each section.

We, as the authors, have thoroughly enjoyed writing this book, and we hope you will also enjoy to the same extent and find it beneficial for your studies as well as in various applications.

Yeliz Karaca and Carlo Cattani

https://doi.org/10.1515/9783110496369-201

Acknowledgment

The authors are deeply indebted to their colleagues and institutions (University of Massachusetts (YK) and Tuscia University (CC)) that have supported this work during the writing process. This book is the result of an impressive work that has lasted for many years. By intensively working in this field, both authors have had the opportunity to discuss the results on the algorithms and applications with colleagues, scholars and students who are involved in the authors' research area and have attended various seminars and conferences offered by the authors.

In particular, the authors would like to express their gratitude to Professor Majaz Moonis (University of Massachusetts) for providing the psychology data and his comments on the interpretation of the obtained results. The authors are also very grateful to Professor Rana Karabudak (Hacettepe University) for providing the neurological data related to the multiple sclerosis patients and the confirmation she has provided with regard to the results obtained by the analyses given in this book.

The authors would like to offer their special thanks to Professor Abul Hasan Siddiqi (Sharda University), Professor Ethem Tolga (Galatasaray University), Professor Günter Leugering (Friedrich-Alexander-University of Erlangen-Nürnberg), Professor Kemal Güven Gülen (Namık Kemal University), Professor Luis Manuel Sánchez Ruiz (Polytechnic University of Valencia) and Professor Majaz Moonis (University of Massachusetts) for their valuable feedback. The authors would like to thank Ahu Dereli Dursun, Boriana Doganova and Didem Sözen Şahiner for their contribution with respect to health communication and health psychology.

The authors would like to extend their gratitude to their families. YK would like to express her deepest gratitude to her mother, Fahriye Karaca; her father, Emin Karaca; her brother, Mehmet Karaca; and his family who have always believed in her, showed their unconditional love and offered any kinds of support all the way through. CC shares his feelings of love to his family: Maria Cristina Ferrarese, Piercarlo and Gabriele Cattani.

The authors would also like to express their gratitude to the editors of De Gruyter: Apostolos Damialis, Leonardo Milla and Lukas Lehmann for their assistance and efforts.

Yeliz Karaca and Carlo Cattani

https://doi.org/10.1515/9783110496369-202

Contents

1 Introduction

In mathematics and computer science, an *algorithm* is defined as an unambiguous specification of how to solve a problem, based on a sequence of logical-numerical deductions. Algorithm, in fact, is a general name given to the systematic method of any kind of numerical calculation or a route drawn for solving a problem or achieving a goal utilized in the mentioned fields. Tasks such as calculation, data processing and automated reasoning can be realized through algorithms. Algorithm, as a concept, has been around for centuries. The formation of the modern algorithm started with endeavors exerted for the solution of what David Hilbert (1928) called *Entscheidungsproblem* (decision problem). The following formalizations of the concept were identified as efforts for the definition of "effective calculability" or "effective method" with Gödel–Herbrand–Kleene recursive functions (1930, 1934 and 1935), lambda calculus of Alonzo Church (1936), "Formulation 1" by Emil Post (1936) and the Turing machines of Alan Turing (1936–37 and 1939). Since then, particularly in the twentieth century, there has been a growing interest in data analysis algorithms as well as their applications to interdisciplinary various datasets.

Data analysis can be defined as a process of collecting raw data and converting it into information that would prove to be useful for the users in their decision-making processes. Data collection is performed and data analysis is done for the purpose of answering questions, testing hypotheses or refuting theories. According to the statistician John Tukey (1961), data analysis is defined as the set of (1) procedures for analyzing data, (2) techniques for interpreting the results of these procedures and (3) methods for planning the gathering of data so that one can render its analysis more accurate and also much easier. It also comprises the entire mechanism and outcomes of (mathematical) statistics, which are applicable to the analyzing of data. Numerous ways exist for the classification of algorithms and each of them has its own merits.

Accordingly, knowledge itself turns into power when it is processed, analyzed and interpreted in a proper and accurate way. With this key motive in mind, our general aim in this book is to ensure the integration of relevant findings in an interdisciplinary approach, discussing various relevant methods, thus putting forth a common approach for both problems and solutions. The main aim of this book is to provide the readers with core skills regarding data analysis in interdisciplinary studies. Data analysis is characterized by three typical features: (1) algorithms for classification, clustering, association analysis, modeling, data visualization as well as singling out the singularities; (2) computer algorithms' source codes for conducting data analysis; and (3) specific fields (economics, physics, medicine, psychology, etc.) where the data are collected.

This book will help the readers establish a bridge from equations to algorithms' source codes and from the interpretation of results to draw meaningful information

https://doi.org/10.1515/9783110496369-001

about data and the process they represent. As the algorithms are developed further, it will be possible to grasp the significance of having a variety of variables. Moreover, it will be showing how to use the obtained results of data analysis for the forecasting of future developments, diagnosis and prediction in the field of medicine and related fields. In this way, we will present how knowledge merges with applications.

With this concern in mind, the book will be guiding for interdisciplinary studies to be carried out by those who are engaged in the fields of mathematics, statistics, economics, medicine, engineering, neuroengineering, computer science, neurology, cognitive sciences and psychiatry and so on.

In this book, we will analyze in detail important algorithms of data analysis and classification. We will discuss the contribution gained through linear model and multilinear model, decision trees, naive Bayesian classifier, support vector machines, *k*-nearest neighbor and artificial neural network (ANN) algorithms. Besides these, the book will also include *fractal and multifractal methods with ANN algorithm*.

The main goal of this book is to provide the readers with core skills regarding data analysis in interdisciplinary datasets. The second goal is to analyze each of the main components of data analysis:
- Application of algorithms to real dataset and synthetic dataset
- Specific application of data analysis algorithm in interdisciplinary datasets
- Detailed description of general concepts for extracting knowledge from data, which undergird the wide-ranging array of datasets and application algorithms

Accordingly, each component has adequate resources so that data analysis can be developed through algorithms. This comprehensive collection is organized into three parts:
- Classification of real dataset and synthetic dataset by algorithms
- Singling out singularities features by fractals and multifractals for real dataset and synthetic datasets
- Achieving high accuracy rate for classification of singled out singularities features by ANN algorithm (learning vector quantization algorithm is one of the ANN algorithms).

Moreover, we aim to coalesce three scientific endeavors and pave a way for providing direction for future applications to
- real dataset and synthetic datasets,
- fractals and multifractals for singled out singularities data as obtained from real datasets and synthetic datasets and
- data analysis algorithms for the classification of datasets.

Main objectives are as follows:

1.1 Objectives

Our book intends to enhance knowledge and facilitate learning, by using linear model and multilinear model, decision trees, naive Bayesian classifier, support vector machines, *k*-nearest neighbor, ANN algorithms as well as fractal and multifractal methods with ANN with the following goals:

- Understand what data analysis means and how data analysis can be employed to solve real problems through the use of computational mathematics
- Recognize whether data analysis solution with algorithm is a feasible alternative for a specific problem
- Draw inferences on the results of a given algorithm through discovery process
- Apply relevant mathematical rules and statistical techniques to evaluate the results of a given algorithm
- Recognize several different computational mathematic techniques for data analysis strategies and optimize the results by selecting the most appropriate strategy
- Develop a comprehensive understanding of how different data analysis techniques build models to solve problems related to decision-making, classification and selection of the more significant critical attributes from datasets and so on
- Understand the types of problems that can be solved by combining an expert systems problem solving algorithm approach and a data analysis strategy
- Develop a general awareness about the structure of a dataset and how a dataset can be used to enhance opportunities related to different fields which include but are not limited to psychiatry, neurology (radiology) as well as economy
- Understand how data analysis through computational mathematics can be applied to algorithms via concrete examples whose procedures are explained in depth
- Handle independent variables that have direct correlation with dependent variable
- Learn how to use a decision tree to be able to design a rule-based system
- Calculate the probability of which class the samples with certain attributes in dataset belong to
- Calculate which training samples the smallest k unit belongs to among the distance vector obtained
- Specify significant singled out singularities in data
- Know how to implement codes and use them in accordance with computational mathematical principles

1.2 Intended audience

Our intended audience are undergraduate, graduate, postgraduate students as well as academics and scholars; however, it also encompasses a wider range of readers who specialize or are interested in the applications of data analysis to real-world problems concerning various fields, such as engineering, medical studies, mathematics, physics, social sciences and economics. The purpose of the book is to provide the readers with the mathematical foundations for some of the main computational approaches to data analysis, decision-making, classification and selecting the significant critical attributes. These include techniques and methods for numerical solution of systems of linear and nonlinear algorithms. This requires making connections between techniques of numerical analysis and algorithms. The content of the book focuses on presenting the main algorithmic approaches and the underlying mathematical concepts, with particular attention given to the implementation aspects. Hence, use of typical mathematical environments, Matlab and available solvers/libraries, is experimented throughout the chapters.

In writing this text, we directed our attention toward three groups of individuals:

- **Academics** who wish to teach a unit and conduct a workshop or an entire course on essential computational mathematical approaches. They will be able to gain insight into the importance of datasets so that they can decide on which algorithm to use in order to handle the available datasets. Thus, they will have a sort of guided roadmap in their hands as to solving the relevant problems.
- **Students** who wish to learn more about essential computational mathematical approaches to datasets, learn how to write an algorithm and thus acquire practical experience with regard to using their own datasets.
- **Professionals** who need to understand how essentials of computational mathematics can be applied to solve issues related to their business by learning more about the attributes of their data, being able to interpret more easily and meaningfully as well as selecting out the most critical attributes, which will help them to sort out their problems and guide them for future direction.

1.3 Features of the chapters

In writing this book, we have adopted the approach of learning by doing. Our book can present a unique kind of teaching and learning experience. Our view is supported by several features found within the text as the following:

- **Plain and detailed examples.** We provide simple but detailed examples and applications of how the various algorithm techniques can be used to build the models.
- **Overall tutorial style.** The book offers flowcharts, algorithms and examples from Chapter 4 onward so that the readers can follow the instructions and carry out their applications on their own.
- **Algorithm sessions.** Algorithm chapters allow readers to work through the steps of data analysis, solving a problem process with the provided mathematical

computations. Each chapter is specially highlighted for easy differentiation from regular ordinary texts.
- **Datasets for data analysis.** A variety of datasets from medicine and economy are available for computational mathematics-oriented data analysis.
- **Web site for data analysis.** Links to several web sites that include dataset are provided as
  UCI Machine Learning Repository
  (http://archive.ics.uci.edu/ml/index.php)
  Kaggle the Home of Data Science & Machine Learning
  (http://kaggle.com)
  The World Bank Data Bank
  (http://data.worldbank.org)
- **Data analysis tools.** All the applications in the book have been carried out by Matlab software.
- **Key term definitions.** Each chapter introduces several key terms that are critical to understanding the concepts. A list of definitions for these terms is provided at the head of each chapter.
- **Each chapter accompanied by some examples as well as their applications.** Each chapter includes examples related to the subject matter being handled and applications that are relevant to the datasets being handled. These are grouped into two categories – computational questions and medical datasets as well as economy dataset applications.
  - *Computational questions* have a mathematical flavor in that they require the reader to perform one or several calculations. Many of the computational questions are appropriate for challenging issues that address the students with advanced level.
  - *Medical datasets and economy dataset applications* discuss advanced methods for classification, learning algorithms (linear model and multilinear model, decision trees algorithms (ID3, C4.5 and CART), naive Bayesian, support vector machines, *k*-nearest neighbor, ANN) as well as multifractal methods with the application of LVQ algorithm which is one of the ANN algorithms.

1.4 Content of the chapters

Medical datasets and economy dataset examples and applications with several algorithms' applications are explained concisely and analyzed in this book.

When data analysis with algorithms is not a viable choice, other options for creating useful decision-making models may be available based on the scheme in Figure 1.1.

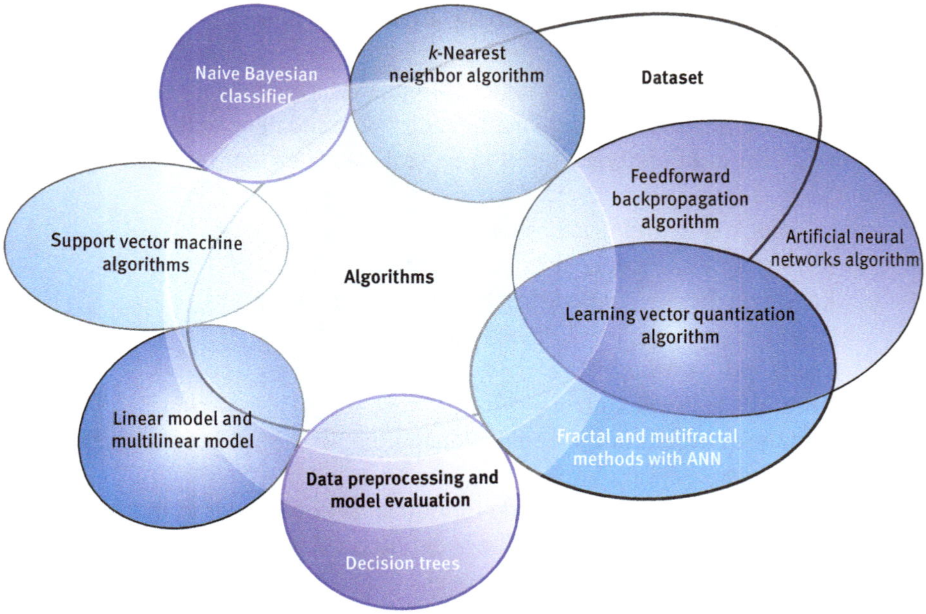

Figure 1.1: Content units in the chapters.

A brief description of the contents in each chapter is as follows:
- **Chapter 2** offers an overview of all aspects of the processes related to dataset as well as basic statistical descriptions of dataset along with relevant examples. It also addresses explanations of attributes of datasets to be handled in the book.
- **Chapter 3** introduces data preprocessing and model evaluation. It first introduces the concept of data quality and then discusses methods for data cleaning, data integration, data reduction and data transformation.
- **Chapter 4** provides detailed information on definition of algorithm. It also elaborates on the following concepts: *flow chart, variables, loop*, and *expressions* for the basic concepts related to algorithms.
- **Chapter 5** presents some basic concepts regarding linear model and multilinear model with details on our datasets (medicine and economics).
- **Chapter 6** provides basic concepts and methods for classification of decision tree induction, including ID3, C4.5 and CART algorithms with rule-based classification. The implementation of the ID3, C4.5 and CART algorithms' model application to medical datasets and economy dataset.
- **Chapter 7** introduces some basic concepts and methods for probabilistic naive Bayes classifiers on Bayes' theorem with strong (naive) independence assumptions between the features.
- **Chapter 8** explains the principles on support vector machines modeling, introducing the basic concepts and methods for classification in a variety of applications that are linearly separable and linearly inseparable.

- **Chapter 9** elaborates on nonparametric *k*-nearest neighbor as to how it is used for classification and regression. The chapter includes all available cases, classifying new cases based on a similarity measure. It is used for the prediction of the property of a substance in relation to the medical datasets and economy dataset for their most similar compounds.
- **Chapter 10** presents two popular neural network models that are supervised and unsupervised, respectively. This chapter provides a detailed explanation of feedforward backpropagation and LVQ algorithms' network training topologies/structures. Detailed examples have been provided, and applications on these neural network architectures are given to show their weighted connections during training with medical datasets and economy dataset.
- **Chapter 11** emphasizes on the essential concepts pertaining to fractals and multifractals. The most important concept on the fractal methods highlights fractal dimension and diffusion-limited aggregation with the specific applications, which are concerned with medical datasets and economy dataset. Brownian motion analysis, one of the multifractal methods, is handled for the selection of singled out singularities with regard to applications in medical datasets and economy dataset. This chapter deals with the application of Brownian motion Hölder regularity functions (polynomial and exponential) on medical datasets and economy dataset, with the aim of singled out singularities attributes in such datasets. By applying the LVQ algorithm on these singularities attributes as obtained from this chapter, we propose a classification accordingly.

1.5 Text supplements

All the applications in the book have been carried out by Matlab software. Moreover, we will also use the following datasets:

1.5.1 Datasets

We focus on issues related to economy dataset and medical datasets. There are three different datasets for this focus, which are made up of two synthetic datasets and one real dataset.

1.5.1.1 Economy dataset

Economy (U.N.I.S.) dataset. This dataset contains real data about economies of the following countries: The United States, New Zealand, Italy and Sweden collected from 1960 to 2015. The dataset has both numeric and nominal attributes.

The dataset has been retrieved from http://databank.worldbank.org/data/home.aspx. Databank is an online resource that provides quick and simple access to collections of time series real data.

1.5.1.2 Medical datasets

The multiple sclerosis (MS) dataset. MS dataset holds magnetic resonance imaging and Expanded Disability Status Scale information about relapsing remitting MS, secondary progressive MS, primary progressive MS patients and healthy individuals. MS dataset is obtained from Hacettepe University Medical Faculty Neurology Department, Ankara, Turkey. Data for the clinically definite MS patients were provided by Rana Karabudak, Professor (MD), the Director of Neuroimmunology Unit in Neurology Department.

In order to obtain synthetic MS dataset, classification and regression trees algorithm was applied to the real MS dataset. There are two ways to deal with synthetic dataset from real dataset: one is to develop an algorithm to obtain synthetic dataset from real dataset and the other is to use the available software (SynTReN, R-package, ToXene, etc.).

Clinical psychology dataset (revised Wechsler Adults Intelligence Scale [WAIS-R]). WAIS-R test contains the analysis of the adults' intelligence. Real WAIS-R dataset is obtained through Majaz Moonis, Professor (M.D.) at the University of Massachusetts (UMASS), Department of Neurology and Psychiatry, of which he is the Stroke Prevention Clinic Director.

In order to obtain synthetic WAIS-R dataset, classification and regression trees algorithm was applied to the real WAIS-R dataset. There are two ways to deal with synthetic dataset from real dataset: one is to develop an algorithm to obtain synthetic dataset from real dataset and other is to use the available software (SynTReN, R-package, ToXene, etc.).

2 Dataset

This chapter covers the following items:
- Defining dataset, big data and data science
- Recognizing different types of data and attributes
- Working through examples

In this chapter, we will discuss the main properties of datasets as a collection of data. In a more general abstract sense with the word "data," we intend to represent a large homogeneous collection of quantitative or qualitative objects that could be recorded and stored into a digital file, thus converting into a large collection of bits. Some examples of data could be time series, numerical sequences, images, DNA sequences, large sets of primes, big datasets and so on. Time series and signals are the most popular examples of data and they could be found in any kind of human and natural activities such as neural activity, stock exchange rates, temperature changes in a year, heart blood pressure during the day and so on. Such kind of data could be processed aiming at discovering some correlations, patterns, trends, and whatever could be useful, for example, to predict the future values of data. When the processing of data is impossible or very difficult to handle, then these data are more appropriately called big data. In the following section, we will discuss how to classify data, by giving the main attributes and providing the basic element of descriptive statistics for analyzing data.

The main task with data is to discover some information from them. It is a strong temptation for anyone to jump into [data] mining directly. However, before doing such a leap, at the initial stage, we have to get the data ready for processing.

We can roughly distinguish two fundamental steps in data manipulation as in Figure 2.1:

1. Data *preprocessing*, which consists of one or several actions such as cleaning, integration, eduction and transformation, in such a way to prepare the data for the next step.
2. Data *processing*, also known as data analysis or data mining, where the data are analyzed by some statistical or numerical algorithms, in order to single out some features, patterns, singular values and so on.

At each step, there are also some more models and algorithms. In this book we limit ourselves to some general definition on preprocessing and on the methods from statistical model (linear–multilinear model), classification methods (decision trees (ID3, C4.5 and Classification and Regression Trees [CART]), Naïve Bayes, support

https://doi.org/10.1515/9783110496369-002

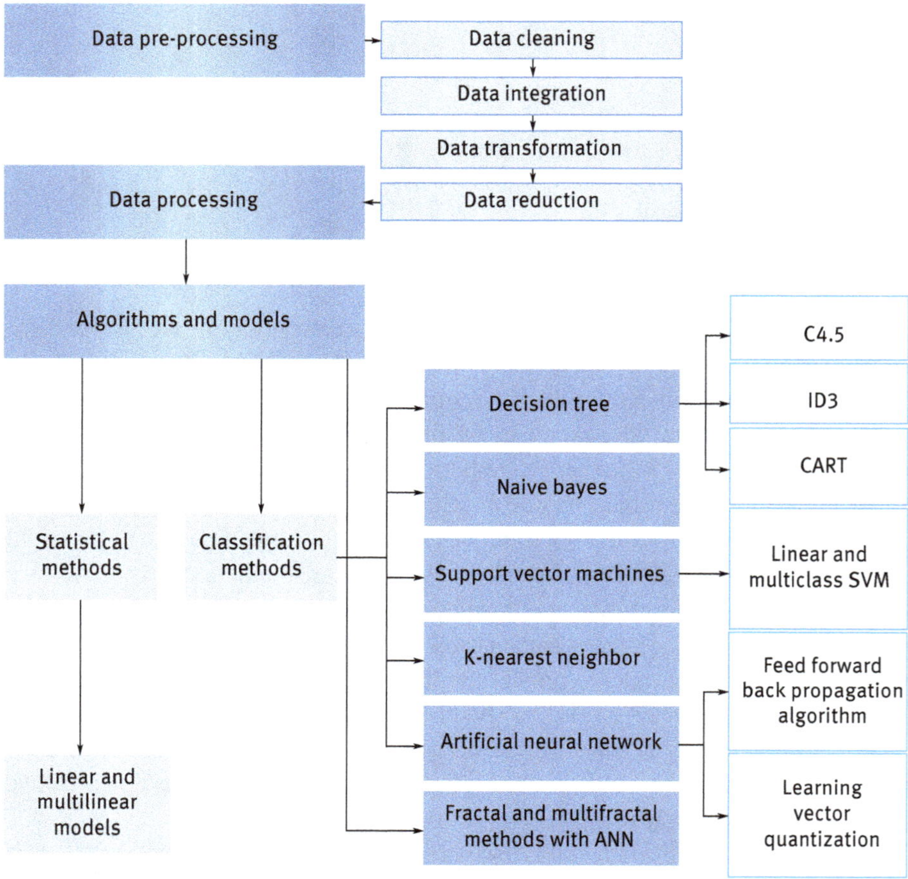

Figure 2.1: Data analysis process.

vector machines (linear, nonlinear), *k*-nearest neighbor, artificial neural network (feed forward back propagation, learning vector quantization) theory to handle and analyze data (at the data analysis step) as well as fractal and multifractal methods with ANN (LVQ).

This process involves taking a deeper insight at the attributes and the data values. Data in the real world tend to be noisy or enormous in volume (often denoted in several gigabytes or more). There are also cases when data may stem from a mass of heterogeneous sources.

This chapter deals with the main algorithms used to organize data. Knowledge regarding your data is beneficial for data preprocessing (see Chapter 3). This preliminary task is a fundamental step for the following data mining process. You may have to know the following for the process:

– What are the sorts of *attributes* that constitute the data?

- What are the type of values each attribute owns?
- Which of the attributes are continuous valued or discrete?
- How do the data *appear*?
- How is the distribution of the values?
- Do some ways exist so that we can visualize data in order to get a better sense of it overall?
- Is it possible to detect any outliers?
- Is it possible that we can measure the similarity of some data objects with regard to others?

Gaining such insight into the data will help us with further analysis. *On the basis of these, we can learn about our data that will prove to be beneficial for data preprocessing.* We begin in Section 2.1 by *this premise*: "*If our data is of enormous amount, we use big data.*" By doing so, we also have to study the various types of attributes, including nominal, binary, ordinal and numeric attributes. Basic *statistical descriptions* can be used to learn more about the values of attributes that are described in Section 2.5. As for *temperature* attribute, for instance, we can determine its *mean* (average value), *median* (middle value) and *mode* (most common value), which are the *measures of central tendency*. This gives us an idea of the "middle" or center of distribution.

If we know these basic statistics regarding each attribute it facilitates filling in missing values, spot outliers and smooth noisy values in data preprocessing. Knowledge of attributes and attribute values can also be of assistance in settling inconsistencies in data integration. Through plotting measures of central tendency, we can see whether the data are skewed or symmetric. Quantile plots, histograms and scatter plots are other graphic displays of basic statistical descriptions. All of these displays prove to be useful in data preprocessing.

Data visualization offers many additional affordances like techniques for viewing data through graphical means. These can help us to identify relations, trends and biases "hidden" in unstructured datasets. Data visualization techniques are described in Section 2.5.

We may also need to examine how *similar* or *dissimilar* data objects are. Suppose that we have a database in which the data objects are patients classified in different subgroups. Such information can allow us to find clusters of like patients within the dataset. In addition, economy data involve the classification of countries in line with different attributes.

In summary, by the end of this chapter, you will have a better understanding as to the different attribute types and basic statistical measures, which help you with the central tendency and dispersion of attribute data. In addition, the techniques to visualize attribute distributions and related computations will be clearer through explanations and examples.

2.1 Big data and data science

Definition 2.1.1. Big Data is an extensive term for any collection of large datasets, usually billions of values. They are so large or complex that it becomes a challenge to process them by employing conventional data management techniques. By these conventional techniques, we mean relational database management systems (RDBMS). In RDBMS, which is substantially a collection of tables, a unique name is assigned to each item. Consisting of a set of attributes (like columns or fields), each table normally stores a large set of tuples (records or rows). Each tuple in a relational table signifies an object that is characterized by a unique key. In addition, it is defined by a set of attribute values. While one is mining relational databases, he/she can proceed by pursuing trends or data patterns [1]. For example, RDBMS, an extensively adopted technique for big data, has been considered as a one-size-fits-all solution.

The data are stored on relation (table); the relation in a relational database is similar to a table where the column represents attributes, and data are stored in rows (Table 2.1). ID is a unique key, which is underlined in Table 2.1.

Table 2.1: RDBMS table example regarding New York Stock Exchange (NYSE) fundamentals data.

Attributes

Data

ID	Ticker symbol	Period ending	Accounts payable
0	AAL	2012-12-31	3068000000,0
1	AAL	2013-12-31	4975000000,0
2	AAL	2014-12-31	4668000000,0

Data science involves employing methods to analyze massive amounts of data and extract the knowledge they encompassed. We can imagine that the relationship between data science and big data is similar to that of raw material and manufacturing plant. Data science and big data have originated from the field of statistics and traditional data management. However, for the time being, they are seen as distinctive disciplines.

The characteristics of big data are often referred to as the three *Vs* as follows [2]:
- *Volume* – This refers to how much data are there.
- *Variety* – This refers to how diverse different types of data are.
- *Velocity* – How fast new data are generated.

These characteristics are often supplemented with a fourth *V* that refers to
- *Veracity*, meaning how accurate is the data.

These four properties make big data different from the data existing in conventional data management tools. As a result, the challenges brought by big data are seen in almost every aspect from data capture to curation, storage to search as well as other aspects including but not limited to sharing, transferring and visualization. Moreover, big data requires the employment of specialized techniques for the extraction of the insights [3, 4].

Following a brief definition of big data and what it requires, let us now have a look at the benefits and uses of big data and data science.

Data science and big data are terms used extensively both in commercial and noncommercial settings. There are a number of cases with big data. The examples we offer in this book will elucidate such possibilities. Commercial companies operating in different areas use data science and big data so as to know more about their staff, customers, processes, goods and services. Many companies use data science to be able to offer a better user experience to their customers. They also use big data for customization.

Let us have a look at the following example to relate the information about aforementioned big data, for example, NYSE (https://www.nyse.com/data/transactions-statistics-data-library) [5]. Big dataset is an area for fundamental and technical analysis. In this link, dataset includes fundamentals.csv file. Fundamentals file show which company has the biggest chance of going out of business, which company is undervalued, what the return on *Investment* is metrics extracted from annual sec 10K fillings (from 2012 to 2016), should suffice to derive most popular fundamental indicators as can be seen from Table 2.2.

Let us have a look at the $4V$ stage for the NYSE data: investments of companies are continuously tracked by NYSE. Thus, this enables related parties to bring the best deals with fundamental indicators for going out of business. Fundamentals involve getting to grips with the following properties:

- *Volume*: Thousand of companies are involved.
- *Velocity*: Companies, company information and prices undervalued.
- *Variety*: Investments: futures, forwards, swaps and so on.
- *Veracity*: Going out of business (refers to the biases, noise and abnormality in data).

In the NYSE, there is big data with a dimension of $1{,}781 \times 79$ (1,781 companies, 79 attributes). There is an evaluation of 1,781 firms as to their status of bankruptcy (going out of business) based on 79 attributes (*ticker symbol, period ending, accounts payable, add income/expense items, after tax ROE (Return On Equity), capital expenditures, capital surplus, cash ratio* and so on). Now, let us analyze the example in Excel table (Table 2.2) regarding the case of NYSE. The data have been selected from NYSE (https://www.nyse.com/data/transactions-statistics-data-library).

If we are to work with big data as in Table 2.2, the first thing we have to do is to understand the dataset. In order to understand the dataset, it is important that we

Table 2.2: NYSE fundamentals data as depicted in RDBMS table (sample data breakdown).

ID	Ticker symbol	Period ending	Accounts payable	Accounts receivable	Additional income/ expense items	After tax ROE	Capital expenditures	Capital surplus	Cash ratio
0	AAL	2012-12-31	3068000000,0	−222000000,0	−1961000000,0	23.0	−1888000000,0	4695000000,0	53.0
1	AAL	2013-12-31	4975000000,0	−93000000,0	−2723000000,0	67.0	−3114000000,0	10592000000,0	75.0
2	AAL	2014-12-31	4668000000,0	−160000000,0	−150000000,0	143.0	−5311000000,0	15135000000,0	60.0
3	AAL	2015-12-31	5102000000,0	352000000,0	−708000000,0	135.0	−6151000000,0	11591000000,0	51.0
4	AAP	2012-12-29	2409453000,0	−89482000,0	600000,0	32.0	−271182000,0	520215000,0	23.0
5	AAP	2013-12-28	2609239000,0	−32428000,0	2698000,0	26.0	−195757000,0	531293000,0	40.0
6	AAP	2015-01-03	3616038000,0	−48209000,0	3092000,0	25.0	−228446000,0	562945000,0	3.0
7	AAP	2016-01-02	3757085000,0	−21476000,0	−7484000,0	19.0	−234747000,0	603332000,0	2.0
8	AAPL	2013-09-28	36223000000,0	−1949000000,0	1156000000,0	30.0	−8165000000,0	0	93.0
9	AAPL	2014-09-27	48649000000,0	−6452000000,0	980000000,0	35.0	−9571000000,0	0	40.0
10	AAPL	2015-09-26	60671000000,0	−3124000000,0	1285000000,0	45.0	−11247000000,0	0	52.0
11	AAPL	2016-09-24	59321000000,0	1044000000,0	1348000000,0	36.0	−1273000000,0	0	85.0
12	ABBV	2012-12-31	5734000000,0	223000000,0	−8000000,0	1507.0	−333000000,0	0	118.0
13	ABBV	2013-12-31	6448000000,0	681000000,0	−54000000,0	92.0	−491000000,0	3671000000,0	144.0
14	ABBV	2014-12-31	6954000000,0	−172000000,0	−651000000,0	102.0	−612000000,0	4194000000,0	74.0
15	ABBV	2015-12-31	8463000000,0	−1076000000,0	−206000000,0	130.0	−532000000,0	13080000000,0	77.0
16	ABC	2013-09-30	14870635000,0	−2312518000,0	−44000,0	19.0	−202450000,0	2360992000,0	8.0
17	ABC	2014-09-30	17250160000,0	−938286000,0	−28594000,0	14.0	−264457000,0	2749185000,0	10.0
18	ABC	2015-09-30	21578227000,0	−1478793000,0	−44220000,0	22.0	−231585000,0	3736477000,0	10.0
19	ABC	2016-09-30	24670159000,0	−912724000,0	5048000,0	67.0	−464616000,0	4333001000,0	11.0
20	ABT	2012-12-31	10889000000,0	36000000,0	−1260000000,0	22.0	−1795000000,0	0	114.0

show the relationship between the attributes (ticker symbol, period ending, accounts payable, accounts receivable, etc.) and records (0th row, 1st row, 2nd row, ..., 1,781st row) in the dataset.

Histograms are one of the ways to depict such a relationship. In Figure 2.2, the histogram provides the analyses of attributes that determine bankruptcy statuses of the firms in the NYSE fundamentals big data. The analysis of the selected attributes has been provided separately in some histogram graphs (Figure 2.3).

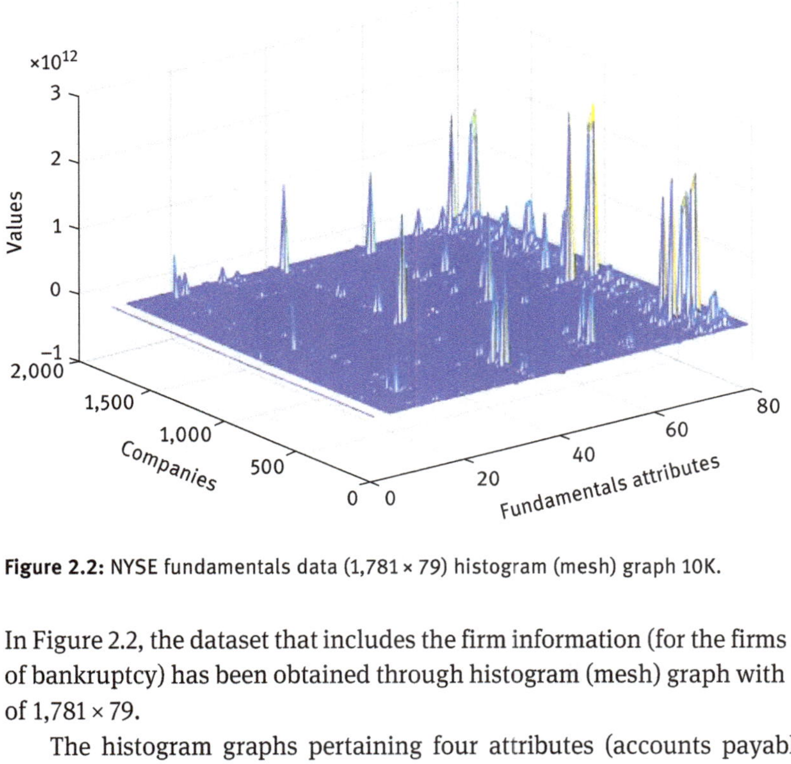

Figure 2.2: NYSE fundamentals data (1,781 × 79) histogram (mesh) graph 10K.

In Figure 2.2, the dataset that includes the firm information (for the firms on the verge of bankruptcy) has been obtained through histogram (mesh) graph with a dimension of 1,781 × 79.

The histogram graphs pertaining four attributes (accounts payable, accounts receivable, capital surplus and after tax ROE) as chosen from 79 attributes that determine the bankruptcy of a total of 1,781 firms are provided in Figure 2.3.

2.2 Data objects, attributes and types of attributes

Definition 2.2.1. Data are defined as a collection of facts. These facts can be words, figures, observations, measurements or even merely descriptions of things.

In some instances, it would be most appropriate to express data values in purely *numerical* or *quantitative terms*, for example, in currencies like in pounds or dollars, or in measurement units like in inches or percentages (measurements values of which are essentially numerical).

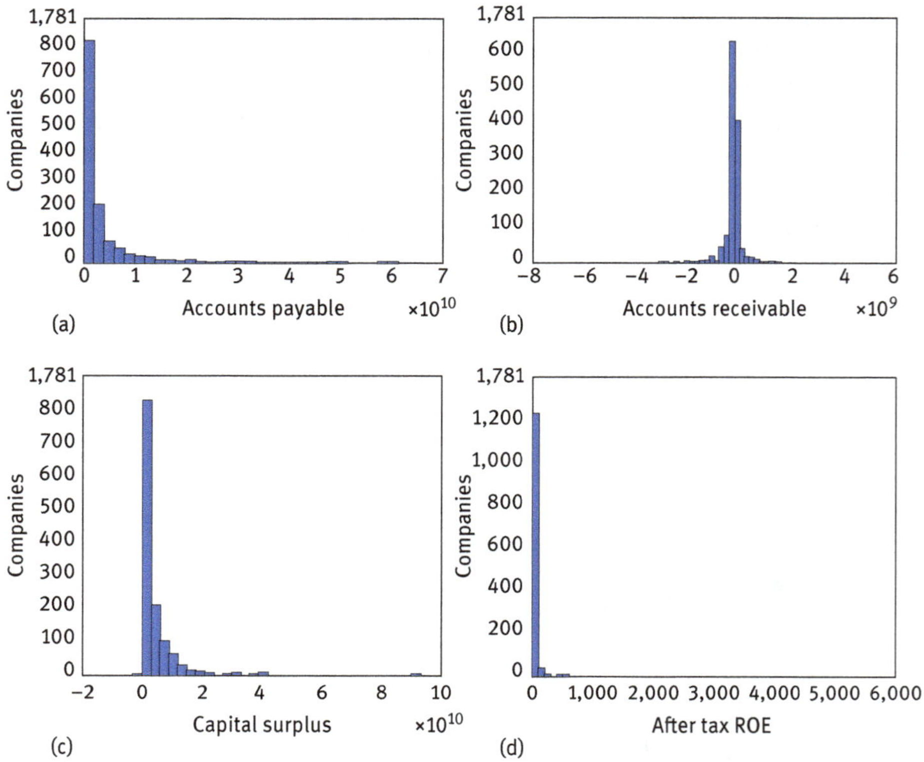

Figure 2.3: NYSE fundamentals attributes histogram graph: (a) accounts payable companies histogram; (b) accounts receivable companies histogram; (c) capital surplus companies histogram and (d) after tax ROE histogram.

In other cases, the observation may signify only the category that an item belongs to. Categorical data are referred to as *qualitative data* (whose data measurement scale is essentially categorical) [6]. For instance, a study might be addressing the class standing-freshman, sophomore, junior, senior or graduate levels of college students. The study may also handle students to be assessed depending on the quality of their education as very poor, poor, fair, good or very good.

Note that even if students are asked to record a number (1–5) to indicate the quality level at which the numbers correspond to a category, the data will still be regarded as qualitative because the numbers are merely codes for the relevant (qualitative) categories.

A *dataset* is a numerable, but not necessarily ordered, collection of *data objects*. A data object represents an entity; for instance, USA, New Zealand, Italy, Sweden (U.N.I.S.) economy dataset, GDP growth, tax payments and deposit interest rate. In medicine, all kind of records about human activity could be represented by datasets of suitable data objects; for instance, the heart beat rate measured by an electrocardiograph has the wave shape with isolated peaks and each R–R wave in an R–R interval could be considered as a data objects. Another example of dataset in medicine is the multiple sclerosis (MS) dataset, where the objects of datasets may include subgroups of MS patients and the subgroup of healthy individuals, magnetic resonance images (MRI), expanded disability status scale (EDSS) scores and other relevant medical findings of the patients. In another example, for instance, clinical psychology dataset may consist of test results of the subjects and demographic features of the subjects such as age, marital status and education level. Data objects, which may also be referred to as samples, examples, objects or data points, are characteristically defined by attributes.

Definition 2.2.2. Attributes are data fields that denote feature of a data objects. In literature, there are instances where other terms are used for attributes depending on the relevant field of study and research. It is possible to see some terms such as feature, variable or dimension. While statisticians prefer to use the term *variable*, professionals in data mining and processing use the term *attribute*.

Definition 2.2.3. Data mining is a branch of computer science and of data analysis, aiming at discovering patterns in datasets. The investigation methods of data mining usually are based on statistics, artificial intelligence and machine learning.

Definition 2.2.4. Data processing is also called information processing is a way to discover meaningful information from data.

Definition 2.2.5. A collection of data depending on a given parameter is called a distribution. If the distribution of data involves only one attribute, it is called univariate. A bivariate distribution, on the other hand, involves two attributes. A set of possible values determine the type of an attribute. For instance, nominal, numeric, binary, ordinal, discrete and continuous are the aspects an attribute can refer to [7, 8].

Let us see some examples of univariate and bivariate distributions. Figure 2.4 shows the graph of univariate distribution for net domestic credit attribute. This is one of the attributes in Italy Economy Dataset Explanation dataset (Figure 2.4(a)). The same dataset is given in Figure 2.4(b), with bivariate distribution graph together

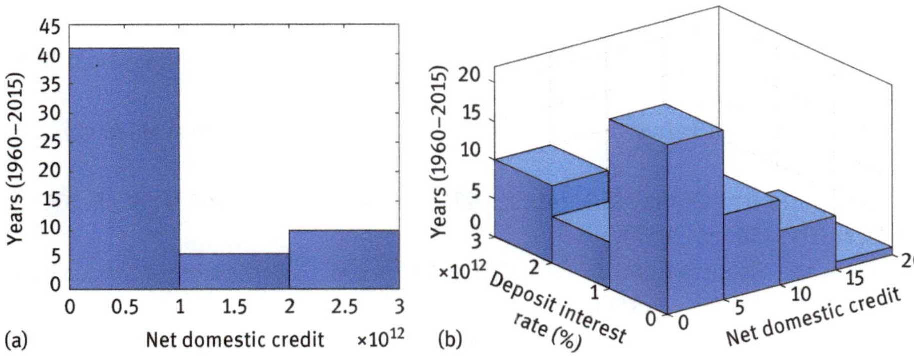

Figure 2.4: (a) Univariate and (b) bivariate distribution graph for Italy Economy Dataset.

with deposit interest rate attributes based on years as chosen from the dataset for two attributes (for the period between 1960 and 2015).

The following sections provide further explanations and basic examples for these attributes.

2.2.1 Nominal attributes and numeric attributes

Nominal refers to names. A *nominal attribute*, therefore, *can be names of things or symbols*. Each value in a nominal attribute represents a code, state or category. For instance, healthy and patient categories in medical research are example of nominal attributes. The education level categories such as graduate of primary school, high school or university are other examples of nominal attributes. In studies regarding MS, subgroups such as relapsing remitting multiple sclerosis (RRMS), secondary progressive multiple sclerosis (SPMS) or primary progressive multiple sclerosis (PPMS) can be listed among nominal attributes. It is important to note that nominal attributes are to be handled distinctively from numeric attributes. They are incommensurable but interrelated. For mathematical operations, a nominal attribute may not be significant. In medical dataset, for example, the schools from which the individuals, on whom Wechsler Adult Intelligence Scale – Revised (WAIS-R) test was administered, graduated (primary school, secondary school, vocational school, bachelor's degree, master's degree, PhD) can be given as examples of nominal attribute. In economy dataset, for example, consumer prices, tax revenue, inflation, interest payments on external debt, deposit interest rate and net domestic credit can be provided to serve as examples of nominal attributes.

Definition 2.2.1.1. Numeric attribute is a measurable quantity; thus, it is quantitative, represented in real or integer values [7, 9, 10]. Numeric attributes can either be interval- or ratio-scaled.

Interval-scaled attribute is measured on a scale that has equal-size units, such as

Celsius temperature (C): (−40 °C, −30 °C, −20 °C, 0 °C, 10 °C, 20 °C, 30 °C, 40 °C, 50 °C, 60 °C, 70 °C, 80 °C).

We can see that the distance from 30 °C to 40 °C is the same as distance from 70 °C to 80°C.

The same set of values converted into Fahrenheit temperature scale is given in the following list:

Fahrenheit temperature (F): (−40 °F, −22 °F, −4 °F, 32 °F, 50 °F, 68 °F, 86 °F, 104 °F, 122 °F, 140 °F, 158 °F, 176 °F).

Thus, the distance from 30 °C to 40 °C [86 °F – 104 °F] is the same distance from 60 °C to 70 °C [140 °F – 158 °F].

Ratio-scaled attribute has inherent zero point. Consider the example of age. The clock starts ticking forward once you are born, but an age of "0" technically means you do not actually exist. Therefore, if a given measurement is ratio scaled, the value can be considered as one that is a multiple or ratio of another value. Moreover, the values are ordered so the difference between values, mean, median and mode can be calculated.

Definition 2.2.1.2. Binary attribute is a nominal attribute that has only two states or categories. Here, 0 typically shows that the attribute is absent and 1 means that the attribute is present. States such as *true* and *false* fall into this category. In medical cases, healthy and unhealthy states pertain to binary attribute type [11–13].

Definition 2.2.1.3. Ordinal attributes are those that yield possible values that enable meaningful order or ranking. A basic example is the one used commonly in clothes size: x-small, small, medium, large and x-large. In psychology, mild and severe refer to ranking a disorder [11–14].

Discrete attributes have either finite set of values or infinite but countable set of values that may or may not be represented as integers [7, 9, 10]. For example, age, marital status and education level each have a finite number of values; therefore, they are discrete. Discrete attributes may have numeric values, for example, age may be between 0 and 120. If an attribute is not discrete, it is *continuous*. Numeric attribute and continuous attribute are usually used interchangeably in literature. However, continuous values are real numbers, while numeric values can be either real numbers of integers. The income level of a person or the time it takes a computer to complete a task can be continuous [15–20].

Definition 2.2.1.4. Data Set (or dataset) is a collection of data. For $i = 0, 1, \ldots, K$ in $x: i_0, i_1, \ldots, i_K; j = 0, 1, \ldots, K$ in $y: j_0, j_1, \ldots, j_L$ $D(K \times L)$ represents a dataset. $x: i.$ is the entry on the line and $y: j.$ is the matrix that represents the attribute in the column.

Figure 2.5 shows the dataset $D(3 \times 4)$ as depicted with row (x) and column (y) numbers.

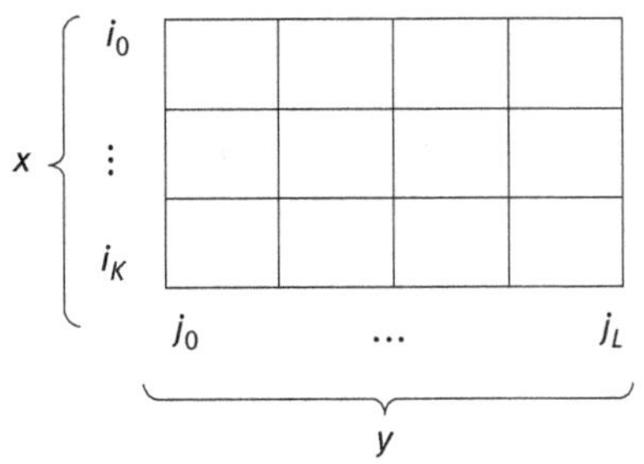

Figure 2.5: Shown in the figure with x row and y column as an example regarding dataset D.

2.3 Basic definitions of real and synthetic data

In this book, there are two main type of datasets identified. These are real dataset and synthetic dataset.

2.3.1 Real dataset

Real dataset refers to specific topic collected for a particular purpose. In addition, the real dataset includes the supporting documentation, definitions, licensing statements and so on. A dataset is a collection of discrete items made up of related data. It is possible to access to this data either individually or in combination. It is also possible that it is managed as an entire sample. Most frequently real dataset corresponds to the contents of a single statistical data matrix in which each column of the table represents a particular attribute and every row corresponds to a given member of the dataset. A dataset is arranged into some sort of data structure. In a database, for instance, a dataset might encompass a collection of business data (sales figures, salaries, names, contact information, etc.). The database can be considered a dataset per se, as can bodies of data within it pertaining to a particular kind of information, such as sales data for a particular corporate department.

In the applications and examples to be addressed in the following sections, dataset from economy field has been applied by having formed real data. Economy dataset, as real data, is selected from World Bank site (you can see http://data.worldbank.org/country/) [21] and used accordingly.

2.3.2 Synthetic dataset

The idea of original fully synthetic data was generated by Rubin in 1993 [16]. Rubin originally designed this with the aim of synthesizing the Decennial Census long-form responses for the short-form households. A year later, in 1994, Fienberg generated the idea of critical refinement. In this refinement he utilized a parametric posterior predictive distribution (rather than a Bayes bootstrap) in order to perform the sampling. They together discovered a solution for how to treat synthetic data partially with missing data [14–17].

Creating synthetic data is a process of data anonymization. In other words, synthetic data are a subset of anonymized, or real, data. It is possible to use synthetic data in different areas to act as a filter for information that would otherwise compromise the confidentiality of particular aspects of the data. The particular aspects generally emerge in the form of patient individual information (i.e., individual ID, age, gender, etc.).

Using synthetic data has been recommended for data analysis applications for the following conveniences it offers:

- it is quick and affordable to produce as much data as needed;
- it can yield accurate labels (labeling that may be impossible or expensive to obtain by hand);

- it is open to modification for the improvement of the model and training;
- it can be used as a substitute for certain real data segments.

In the applications to be addressed in the following sections, datasets from medicine have been applied forming synthetic data by CART algorithm (see Section 6.5) Matlab. There are two ways to deal with generating synthetic dataset from real dataset: one is to develop an algorithm to generate synthetic dataset from real dataset or another may use available software (SynTReN, R-package, ToXene, etc.) [18–20].

The synthetic datasets in our book have been obtained by the CART algorithm (for more details see Section 6.5) application on the real dataset.

For real dataset: $i = 0, 1, \ldots, R$ in $x: i_0, i_1, \ldots, i_R$; $j = 0, 1, \ldots, L$ in $y: j_0, j_1, \ldots, j_L$ $D(R \times L)$ represents a dataset. $x: i.$ is the entry on the line and $y: j.$ is the matrix representing the attribute in the column.

For synthetic dataset, $i = 0, 1, \ldots, K$ in $x: i_0, i_1, \ldots, i_K$; $j = 0, 1, \ldots, K$ in $y: j_0, j_1, \ldots, j_L$ $D(K \times L)$ represent a dataset. $x: i.$ is the entry on the line and $y: j.$ is the matrix representing the attribute in the column.

A brief description of the contents in synthetic dataset is shown in Figure 2.7.

The steps pertaining to the application of CART algorithm (Figure 2.6) to the real dataset $D(K \times L)$ for obtaining the synthetic dataset $D(R \times L)$ are provided as follows.

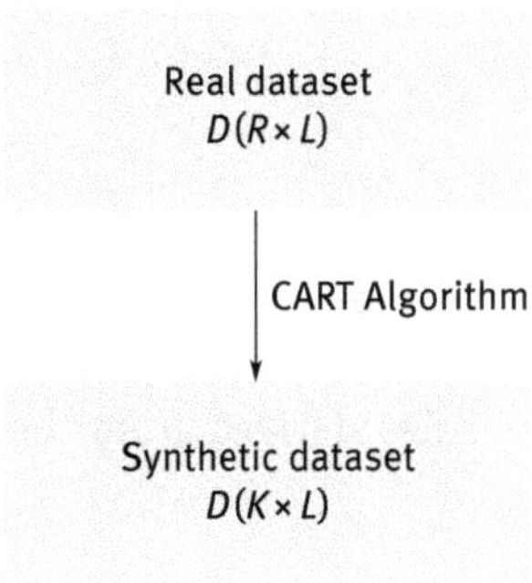

Figure 2.6: Obtaining the synthetic dataset.

The steps of getting the synthetic dataset through the application of general CART algorithm are provided in Figure 2.7.

Steps (1–6) The data belonging to each attribute in the real dataset are ranked from lowest to highest (when the attribute values are same, only one value is included in the ranking). For each attribute in the real dataset, the average of median and the value following the median is calculated. x_{median} represents the median value of the attribute and $x_{\text{median}+1}$ represents the value after the median value (see eq. (2.4(a))).

The average of x_{median} and $x_{\text{median}+1}$ is calculated as $\frac{x_{\text{median}} + x_{\text{median}+1}}{2}$ (see eq. (2.1)).

Algorithm: CART algorithm

Input: Real Data Set $D(R \times L)$// in the real dataset, the line number is with R dimension and the column number is with L dimension.

Output: Synthetic Data Set $D(K \times L)$// in the synthetic dataset formed from real data set, the line number is with K dimension and the column number is with L dimension.

Method:

(1) **while**(j is to L) **do**{
(2) **while**(i is to R) **do**{
(3) //Each attribute in the real dataset is split into two different groups.
(4) //The data belonging to the real dataset are ranked from low to high.
(5) **for each** attribute in the Real Dataset the average of median and the value following the median is calculated.
(6) $\frac{x_{median} + x_{median+1}}{2}$}
(7) //Formation of Decision Tree belonging to Real dataset (through CART algorithm (see more details Section 6.5.)
(8) //the Gini value of the attribute in the real data set is calculated (see more details Section 6.5).
(9) //generating the Decision tree rules belonging to the Real Data set
(10) //forming of the synthetic data set $D(K \times L)$ based on the decision tree rules of real data set
(11) }

Figure 2.7: Obtaining the synthetic dataset from the application of general CART algorithm on the real dataset.

Steps (7–10) The Gini value of each attribute is calculated for the formation of decision tree from the real dataset (see Section 6.5). The lowest Gini value calculated is the root of the decision tree. Synthetic data are formed based on the decision tree obtained from the real dataset and the rules derived from the decision tree.

Let us now explain how the synthetic dataset is generated from real dataset through a smaller scale of sample dataset.

Example 2.1 Dataset $D(10 \times 2)$ consists of two classes (gender as female or male) and two attributes (weight in kilograms and height in centimeters; see Table 2.3) as CART algorithm has been applied and synthetic data need to be generated.

Now we can apply CART algorithm on the sample dataset $D(10 \times 2)$ (see Table 2.3)

Steps (1–6) The data that belong to each attribute in the dataset $D(10 \times 2)$ are ranked from lowest to highest (when the attribute values are the same, only one value is incorporated in the ranking; see Table 2.3). For each attribute in the dataset the average of median and the value following the median is computed. x_{median} represents the

Table 2.3: Dataset $D(10 \times 2)$.

Class	The attributes of the dataset $D(10 \times 2)$	
	Weight	Height
Female	70	170
Female	60	165
Female	55	160
Female	47	167
Female	65	170
Male	90	190
Male	82	180
Male	83	182
Male	85	186
Male	100	185

median value of the attribute and $x_{\text{median}+1}$ represents the value after the median value (see eq. (2.4(a))).

The average of x_{median} and $x_{\text{median}+1}$ is calculated as $\frac{x_{\text{median}}+x_{\text{median}+1}}{2}$ (see eq. (2.1)).

The threshold value calculation for each attribute (column; *j*: 1, 2) in the dataset $D(10 \times 2)$ in Table 2.3 is applied based on the following steps.

In the following, we discuss the steps of the threshold value calculation for the *j*: 1: {*Weight*} attribute.

In Table 2.3. *{Weight}* values (line: *i*: 1,. . ., 10) in *j*: 1 column is ranked from lowest to highest (when the attribute values are the same, only one value of is incorporated in the ranking).

$$\text{Weight}_{\text{sort}} = \{47,\ 55,\ 60,\ 65,\ 70,\ 82,\ 83,\ 85,\ 90,\ 100\}$$

$$\text{Weight}_{\text{median}} = 70$$

$$\text{Weight}_{\text{median}+1} = 82$$

$$\overline{Weight} = \frac{70+82}{2} = 76$$

For "Weight ≥ 76", "Weight <76" two groups are formed.

The following are the steps for the calculation of threshold value for the *j*: 2: {*Height*} attribute.

In Table 2.3, *{Height} values* (line: *i: 1,. . ., 10) in j: 2 column are ranked from lowest to highest.*(When the attribute values are same, only one value is incorporated in the ranking.)

$$\text{Height}_{\text{sort}} = \{160, 165, 167, 170, 180, 182, 185, 186, 190\}$$

$$\text{Height}_{\text{median}} = 180$$

$$\text{Height}_{\text{median}+1} = 182$$

$$\overline{\text{Height}} = \frac{180 + 182}{2} = 181$$

For "Height ≥ 181", "Height <181," two groups are formed.

Steps (7–10) The Gini value of each attribute is calculated for the formation of decision tree of the dataset $D(10 \times 2)$. For each attribute, Gini value calculation is done after being split into two groups, and the root of the tree is found (more details in Section 6.5.)

In Table 2.4, you can see the results derived from the splitting of the binary groups for the attributes ($j = 1, 2$) that belong to the sample real data (female, male class).

Table 2.4: Results derived from the splitting of the binary groups for the attributes that belong to dataset (female, male class).

Class	Weight		Height	
	≥76	<76	≥181	<181
Female	0	5	0	5
Male	5	0	4	1

The results provided in Table 2.4 for each attribute in sample dataset $D(10 \times 2)$ will be used for the Gini calculation.

The calculation of *Gini* (*Weight*) for *j: 1* (see Section 6.5, eq. (6.8)):

$$\text{Gini}_{\text{left}}(\text{Weight}) = 1 - [(0/5)^2 + (5/5)^2] = 0$$

$$\text{Gini}_{\text{right}}(\text{Weight}) = 1 - [(5/5)^2 + (0/5)^2] = 0$$

$$\text{Gini}(\text{Weight}) = (5(0) + 5(0))/10 = 0$$

The calculation of *Gini* (*Height*) value for *j*: 1 (see Section 6.5, eq. (6.8)):

$$\text{Gini}_{\text{left}}(\text{Height}) = 1 - [(0/4)^2 + (4/4)^2] = 0$$

$$\text{Gini}_{\text{right}}(\text{Height}) = 1 - [(5/6)^2 + (1/6)^2] = 0.284$$

The lowest Gini value is *Gini*(*Weight*) = 0. The lowest value in the Gini values is the root of the decision tree. The decision tree obtained from the dataset is shown in Figure 2.8.

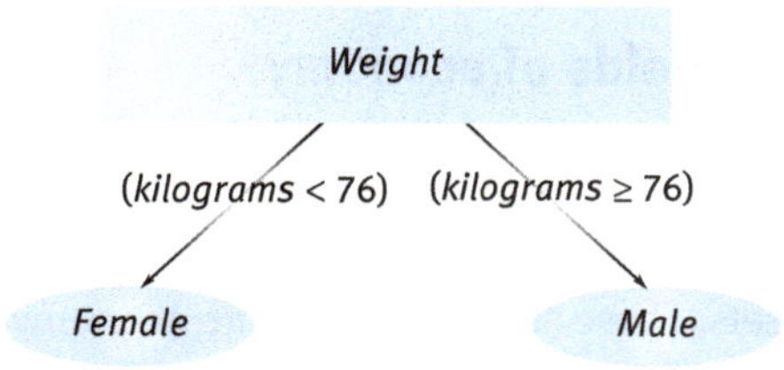

Figure 2.8: The decision tree obtained from dataset *D*(10 × 2).

The decision rule as obtained from dataset in Figure 2.8 is given as Rules 1 and 2:
Rule 1: If *Weight* ≥76, class is *male*.
Rule 2: If *Weight* <76, class is *female*.

In Table 2.5, synthetic dataset *D*(3 × 10) in Rule 1 *Weight* ≥ 76 and in Rule 2 *Weight* <76 (see Figure 2.8) are applied and obtained accordingly.

Table 2.5: Forming the sample synthetic dataset for *Weight* ≥ 76 and *Weight* <76.

Weight	Height
76	180
70	162
110	185

In Table 2.6, synthetic dataset *D*(3 × 2) is obtained by applying Rules 1 and 2 (see Figure 2.8).

Table 2.6: Forming the sample synthetic dataset for *Weight* ≥76, then class is *male* and Rule 2: *Weight* <76, then class is *female*.

Weight	Height	Class
76	180	Male
70	162	Female
110	185	Male

CART algorithm is applied to the real dataset. On applying this, we have received synthetic dataset. Synthetic dataset is composed of two entries from the male class and one entry (row) from the female class (it is possible to increase or reduce the entry number in the synthetic dataset).

2.4 Real and synthetic data from the fields of economy and medicine

In this book, to apply and explain the models described in the following section, we will refer to only three different types of datasets. These are concerned with economy data (USA, New Zealand, Italy, Sweden), medical data (Multiple Sclerosis dataset), clinical psychology data (WAIS-R dataset). The sample datasets taken from these diverse fields have been discussed throughout the book. These can be found in the examples and applications.

2.4.1 Economy data: Economy (U.N.I.S.) dataset

As the first set of real dataset, we will use, in the following sections, some data related to USA (abbreviated by the letter U), New Zealand (N), Italy (I) and Sweden (S) economies for U.N.I.S. countries' economies. Data belonging to their economies from 1960 to 2015 are defined based on the attributes given in Table 2.7 (Economy dataset; http://data.worldbank.org) [21].

The matrix dimension of economy (U.N.I.S.) dataset is defined as follows (see Figure 2.9).

For $i = 0, 1, ..., 228$ in $x: i_0, i_1, ..., i_{228}$; $j = 0, 1, ..., 18$ in $y: j_0, j_1, ..., j_{18}$, $D(228 \times 18)$ represents economy dataset.

In Table 2.7, the economic growth and parameters thereof are explained as to U. N.I.S. countries for the period between 1960 and 2015.

The sample data values that belong to these economies can be presented as in Table 2.8 (see the World Bank website [21]).

2.4.2 MS data and content (neurology and radiology data): MS dataset

MS is a demyelinating disease of central nervous system (CNS). The disease is brought about by an autoimmune reaction that results from a complex interaction of genetic and environment factors [22–25]. MS may affect any site in the CNS; nevertheless, it is possible to recognize six types regarding the disease (Table 2.9).

Table 2.7: U.N.I.S. economies dataset explanation.

Country (between 1960–2015)	Attributes	Data Size
USA New Zealand Italy Sweden } 57 years of enrollment × 4 countries = 228	Years Tax revenue Tax payments Time to import Time to export Deposit interest rate Net domestic credit Inflation, consumer prices Unemployment, youth male Adjusted net national income Merchandise exports attribute Unemployment, youth female Gross value added at factor cost Account at a financial institution, female Account at a financial institution, male Interest payments on external debt, total } 16 Numeric GDP growth GDP per capita attributes } 2 Nominal	228×18

1. *RRMS:* This is the typical form of MS that usually has onset in the late teens or twenties characterized by a severe attack that is followed by complete or incomplete recovery. About 70% of patients with MS experience a relapsing–remitting course [22, 23]. Further attacks may occur at unpredictable intervals, and each of such attacks are followed by increasing disability. Relapsing–remitting pattern tends to turn into secondary progressive form of the disease in the late thirties.
2. *PPMS*: The disease displays a steady deteriorating course that may be interrupted by periods of quiescence without improvement. The rate of progression may vary: This form can end up with death within a few years in the most severe cases. In contrast, the more chronic form of progressive MS resembles the benign form of the disease [22–24].
3. *SPMS*: The relapsing–remitting form of the disease usually develops into SPMS following a changeable period of time [22] (usually this corresponds to the late thirties).
4. *Relapsing progressive MS*: Occasionally it has been observed that patients with progressive form of MS have superimposed relapses with no significant recovery [22].
5. *Benign MS*: The benign form of MS is seen in approximately 20% of the cases. It is particularly improbable that these patients will ever be debilitated by the disease

Table 2.8: An excel table to provide an example of economy (U.N.I.S.) data.

Years	Class (country name)	Unemployment, youth male	Unemployment, youth female	GDP growth	...	Tax payments	Deposit interest rate	...	Net domestic credit
1960	USA	0	0	0	...	0	0	...	3,29E + 11
...	...	...	...	...	...	...	...	...	...
2015	USA	12.729	10.343	0	...	0	0	...	0
1960	New Zealand	0	0	0	...	0	0	...	619,3
...	...	...	...	...	...	...	...	...	...
2015	New Zealand	4.881	5.893	3.576	...	8	3.730	...	0
1960	Italy	0	0	0	...	0	0	...	0
...	...	...	...	...	...	...	...	...	...
2015	Italy	0	0	...	...	14	0	...	2,803E + 12
1960	Sweden	0	0	0	...	0	0	...	400,73
...	...	...	...	...	...	...	...	...	...
2015	Sweden	7.571	7.277	0	...	6	0	...	6,331E + 12

Table 2.9: Six types of MS [23].

1 – Relapsing-Remitting MS (RRMS)
2 – Primary Progressive MS (PPMS)
3 – Secondary Progressive MS (SPMS)
4 – Relapsing Progressive MS (RPMS)
5 – Benign MS
6 – Spinal form of MS

and they can go on leading a full life span, with only occasional minor symptoms. The existence of a benign form of MS emphasizes the significance of recording the date of the first symptoms patients develop with few residual abnormal signs several years after the beginning of the disease. It is important that these patients be informed about the existence of a benign form of MS 10 years after the first record of symptoms, and also about the fact that the benign course will continue in the upcoming years of their lives [22].

6. *Spinal form of MS*: This form of MS manifests symptoms of predominantly involvement of spinal cord from the beginning and keeps up with this .pattern. There may be a clear-cut pattern of relapse and remission initially, followed by the secondary progressive form of the disease following several years, or the manifestation may be seen as one of the steady [22].

One of the most important tools in the diagnosis of MS is MRI. Using this tool, specialists examine the current formation of the plaques that form in the MS disease.

MRI: This is a very important tool that can reveal the inflammated or harmed tissue sites in the diagnosis of MS. This study makes use of the MRI of the patients who have been diagnosed with MS based on Mc Donald criteria [24]. The dataset has been formed by getting the measurement as to the lesions from three sites based on the MRI [25]. These sites are the brain stem, corpus callosum-periventricular region and upper cervical region (see Figure 2.9).

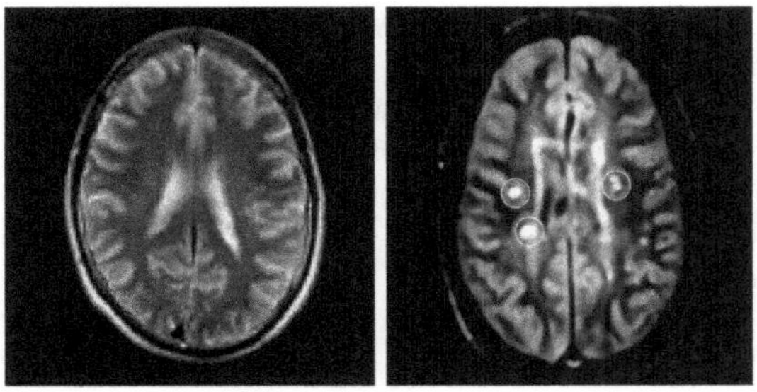

Figure 2.9: A sample MRI image that belongs to a healthy and an MS patient (obtained from Hacettepe University). (a) An MRI image of a healthy subject and (b) an MRI image with lesions indicated.

EDSS: EDSS is a scale that is capable of measuring eight regions, partially of the CNS known as the functional system. This scale measures the freedom of movement first by using the temporary numbness in the face and fingers or visual impairments and walking distance [26]. The ranking of each item from these systems is conducted based on the disability status, and to this functional system, mobility as well as daily life limitations are added. Twenty ranks within EDSS are provided with their corresponding definitions in Table 2.10 [26, 27].

Table 2.10: Description of EDSS [26, 27].

Score	Description
1.0	No disability, minimal signs in one FS
1.5	No disability, minimal signs in more than one FS
2.0	Minimal disability in one FS
2.5	Mild disability in one FS or minimal disability in two FS
3.0	Moderate disability in one FS, or mild disability in three or four FS. No impairment to walking
3.5	Moderate disability in one FS and more than minimal disability in several others. No impairment to walking
4.0	Significant disability but self-sufficient and up and about some 12 h a day. Able to walk without aid or rest for 500 m
4.5	Significant disability but up and about much of the day, able to work a full day; may otherwise have some limitation of full activity or require minimal assistance. Able to walk without aid or rest for 300 m
5.0	Disability severe enough to impair full daily activities and ability to work a full day without special provisions. Able to walk without aid or rest for 200 m
5.5	Disability severe enough to preclude full daily activities. Able to walk without aid or rest for 100 m
6.0	Requires a walking aid-cane, crutch and so on to walk about 100 m with or without resting
6.5	Requires two walking aids-pair of canes, crutches and so on to walk about 20 m without resting
7.0	Unable to walk beyond approximately 5 m even with aid. Essentially restricted to wheelchair; though wheels self in standard wheelchair and transfers alone. Up and about in wheelchair some 12 h a day
7.5	Unable to take more than a few steps. Restricted to wheelchair and may need aid in transferring. Can wheel self but cannot carry on in standard wheelchair for a full day and may require a motorized wheelchair
8.0	Essentially restricted to bed or chair or pushed in wheelchair. May be out of bed itself much of the day. Retains many self-care functions. Generally has effective use of arms
8.5	Essentially restricted to bed much of day. Has some effective use of arms retains some self-care functions
9.0	Confined to bed. Can still communicate and eat
9.5	Confined to bed and totally dependent. Unable to communicate effectively or eat/swallow
10.0	Death due to MS

Now let us provide definitions for the MS dataset (synthetic MS data) with details, which are used in the applications of the algorithms in the following sections.

2.4.2.1 Obtaining synthetic MS dataset

The MS dataset consist of EDSS and MRI data belonging the MS dataset (139 × 112) which constitute 120 MS patients (with RRMS, SPMS, PPMS subgroups) and 19 healthy individuals. Level of disability in MS patients was determined by using the Extended Disability Status Scale (EDSS). Besides EDSS, MRIs of the patients were also used. Magnetic resonance images are obtained with a 1.5 Tesla device (Magnetom, Siemens Medical Systems, Erlangen, Germany). The brainstem, corpus callosum-periventricular region, including the upper cervical lesions in the three regions were included in the information. The MRIs have been evaluated by radiologists. MRIs of the patients were obtained from Hacettepe University Radiology Department and Primer Magnetic Resonance Imaging Center. The MS data have been obtained from Rana Karabudak, Professor (MD), the Director of Neuroimmunology Unit in Neurology Department, Ankara, Turkey.

A brief description of the contents in synthetic MS dataset is shown in Figure 2.10.

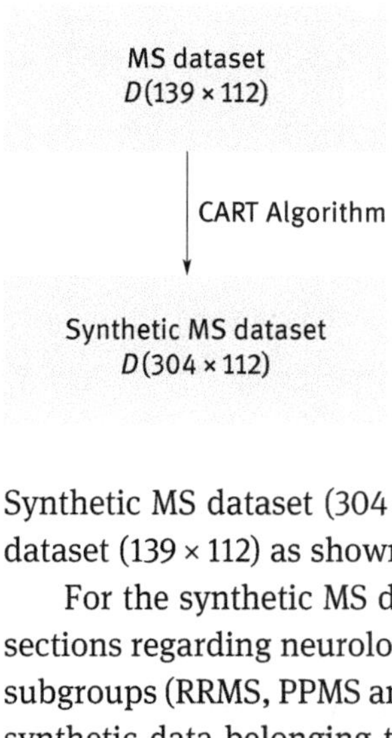

Figure 2.10: Obtaining the synthetic MS dataset.

Synthetic MS dataset (304 × 112) is obtained by applying CART algorithm to the MS dataset (139 × 112) as shown in Figure 2.10.

For the synthetic MS dataset $D(304 \times 12)$ addressed in the following application sections regarding neurology and radiology data, there are three clinical definite MS subgroups (RRMS, PPMS and SPMS) and a healthy group to be handled [25]. From the synthetic data belonging to these patients, the following subgroups of 76 patients with RRMS, 76 with PPMS, 76 with SPMS and 76 healthy individuals have been formed homogenously. Dataset is composed of MRI for the three sites of the brain with lesion size and lesion count and EDSS belonging to the MS patients.

Table 2.11 lists the relevant data for MRI for the three sites of the brain with lesion size and lesion count and EDSS belonging to the MS patients.

The matrix dimension of the MS dataset is defined as follows (see Figure 2.10).

For $i = 0, 1, ..., 304$ in $x: i_0, i_1, ..., i_{304}$; $j = 0, 1, ..., 112$ in $y: j_0, j_1, ..., j_{112}$, $D(304 \times 112)$ represents synthetic MS dataset.

Table 2.11: Synthetic MS dataset explanation.

Number of Subjects		*Attributes*			*Data Size*
76 RRMS 76 PPMS 76 SPMS 76 Healthy	304 individuals	from	**MRI** brain stem region corpus callosum - periventricular region upper cervical	Lesion size Lesion count → 111 numeric attributes	304×112
			EDSS score	1 nominal attribute	

The sample data values of synthetic MS dataset belonging to the individuals are listed in Table 2.12.

The first region represents the brain stem lesion count (MRI 1) and lesion size, while the second region represents corpus callosum lesion count (MRI 2) and lesion size. The third region represents upper cervical region lesion count (MRI 3) and lesion size. The number of lesions belonging to the brain stem is (0 mm, 1 mm, ..., 16 mm, ...) and for corpus callosum the number of lesions is (0 mm, 2.5 mm, ..., 8 mm, ..., 27 mm, ...). The number of lesions for the upper cervical region is (0 mm, 2.5 mm, ..., 8 mm, ..., 15 mm, ...).

CART algorithm has been used to obtain the synthetic MS dataset $D(139 \times 121)$ from MS dataset $D(304 \times 121)$.

Let us now explain how synthetic MS dataset $D(304 \times 112)$ is generated from MS dataset through a smaller scale of sample dataset by using CART algorithm shown in Figure 2.7 (for more details see Section 6.5).

Example 2.2 Sample MS dataset $D(20 \times 5)$ (see Table 2.13) is chosen from MS dataset $D(304 \times 112)$. CART algorithm is applied and sample synthetic dataset is derived. The steps for this process are provided in detail.

Let us apply the CART algorithm to the sample MS dataset below $D(20 \times 5)$ (see Table 2.14).

The steps of obtaining the sample synthetic MS dataset through the application of CART algorithm on the sample MS dataset $D(20 \times 5)$ are shown in Figure 2.7 (for more details see Section 6.5). The steps for the algorithm are given in detail.

Steps (1–6) The data belonging to each attribute in the sample MS dataset $D(20\times 5)$ (see Table 2.14) are ranked from lowest to highest (when the attribute values are same, only the one value is included in the ranking). For each attribute in the sample MS dataset, the average of median and the value following the median are calculated. x_{median} represents the median value of the attribute and $x_{\text{median}+1}$ represents the value after the median value (see eq. (2.4(a))).

The average of x_{median} and $x_{\text{median}+1}$ is calculated as $\frac{x_{\text{median}}+x_{\text{median}+1}}{2}$ (see eq. (2.1)). The threshold value calculation for each attribute (column; j: 1,2...,6) in the sample MS dataset $D(20 \times 5)$ listed in Table 2.7 is applied according to the following steps.

Table 2.12: An excel table to provide an example of synthetic MS data.

Individual (ID)	MS subgroup identified	EDSS	MRI of MS patients, regarding the three sides of the brain																
			First region						Second region					Third region					
			MRI 1	0 mm	...	1.5 mm	2 mm	...	MRI 2	0 mm	1 mm	...	...	MRI 3	0 mm	1 mm	...	15 mm	...
1	RRMS	6	10	0	...	0	0	...	17	0	0	...	...	0	0	0	...	0	...
2	RRMS	3.5	6	0	...	0	0	...	11	0	0	...	...	0	0	0	...	0	...
...	...	...	...	...	...	...	...	...	...	...	...	...	...	...	...	...	...	...	...
77	PPMS	5	5	0	...	0	0	...	11	0	0	...	...	0	0	0	...	0	...
78	PPMS	4.5	5	0	...	0	1	...	19	0	0	...	...	2	0	0	...	0	...
...	...	...	...	...	...	...	...	...	...	...	...	...	...	...	...	...	...	...	...
152	SPMS	7	9	0	...	0	0	...	13	0	0	...	...	4	0	0	...	0	...
153	SPMS	2.5	5	0	...	0	1	...	13	0	0	...	...	0	0	0	...	0	...
...	...	...	...	...	...	...	...	...	...	...	...	...	...	...	...	...	...	...	...
238	Healthy	0	0	0	...	0	...	...	0	0	...	...	...	0	0	0	...	0	...
239	Healthy	0	0	0	...	0	0	...	0	0	...	...	...	0	0	0	...	0	...
...	...	...	...	...	...	...	...	...	...	...	...	...	...	...	...	...	...	...	...
304	Healthy	0	0	0	...	0	...	...	0	0	...	...	...	0	0	0	...	0	...

Table 2.13: Sample MS dataset $D(20 \times 5)$ chosen from the MS dataset.

Class	Attributes of sample MS dataset $D(20 \times 5)$				
	EDSS	MRI 1	First region lesion size is 1 mm	MRI 2	Second region lesion size is 1 mm
RRMS	6	10	0	17	3
RRMS	3.5	6	1	11	2
SPMS	7	9	2	13	0
SPMS	7	5	0	11	0
RRMS	4	0	0	7	0
SPMS	6.5	1	0	6	0
SPMS	6.5	5	0	13	0
RRMS	4	8	1	9	1
RRMS	3.5	2	0	7	0
RRMS	5.5	3	0	11	0
RRMS	5	9	0	14	0
SPMS	6	9	0	9	0
SPMS	6	4	1	12	2
SPMS	6.5	4	0	10	0
RRMS	5.5	7	1	12	0
SPMS	7.5	8	0	9	1
SPMS	8	9	0	13	0
SPMS	6.5	0	0	0	0
SPMS	6.5	9	1	6	0
RRMS	3	4	0	9	0

Table 2.14: Results derived from the splitting of the binary groups for the attributes that belong to sample MS dataset (RRMS and SPMS subgroups).

Class	EDSS		MRI 1		First region lesion size is 1 mm		MRI 2		Second region lesion size is 1 mm	
	≥5.75	<5.75	≥5	<5	≥0.5	<0.5	≥10.5	<10.5	≥1.5	<1.5
RRMS	1	8	5	4	3	6	5	4	2	7
SPMS	11	0	7	4	3	8	5	6	1	10

The steps of the threshold value calculation for the *j*: 1: {*EDSS*} attribute are as follows:

In Table 2.13, {*EDSS*} values (line: *i*: 1,..., 20) in *j: 1 column* are ranked from lowest to highest (when the attribute values are same, only one value is included in the ranking).

$$\text{EDSS}_{\text{sort}}=\{3, 3.5, 4, 5, 5.5, 6, 6.5, 7, 7.5, 8\}$$

$$\text{EDSS}_{\text{median}} = 5.5 \qquad \text{(see eq. (2.4(a)))}$$

$$\text{EDSS}_{\text{median}+1} = 6$$

$$\overline{\text{EDSS}} = \frac{5.5+6}{2} = 5.75 \qquad \text{(see eq. (2.1))}$$

Two groups are formed for "EDSS ≥ 5.75" and "EDSS < 5.75."

The steps for calculating the threshold value of *j*: 2: {*MRI 1*} attribute are as follows:

In Table 2.13, {*MRI 1*} values (line: *i*: 1,..., 20) in *j*: 2 column are ranked from lowest to highest. (When the attribute values are same, only one value is included in the ranking.)

$$\text{MRI } 1_{(\text{sort})} = \{0, 1, 2, 3, 4, 5, 6, 7, 8, 9, 10\}$$

$$\text{MRI } 1_{\text{median}} = 5$$

$$\text{MRI } 1_{\text{median}+1} = 6$$

$$\overline{\text{MRI } 1} = \frac{5+6}{2} = 5.5$$

For "MRI 1 ≥ 5.5" and "MRI 1 < 5.5," two groups are formed.

The steps for calculating the threshold value of *j*: 3: {*First region lesion size is 1 mm*} attribute are as follows:

In Table 2.13, {*First region lesion size is 1 mm*} values (line: *i*: 1,..., 20) in *j*: 3 column are ranked from lowest to highest (when the attribute values are same, only one value is included in the ranking).

$$\text{First region lesion size is 1 mm}_{\text{sort}} = \{0, 1, 2\}$$

$$\text{First region lesion size is 1 mm}_{\text{median}} = 0$$

$$\text{First region lesion size is 1 mm}_{\text{median}+1} = 1$$

$$\overline{\text{First region lesion size is 1 mm}} = \frac{0+1}{2} = 0.5$$

"First region lesion size is 1 mm ≥ 0.5" and "First region lesion size is 1 mm < 0.5," and for these, two groups are formed.

The steps for calculating the threshold value of *j*: 4: {*MRI 2*} attribute are as follows:

In Table 2.13, *{MRI 2}* values (line: *i*: 1,..., 20) in *j*: 4 column are ranked from lowest to highest. (When the attribute values are same, only one value is included in the ranking.)

$$\text{MRI } 2_{\text{sort}} = \{0, 6, 7, 9, 10, 11, 12, 13, 14, 17\}$$

$$\text{MRI } 2_{\text{median}} = 10$$

$$\text{MRI } 2_{\text{median}+1} = 11$$

$$\overline{\text{MRI } 2} = \frac{10 + 11}{2} = 10.5$$

"MRI 2 ≥ 10.5" and "MRI 2 < 10.5." For these, two groups are formed.

The steps for calculating the threshold value of *j*: 5: {*Second region lesion size is 1 mm*} attribute are as follows:

In Table 2.13, *{Second region lesion size is 1 mm}* values (line: *i*: 1,..., 20) in *j*: 5 column are ranked from lowest to highest. (When the attribute values are the same, only one value is included in the ranking.)

$$\text{Second region lesion size is 1 mm}_{\text{sort}}: \{0, 1, 2, 3\}$$

$$\text{Second region lesion size is 1 mm}_{\text{median}} = 1$$

$$\text{Second region lesion size is 1 mm}_{\text{median}+1} = 2$$

$$\overline{\text{Second region lesion size is 1 mm}} = \frac{1 + 2}{2} = 1.5$$

"Second region lesion size is 1 mm ≥ 1.5" and "Second region lesion size is 1 mm < 1.5." For these, two groups are formed.

Steps (7–10) The Gini value of each attribute is calculated for the formation of decision tree from the sample MS dataset $D(20 \times 5)$. After all the attributes are split into two groups (see Table 2.14), calculation for Gini value for each attribute is conducted and the root of the tree is found (for more details see Section 6.5).

Table 2.14 lists the results derived from the splitting of the binary groups for the attributes ($j = 1, 2, \ldots, 6$) belonging to sample MS dataset (RRMS and SPMS subgroups).

The results in Table 2.14 for each attribute in sample MS dataset $D(20 \times 6)$ will be used in the Gini calculation.

Calculation of *Gini (EDSS)* value for *j*: 1 (see Section 6.5; eq. (6.8)) is as follows.

$$\text{Gini}_{\text{left}}(\text{EDSS}) = 1 - [(1/12)^2 + (11/12)^2] = 0.17$$

$$\text{Gini}_{\text{right}}(\text{EDSS}) = 1 - [(8/8)^2 + (0/8)^2] = 0$$

$$\text{Gini}(\text{EDSS}) = (12(0.17) + 8(0))/20 = 0.102$$

Calculation of *Gini (MRI 1)* value for *j*: 2 (see Section 6.5; eq. (6.8)) is as follows:

$$\text{Gini}_{\text{left}}(\text{MRI 1}) = 1 - [(5/12)^2 + (7/12)^2] = 0.51$$

$$\text{Gini}_{\text{right}}(\text{MRI 1}) = 1 - [(4/8)^2 + (4/8)^2] = 0.5$$

$$\text{Gini}(\text{MRI 1}) = (12(0.51) + 8(0.5))/20 = 0.506$$

Calculation of *Gini (First region lesion size is 1 mm)* value for *j*: 3 (see Section 6.5; eq. (6.8)) is as follows:

$$\text{Gini}_{\text{left}}(\text{First region lesion size is 1 mm}) = 1 - [(3/6)^2 + (3/6)^2] = 0.5$$

$$\text{Gini}_{\text{right}}(\text{First region lesion size is 1 mm}) = 1 - [(6/14)^2 + (8/14)^2] = 0.51$$

$$\text{Gini}(\text{First region lesion size is 1 mm}) = (6(0.5) + 14(0.51))/20 = 0.507$$

Calculation of *Gini* (*MRI 2*) value for *j*: 4 (see Section 6.5; eq. (6.8)) is as follows:

$$\text{Gini}_{\text{left}}(\text{MRI 2}) = 1 - [(5/10)^2 + (5/10)^2] = 0.5$$

$$\text{Gini}_{\text{right}}(\text{MRI 2}) = 1 - [(4/10)^2 + (6/10)^2] = 0.48$$

$$\text{Gini}(\text{MRI 2}) = (10(0.5) + 10(0.48))/20 = 0.49$$

Calculation of *Gini* (*Second region lesion size is 1 mm*) value for *j*: 5 (see Section 6.5; eq. (6.8)) is as follows:

$$\text{Gini}_{\text{left}}(\text{Second region lesion size is 1 mm}) = 1 - [(2/3)^2 + (1/3)^2] = 0.47$$

$$\text{Gini}_{\text{right}}(\text{Second region lesion size is 1 mm}) = 1 - [(7/17)^2 + (10/17)^2] = 0.51$$

$$\text{Gini}(\text{Second region lesion size is 1 mm}) = (3(0.47) + 17(0.51))/20 = 0.504$$

The lowest Gini values are Gini(EDSS) = 0.102. The lowest value in the Gini value is the root of the decision tree. The decision tree obtained from the sample MS dataset is shown in Figure 2.11.

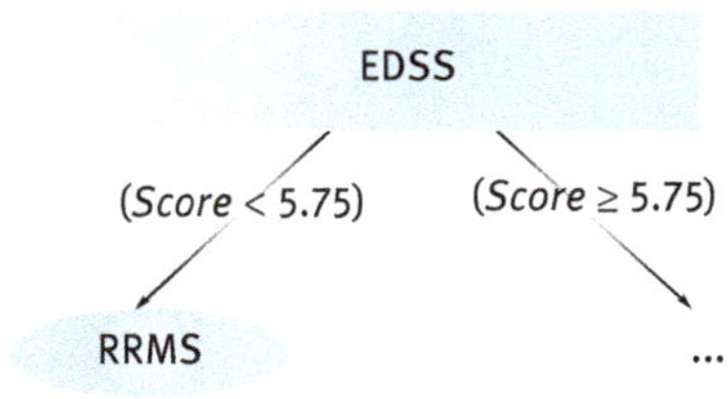

Figure 2.11: Decision tree obtained from sample MS dataset $D(20 \times 5)$.

The decision rule as obtained from sample MS dataset shown in Figure 2.11 is given in Rule 1:

Rule 1: If EDSS ≤ 5.75, the class is *RRMS*.

In Table 2.15, sample synthetic MS dataset $D(8 \times 6)$ in Rule 1 EDSS ≤ 5.75 (see Figure 2.11) is applied and obtained.

Table 2.15: Forming the sample synthetic MS dataset for EDSS ≤ 5.75.

EDSS	MRI 1	First region lesion size is 1 mm	MRI 2	Second region lesion size is 1 mm
3.5	6	1	11	0
4	0	0	7	0
4	8	1	9	1
3.5	2	0	7	0
5.5	3	0	11	0
5	9	0	14	0
5.5	7	1	12	0
3	4	0	9	0

Table 2.15 lists sample synthetic MS dataset $D(8 \times 6)$; *if* Rule 1 EDSS is ≤ 5.75 (see Figure 2.14), the class is *RRMS* (see Figure 2.11), which is applied and obtained.

In brief, sample MS dataset is chosen randomly from the MS dataset. CART algorithm is applied to the MS dataset. By applying this, we got the synthetic sample dataset. Sample synthetic dataset is composed of eight RRMS patients' data (see Table 2.16; it is possible to increase or reduce the entry number in the synthetic MS dataset).

2.4.3 Clinical psychology data: WAIS-R dataset

Mental activity and/or mental disorder is a challenging topic in clinical psychology. There are several methods to analyze the cognitive ability of adults and to detect

Table 2.16: If sample synthetic MS dataset is EDSS ≤ 5.75, the class *RRMS* is formed.

EDSS	MRI 1	First region lesion size is 1 mm	MRI 2	Second Region lesion size is 1 mm	Class
3.5	6	1	11	0	RRMS
4	0	0	7	0	RRMS
4	8	1	9	1	RRMS
3.5	2	0	7	0	RRMS
5.5	3	0	11	0	RRMS
5	9	0	14	0	RRMS
5.5	7	1	12	0	RRMS
3	4	0	9	0	RRMS

functions that they are not able to perform well. Among them one of the most expedient methods is the so-called WAIS given by David Wechsler in 1955 [28], following his preliminary work of 1939 [28–32]. The WAIS-R is a general intelligence test. It is defined as the global capacity of the individual to think rationally, to act in a purposeful way and to manage his/her environment in an effective manner. In line with this definition of intelligence as an aggregate of mental abilities or aptitudes, the WAIS-R includes 11 subtests that are divided into two parts, which are the verbal and the performance parts (Table 2.18 provides the details). The WAIS-R includes six verbal subtests and five performance subtests. The verbal tests are named as follows: information, comprehension, arithmetic, digit span, similarities and vocabulary. The performance subtests are as follows: picture arrangement, picture completion, block design, object assembly and digit symbol. The scores from this test are a verbal IQ, a performance IQ and a full-scale IQ (DM). The DM is a standard score, with a mean of 100 and a standard deviation of approximately 15. The parameters administered on the individuals on WAIS-R test are shown in Figure 2.12.

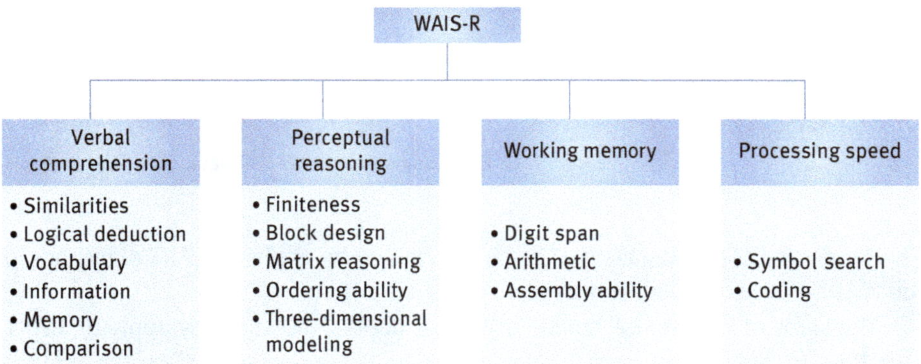

Figure 2.12: WAIS-R test parameters.

Table 2.17 provides elaborate explanations on the WAIS-R test parameters applied on the individuals. The parameters are the index, task, description and proposed abilities measured.

WAIS-R is used for the clinical psychology evaluation of the mental activity in individuals [30]. The attributes provided in Table 2.7 are determinants in making a decision whether the subjects are healthy or afflicted by the mental disorder [31, 32].

Now, let us explain in detail the synthetic WAIS-R data to be used in the algorithm applications for the next sections.

2.4.3.1 Obtaining the synthetic WAIS-R dataset

WAIS-R test contains the analysis of the adults' intelligence. WAIS-R dataset (600 × 21) is obtained from Majaz Moonis, Professor (M.D.) at the University of Massachusetts (UMASS), Department of Neurology and Psychiatry, Stroke Prevention Clinic Director, Worcester, USA.) In order to get synthetic WAIS-R dataset, CART algorithm was applied to the WAIS-R dataset. This test was administered on a total of 400 individuals, where 200 of these subjects were healthy, and the remaining 200 were afflicted by some disorder. The synthetic data have been formed in a homogenous manner. The detailed explanations about the WAIS-R dataset are provided in Table 2.18.

Synthetic WAIS-R dataset was obtained by applying CART algorithm to the WAIS-R dataset. A brief description of the contents in synthetic WAIS-R dataset is shown in Figure 2.13.

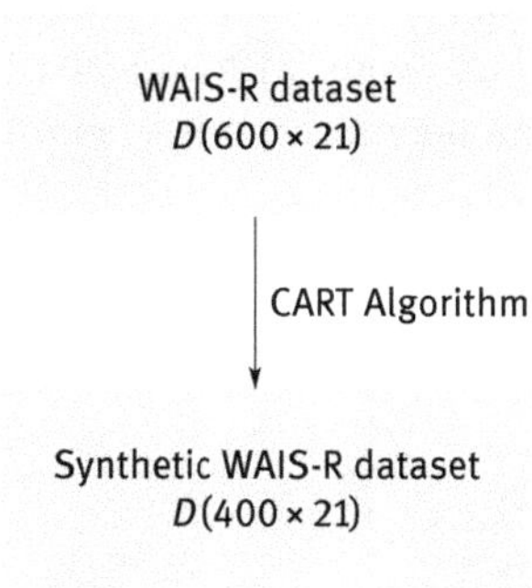

Figure 2.13: Obtaining synthetic WAIS-R data.

The synthetic WAIS-R dataset (400 × 21) is obtained by applying CART algorithm to the WAIS-R dataset (600 × 21) as shown in Figure 2.13.

The steps of getting the synthetic WAIS-R dataset (400 × 21) by applying CART algorithm (as shown in Figure 2.7; for more details see Section 6.5) on the WAIS-R dataset (600 × 21) are shown in Figure 2.13.

Table 2.17: The detailed descriptions of the WAIS-R test parameters [31, 32].

Index	Task	Description	Proposed abilities measured
Verbal comprehension	**Similarities**	Individual to indicate in what are similar two elements	It measures the ability of concrete and abstract thought and implies the ability to distinguish between the common features substantial ones from those irrelevant
	Vocabulary	Consists of a list of 35 nouns or verbs, arranged in order of increasing difficulty, of which the person must explain the meaning	This subtest assesses the degree of verbal learning accumulated from the subject, the extent of his knowledge and the quality of his reasoning
	Information	Comprised 29 general information questions	Derives from the Army Alpha battery and was entered in the scale to assess the extent of knowledge and interests of the subject
	Memory	Measuring its ability to absorb and withhold the information from the surrounding environment and remind them constantly	It is a good index of the qualities of the subject's intellectual ambitions, his degree of mental curiosity, understanding, thinking and long-term memory
	Comparison and comprehension	Replaced items	Revised scoring rules, added sample responses
	Block design	Consists in to re-create nine geometrical figure complexity increasing combining cubes of colored surfaces in white and red	It measures the ability to perceive, divide mentally into parts according to a new organization and reproduce abstract patterns
	Matrix reasoning	Retained 2 of 4 item types, added sample item for each type	Clarified 30-second guideline

Table 2.17 (continued)

Index	Task	Description	Proposed abilities measured
Working memory	**Digital span**	Divided into two parts, called direct and reverse repetition, in the first a series of digits is marked to person by increasing length and he has to repeat them in the same order in which they were handed down, while in the second he must repeat them in the opposite direction	The test evaluates the ability of the person to keep vigilant attention and direct it and keep it focused on desired tasks by inhibiting competing activities
	Arithmetic	Problems similar to those taught at the elementary and middle school; must be solved in mind, without the help of pencil and paper	It's necessary mental alertness and attention span and concentration
	Assembly ability	Four family figures divided into pieces. There are additional scores for speed	It reflects the ability to organize items available in facilities equipped with meaning
Processing speed	**Symbol search**	This subtest measures the skill to learn visuo-motor schemas by repetitive experiences and it indicates the degree of flexibility used in new learning situations	The test requires visuomotor dexterity, coordination of movements, speed and accuracy and it indicates the skill of immediate visual recall of informations
	Coding	Using a paper key, the examinee copies symbols that are paired with numbers within a specified time limit	This subtest measures processing speed, short-term visual memory, learning ability, psychomotor speed, visual perception, visual motor coordination, and visual scanning ability

Steps (1–6) The data belonging to each attribute in the WAIS-R dataset are ranked from lowest to highest (when the attribute values are the same, only one value is included in the ranking). For each attribute in the WAIS-R dataset, the average lof median and the value following the median are calculated, where x_{median} represents the median value of the attribute and $x_{median+1}$represents the value after the median value (see eq. (2.4(a))).

The average of x_{median} and $x_{median+1}$ is calculated as $\frac{x_{median}+x_{median+1}}{2}$ (see eq. (2.1)).

Steps (7–10) The Gini value of each attribute is calculated for the formation of decision tree from the dataset (see Section 6.5). The lowest Gini value calculated is the root of the decision tree. The WAIS-R dataset is $D(600 \times 21)$, and on the basis of the decision tree obtained and the rules derived from this decision tree, the synthetic WAIS-R dataset $D(400 \times 21)$ is obtained. In our study, the synthetic WAIS-R dataset $D(400 \times 21)$ formed by the application of CART algorithm on the WAIS-R dataset is $D(400 \times 1)$.

Table 2.18: Synthetic WAIS-R dataset explanation.

Number of subjects		Attributes		Data size
200 Patient 200 Healthy	400 individuals	School education (primary school, junior high school, vocational school, bachelor, master's degree, PhD)	2 Nominal attribute	400 × 21
		Gender		
		Age	19 Numeric Attribute	
		Memory		
		Similarity		
		Arithmetic		
		Vocabulary		
		Information		
		Comparison		
		Ordering ability		
		Assembly ability		
		Logical deduction		
		Finiteness or Closure		
		Three-dimensional modeling		
		Verbal information total (VIT)		
		Quantity information total (QIT)		
		Verbal information volume (VIV)		
		Quantity information verbal (QIV)		
		Full-scale intelligence quotient (DM)		
		Performance information verbal (VIP)		
		Quantity Information Performance (QIP)		

The matrix dimension of the WAIS-R dataset is defined as follows (see Figure 2.5): For $i = 0,1,2,...,400$ in x: $i_0, i_1, \ldots, i_{400}$; $j = 0, 1, \ldots, 21$ in y: $j_0, j_1, \ldots, j_{21}$ $D(400 \times 21)$ represents synthetic WAIS-R dataset.

The sample data values of synthetic WAIS-R dataset is listed in Table 2.19.

CART algorithm has been used to obtain the synthetic WAIS-R dataset $D(400 \times 21)$ from WAIS-R dataset $D(600 \times 21)$.

Let us now explain how the synthetic WAIS-R dataset $D(400 \times 21)$ is generated from WAIS-R dataset through a smaller scale of sample dataset by using CART algorithm as shown in Figure 2.7 (for more details see Section 6.5).

Example 2.3 Sample WAIS-R dataset $D(10 \times 10)$ (see Table 2.20) is chosen from WAIS-R dataset $D(600 \times 21)$. CART algorithm is applied and sample synthetic dataset is derived. The steps for this process are provided in detail.

Let us apply CART algorithm on the sample WAIS-R dataset $D(10 \times 10)$ (see Table 2.15).

Steps (1–6) The data belonging to each attribute in the WAIS-R dataset $D(10 \times 10)$ are ranked from lowest to highest (when the attribute values are same, only one value is included in the ranking; see Table 2.20). For each attribute in the WAIS-R dataset, the average of median and the value following the median is calculated. x_{median} represents the median value of the attribute and $x_{\text{median}+1}$ represents the value following the median value (see eq. (2.4(a))).

The average of x_{median} and $x_{\text{median}+1}$ is calculated as $\frac{x_{\text{median}} + x_{\text{median}+1}}{2}$ (see eq. (2.1)).

The calculation of threshold value for each attribute (column; j: 1,2...,10) in the sample WAIS-R dataset $D(10 \times 10)$ listed in Table 2.20 is applied according to the following steps.

The steps of the threshold value calculation for the j: 1: {*Age*} attribute are as follows.

In Table 2.20, {Age} values (line: i: 1,...,10) in j: 1 column (see Table 2.20) are ranked from lowest to highest. (When the attribute values are same, only one value is included in the ranking.)

$$\text{Age}_{\text{sort}} = \{17,\ 22,\ 23,\ 44,\ 53,\ 55,\ 56,\ 62,\ 63,\ 71\}$$

$$\text{Age}_{\text{median}} = 53$$

$$\text{Age}_{\text{median}+1} = 55$$

$$\overline{\text{Age}} = \frac{53 + 55}{2} = 54$$

For "Age ≥ 54" and "Age < 54," two groups are formed.

The steps of the threshold value calculation for the j: 2: {*Information*} attribute are as follows.

Table 2.19: An excel table to provide an example of synthetic WAIS-R dataset $D(400 \times 21)$.

Individual (ID)	Class	School Education	Gender	Age	…	Memory	Ordering Ability	Dimensional Ability	…	VIP	DM
1	Patient	1	Female	17	…	9	11	9	…	1.76	0.285
2	Patient	0	Female	24	…	7	14	8	…	2.48	−0.12
…	…	…	…	…	…	…	…	…	…	…	…
195	Patient	2	Female	27	…	9	7	8	…	2.32	0.00
196	Patient	1	Female	18	…	11	5	4	…	0.64	−0.26
…	…	…	…	…	…	…	…	…	…	…	…
201	Healthy	2	Male	60	…	6	11	6	…	1.92	−0.26
202	Healthy	3	Male	66	…	8	13	14	…	2.16	0.00
…	…	…	…	…	…	…	…	…	…	…	…
400	Healthy	4	Male	57	…	16	14	6	…	1.52	0.13

Table 2.20: Sample WAIS-R dataset chosen from WAIS-R dataset (with a dimension of $D(10 \times 10)$).

Class	The attributes of the Sample WAIS-R Data set D(10 × 10)									
	Age	**Information**	**Memory**	**Logical Deduction**	**Ordering ability**	**3 Dimensional Modeling**	**VIV**	**VIP**	**VIT**	**DM**
Patient	17	10	9	9	11	9	1.16	1.76	1.46	0.285
Patient	53	3	11	9	13	10	1.5	1.84	1.73	−0.21
Patient	44	2	3	6	8	5	2.1	1.52	1.86	0.07
Patient	23	7	6	9	9	10	1.25	1.56	1.45	−0.11
Patient	22	7	6	6	11	6	1	1.92	1.74	0.06
Patient	56	7	12	8	11	9	1.83	2.8	1.98	0.00
Healthy	62	12	16	15	10	12	1.5	1.68	2.72	0.00
Healthy	63	15	16	16	11	15	1.2	1.84	2.14	0.00
Healthy	71	12	13	9	11	8	0.83	1.92	2.3	0.00
Healthy	55	14	7	13	13	14	2.8	1.52	2.2	0.10

In Table 2.20, {Information} values (line: i: 1,. . ., 10) in *j: 2 column* (see Table 2.20) are ranked from lowest to highest. (When the attribute values are same, only one value is included in the ranking.)

$$\text{Information}_{\text{sort}} = \{2, 3, 7, 10, 12, 14, 15\}$$

$$\text{Information}_{\text{median}} = 10$$

$$\text{Information}_{\text{median}+1} = 10$$

$$\overline{\text{Information}} = \frac{10+12}{2} = 11$$

For “Information ≥ 11” and “Information < 11,” two groups are identified.

The steps of the threshold value calculation for the j: 3: {*Memory*} attribute are as follows.

In Table 2.20, {Memory} values (line: i: 1,. . ., 10) in j: 3 column are ranked from lowest to highest. (When the attribute values are same, only one value is included in the ranking.)

$$\text{Memory}_{\text{sort}} = \{3, 6, 7, 9, 11, 12, 13, 16\}$$

$$\text{Memory}_{\text{median}} = 9$$

$$\text{Memory}_{\text{median}+1} = 11$$

$$\overline{\text{Memory}} = \frac{9+11}{2} = 10$$

For “Memory ≥ 10” and “Memory < 10,” two groups are formed.

The steps of the threshold value calculation for the j: 4: {*Logical Deduction*} attribute are as follows.

In Table 2.20, {Logical deduction} values (line: i: 1,. . ., 10) in j: 4 column are ranked from lowest to highest. (When the attribute values are same, only one value is included in the ranking.)

$$\text{Logical deduction}_{\text{sort}} = \{6, 8, 9, 13, 15, 16\}$$

$$\text{Logical deduction}_{\text{median}} = 9$$

$$\text{Logical deduction}_{\text{median}+1} = 13$$

$$\overline{\text{Logical deduction}} = \frac{9+13}{2} = 11$$

For "Logical deduction ≥ 11" and "Logical deduction < 11," two groups are formed.

The steps of the threshold value calculation for the *j*: 5: {*Ordering ability*} attribute are as follows.

In Table 2.20, {Ordering ability} values (line: *i*: 1,. . ., 10) in *j*= 5 column are ranked from lowest to highest. (When the attribute values are same, only one value is included in the ranking.)

$$\text{Ordering ability}_{\text{sort}}: \{8, 9, 10, 11, 13\}$$

$$\text{Ordering ability}_{\text{median}} = 10$$

$$\text{Ordering ability}_{\text{median}+1} = 11$$

$$\overline{\text{Ordering ability}} = \frac{10+11}{2} = 10.50$$

For "Ordering ability ≥ 10.50" and "Ordering ability < 10.50," two groups are formed.

The steps of the threshold value calculation for the *j*: 6: {*Three-dimensional modeling*} attribute are as follows.

In Table 2.20, {Three-dimensional modeling} values (line: *i*: 1,. . ., 10) in *j*: 6 column are ranked from lowest to highest. (When the attribute values are same, only one value is included in the ranking.)

$$\text{Three}-\text{dimensional modeling}_{\text{sort}}: \{5,\ 6,\ 8,\ 9,\ 10,\ 12,\ 14,\ 15\}$$

$$\text{Three}-\text{dimensional modeling}_{\text{median}} = 9$$

$$\text{Three}-\text{dimensional modeling}_{\text{median}+1} = 10$$

$$\overline{\text{Three}-\text{dimensional modeling}} = \frac{9+10}{2} = 9.50$$

For "Three-dimensional modeling ≥ 9.50" and "Three-dimensional modeling < 9.50," two groups are formed.

The steps of the threshold value calculation for the *j*: 7: {*VIV*} attribute are as follows.

In Table 2.20, {VIV} values (line: *i*: 1,. . ., 10) in *j*: 7 column are ranked from lowest to highest. (When the attribute values are same, only one value is included in the ranking.)

$$\text{VIV}_{\text{sort}}: \{0.83,\ 1,\ 1.16,\ 1.20,\ 1.25,\ 1.50,\ 1.83,\ 2.10,\ 2.80\}$$

$$\text{VIV}_{\text{median}} = 1.25$$

$$\text{VIV}_{\text{median}+1} = 1.50$$

$$\overline{\text{VIV}} = \frac{1.25 + 1.50}{2} = 1.30$$

For "VIV ≥ 1.30" and "VIV < 1.30," two groups are formed.

The steps of the threshold value calculation for the *j*: 8: {*VIP*} attribute are as follows.

In Table 2.20, {VIP} values (line: *i*: 1,. . ., 10) in *j*: 8 column are ranked from lowest to highest. (When the attribute values are same, only one value is included in the ranking).

$$\text{VIV}_{\text{sort}}\text{: } \{1.52,\ 1.56,\ 1.68, 1.76, 1.84,\ 1.92,\ 2.80\}$$

$$\text{VIP}_{\text{median}} = 1.7$$

$$\text{VIP}_{\text{median}+1} = 1.84$$

$$\overline{\text{VIP}} = \frac{1.76 + 1.84}{2} = 1.80$$

For "VIP ≥ 1.80" and "VIP < 1.80," two groups are formed.

The steps of the threshold value calculation for the *j*: 9: {*VIT*} attribute are as follows.

In Table 2.20, {VIT} values (line: *i*: 1,. . ., 10) in *j*: 9 column are ranked from lowest to highest. (When the attribute values are same, only one is included in the ranking.)

$$\text{VIT}_{\text{sort}} = \{1.45,\ 1.46,\ 1.73,\ 1.74,\ 1.86,\ 1.98,\ 2.14,\ 2.20,\ 2.30,\ 2.72\}$$

$$\text{VIT}_{\text{median}} = 1.86$$

$$\text{VIT}_{\text{median}+1} = 1.98$$

$$\overline{\text{VIT}} = \frac{1.86 + 1.98}{2} = 1.90$$

For "VIT ≥ 1.90" and "VIT < 1.90," two groups are formed.

The steps of the threshold value calculation for the *j*: 10: {*DM*} attribute are as follows.

In Table 2.20, {DM} values (line: *i*: 1,. . ., 10) in *j*: 10 column are ranked from lowest to highest. (When the attribute values are same, only one value is included in the ranking.)

$$\text{DM}_{\text{sort}} = \{-0.21,\ -0.11,\ 0,\ 0.06,\ 0.07,\ 0.10,\ 0.285\}$$

$$\mathrm{DM}_{\mathrm{median}} = 0.06$$

$$\mathrm{DM}_{\mathrm{median}+1} = 0.07$$

$$\overline{\mathrm{DM}} = \frac{0.06 + 0.07}{2} = 0.065$$

For "DM ≥ 0.065" and "DM < 0.065," two groups are formed.

Steps (7–10) The Gini value of each attribute is calculated for the formation of decision tree of the sample WAIS-R dataset $D(10 \times 10)$. After being split into two groups, for each attribute, Gini value calculation is done and the root of the tree is found (for more details see Section 6.5).

Table 2.21 lists the results derived from the splitting of the binary groups for the attributes ($j = 1, 2, \ldots, 6$) belonging to sample WAIS-R dataset (patient and healthy class).

The results provided in Table 2.17 for each attribute in the sample WAIS-R dataset $D(10 \times 10)$ will be used for the Gini calculation.

The calculation of *Gini(Age)* for *j*: 1 (see Section 6.5; eq. (6.8)) is as follows:

$$\mathrm{Gini}_{\mathrm{left}}(\mathrm{Age}) = 1 - [(1/5)^2 + (4/5)^2] = 0.32$$

$$\mathrm{Gini}_{\mathrm{left}}(\mathrm{Age}) = 1 - [(1/5)^2 + (4/5)^2] = 0.32$$

$$\mathrm{Gini}(\mathrm{Age}) = (5(0.32) + 5(0))/10) = 0.16$$

The calculation of *Gini(Information)* value for *j*: 1 (see Section 6.5; eq. (6.8)) is as follows:

$$\mathrm{Gini}_{\mathrm{left}}(\mathrm{Information}) = 1 - [(0/4)^2 + (4/4)^2] = 0$$

$$\mathrm{Gini}_{\mathrm{right}}(\mathrm{Information}) = 1 - [(6/6)^2 + (0/6)^2] = 0$$

$$\mathrm{Gini}(\mathrm{Information}) = (4(0) + 6(0))/10 = 0$$

The calculation of *Gini(Memory)* value for *j*: 3 (see Section 6.5; eq. (6.8)) is as follows:

$$\mathrm{Gini}_{\mathrm{left}}(\mathrm{Memory}) = 1 - [(2/5)^2 + (3/5)^2] = 0.48$$

$$\mathrm{Gini}_{\mathrm{right}}(\mathrm{Memory}) = 1 - [(4/5)^2 + (1/5)^2] = 0.32$$

$$\mathrm{Gini}(\mathrm{Memory}) = (5(0.48) + 5(0.32))/10 = 0.4$$

The calculation of *Gini(Logical deduction)* value for *j*: 4 (see Section 6.5; eq. (6.8)) is as follows:

$$\mathrm{Gini}_{\mathrm{left}}(\mathrm{Logical\ deduction}) = 1 - [(0/3)^2 + (3/3)^2] = 0$$

Table 2.21: Results derived from the splitting of the binary groups for the attributes that belong to sample WAIS-R dataset (patient and healthy class).

Class	Age		Information		Memory		Logical Deduction		Ordering Ability		3 Dimensional Modeling		VIV		VIP		VIT		DM	
	≥54	<54	≥11	<11	≥10	<10	≥11	<11	≥10.5	<10.5	≥9.5	<9.5	≥1.3	<1.3	≥1.8	<1.8	≥1.9	<1.9	≥0.065	<0.065
Patient	1	5	0	6	2	4	0	6	4	2	2	4	3	3	3	3	1	5	2	4
Healthy	4	0	4	0	3	1	3	1	3	1	3	1	2	2	2	2	4	0	1	3

$$\text{Gini}_{\text{right}}(\text{Logical deduction}) = 1 - [(6/7)^2 + (1/7)^2] = 0.26$$

$$\text{Gini}(\text{Logical deduction}) = (3(0) + 7(0.26))/10 = 0.182$$

The calculation of *Gini(Ordering ability)* value for *j*: 5 (see Section 6.5; eq. (6.8)) is as follows:

$$\text{Gini}_{\text{left}}(\text{Ordering ability}) = 1 - [(4/7)^2 + (3/7)^2] = 0.5$$

$$\text{Gini}_{\text{right}}(\text{Ordering ability}) = 1 - [(2/3)^2 + (1/3)^2] = 0.46$$

$$\text{Gini}(\text{Ordering ability}) = (7(0.5) - 3(0.46))/10 = 0.49$$

The calculation of *Gini(Three-dimensional modeling)* value for *j*: 6 (see Section 6.5; eq. (6.8)) is as follows:

$$\text{Gini}_{\text{left}}(\text{Three - dimensional modeling}) = 1 - [(2/5)^2 + (3/5)^2] = 0.48$$

$$\text{Gini}_{\text{right}}(\text{Three - dimensional modeling}) = 1 - [(4/5)^2 + (1/5)^2] = 0.32$$

$$\text{Gini}(\text{Three - dimensional modeling}) = (5(0.48) + 5(0.32))/10 = 0.4$$

The calculation of *Gini(VIV)* value for *j*: 7 (see Section 6.5 eq. (6.8)).

$$\text{Gini}_{\text{left}}(\text{VIV}) = 1 - [(3/5)^2 + (2/5)^2] = 0.48$$

$$\text{Gini}_{\text{right}}(\text{VIV}) = 1 - [(3/5)^2 + (2/5)^2] = 0.48$$

$$\text{Gini}(\text{VIV}) = (5(0.48) + 5(0.48))/10 = 0.48$$

The calculation of *Gini(VIP)* value for *j*: 8 (see Section 6.5; eq. (6.8)) is as follows:

$$\text{Gini}_{\text{left}}(\text{VIP}) = 1 - [(3/5)^2 + (2/5)^2] = 0.48$$

$$\text{Gini}_{\text{right}}(\text{VIP}) = 1 - [(3/5)^2 + (2/5)^2] = 0.48$$

$$\text{Gini}(\text{VIP}) = (5(0.48) + 5(0.48))/10 = 0.48$$

The calculation of *Gini(VIT)* value for *j*: 9 (see Section 6.5; eq. (6.8)) is as follows:

$$\text{Gini}_{\text{right}}(\text{VIT}) = 1 - [(5/5)^2 + (0/5)^2] = 0$$

$$\text{Gini}_{\text{left}}(\text{VIT}) = 1 - [(1/5)^2 + (4/5)^2] = 0.32$$

$$\text{Gini}(\text{VIT}) = (5(0.32) + 5(0))/10 = 0.16$$

The calculation of *Gini(DM)* value for *j*: 10 (see Section 6.5; eq. (6.8)) is as follows:

$$\text{Gini}_{\text{left}}(\text{DM}) = 1 - [(2/3)^2 + (1/3)^2] = 0.46$$

$$\text{Gini}_{\text{right}}(\text{DM}) = 1 - [(4/7)^2 + (3/7)^2] = 0.5$$

$$\text{Gini}(\text{DM}) = (3(0.46) + 7(0.5))/10 = 0.49$$

The lowest Gini value is *Gini(Information)* = 0. The lowest value in the Gini values is the root of the decision tree. The decision tree obtained from the Sample WAIS-R dataset is provided in Figure 2.14.

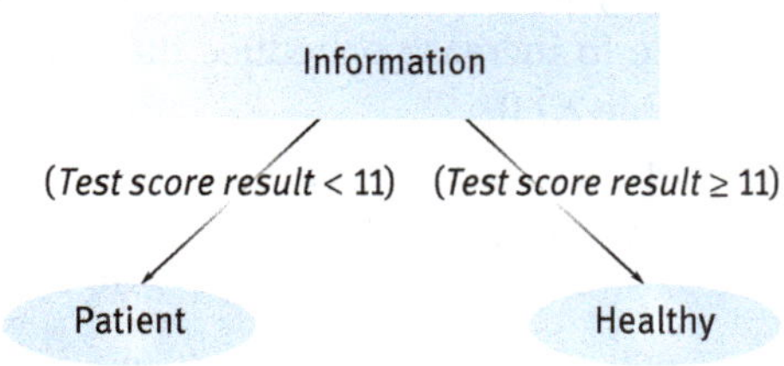

Figure 2.14: The decision tree obtained from sample WAIS-R dataset $D(10 \times 10)$.

The decision rule as obtained from sample WAIS-R dataset in Figure 2.14 is given as Rules 1 and 2:

Rule 1: If *Information* ≥11, class is "*healthy.*"
Rule 2: If *Information* < 11, class is "*patient.*"

Sample synthetic WAIS-R dataset $D(3 \times 10)$ in Rule 1 *Information* ≥ 11 and in Rule 2 *Information* <11 (see Figure 2.14) is applied and obtained accordingly (Table 2.19).

In Table 2.19, sample synthetic WAIS-R dataset $D(3 \times 10)$ is obtained by applying the following: Rules 1 and 2 (see Figure 2.14).

In brief, sample WAIS-R dataset is randomly chosen from the WAIS-R dataset. CART algorithm is applied to the WAIS-R dataset. By applying this, we can get

Table 2.22: Forming the sample synthetic WAIS-R dataset for information ≥ 11 and information <11 as per Rule 1 and Rule 2.

Age	Information	Memory	Logical deduction	Ordering ability	3 Dimensional modeling	VIV	VIP	VIT	DM
24	10	13	9	12	12	1.5	1.04	1.45	0.11
32	12	12	12	13	15	1	2.4	1.63	0.14
20	4	6	2	7	1	0.9	1.84	1.54	-0.06

Table 2.23: Forming the sample synthetic WAIS-R dataset for Rule 1: *Information* ≥11, then class is "*healthy*" and Rule 2: *Information* <11, then class is "*patient as per Table 2.22.*"

Age	Information	Memory	Logical deduction	Ordering ability	3 Dimensional modeling	VIV	VIP	VIT	DM	Class
24	10	13	9	12	12	1.5	1.04	1.45	0.11	Patient
32	12	12	12	13	15	1	2.4	1.63	0.14	Healthy
20	4	6	2	7	1	0.9	1.84	1.54	-0.06	Patient

synthetic sample dataset (see Table 2.22). Sample synthetic dataset is composed of two patients and one healthy data (it is possible to increase or reduce the entry number in the synthetic WAIS-R dataset) (see Table 2.23).

Note that for the following sections, synthetic datasets are used for those in medical area and real datasets are used for those in economy with regard to applications and examples.

2.5 Basic statistical descriptions of data

For a successful preprocessing, it is important to have an overview and general picture of the data. This can be done by some descriptive variables that can be obtained by some elementary or basic statistics. *Mean, median, mode* and midrange are *measures of central tendency*, which measure the location of the middle or center of a data distribution. Apart from the central tendency of the dataset, it is also important to have an idea about the *dispersion of the data*, which refers to how the data spread out. The most frequently used ones in this area are the *range, quartiles* and *interquartile range*; *boxplots*; the *variance* and *standard deviation* of the data. Data summaries and distributions as other display tools include bar charts, quantile plots, quantile–quantile plots and histograms and scatter plots [33–39].

2.5.1 Central tendency: Mean, median and mode

$x_1, x_2, \ldots, x_N$ is the set of N observed values or observations for X. The values here may also be referred to as the dataset (for$D$). If the observations were plotted, where would most of the values fall? This would elicit an idea of the data central tendency. Measures of central tendency comprise the mean, median, mode and midrange [33–39].

The most effective and common numeric measure of the "center" of a set of data is the (arithmetic) mean.

Definition 2.5.1.1. The *arithmetic mean* of this set of values is given as follows:

$$\bar{x} = \frac{\sum_{i=1}^{N} x_i}{N} = \frac{x_1 + x_2 + \ldots + x_N}{N} \tag{2.1}$$

Example 2.4 Suppose that we have the following 10 values $X = \{2.5, 3.5, 3.5, 4, 4, 4.5, 5, 5, 6, 6.5\}$. Using eq. (2.1), the mean is given as:

$$\bar{x} = \frac{2.5 + 3.5 + 3.5 + 4 + 4 + 4.5 + 5 + 5 + 6 + 6.5}{10} = \frac{44.5}{10} = 4.45$$

Sometimes each value x_i in a set may be linked with a weight w_i for $i = 1, \ldots, N$.. Here the weights denote the significance or frequency of occurrence according to their values in the relevant order. In this case, we can have a look at the following definition.

Definition 2.5.1.2. Weighted arithmetic mean or *weighted average*

$$\bar{x} = \frac{\sum_{i=1}^{N} w_i x_i}{\sum_{i=1}^{N} w_i} = \frac{w_1 x_1 + w_2 x_2 + \ldots + w_N x_N}{w_1 + w_2 + \ldots + w_N} \tag{2.2}$$

The mean is the only useful quantity for the description of a dataset, but it does not always offer the optimal way of measuring the center of the data. A significant problem related to the mean is its sensitivity to outlier values, also known as extreme values. It is possible to see cases in which even small extreme values can corrupt the mean. For instance, the mean income for bank employees may be significantly raised by a manager whose salary is extremely high. Similarly, the mean score of a class in a college course in a test could be pulled down slightly by a few very low scores to counterbalance the effect by a small number of outliers. In order to avoid such problems, sometimes the *trimmed mean* is used.

Definition 2.5.1.3. Trimmed mean is defined as the mean obtained after one chops off the values that lie at the high and low extremes. For example, it is possible to arrange the values observed for credits and remove the top and bottom 3% prior to calculating the mean. It should be noted that one is supposed to avoid trimming a big portion (such as 30%) at both ends; such a procedure may bring about a loss of valuable information.

For skewed or asymmetric data, median is known to be a better measure of the center of data.

Definition 2.5.1.4. Median is defined as the middle value in a set of ordered data values. It is the value that separates the higher half of a dataset from the lower half. In the fields of statistics and probability, the median often signify numeric data, and, it is possible to extend the concept of ordinal data. Assume that a given dataset of N values for X attribute is arranged in growing order. If N is odd, the median is the value lying at the middle position of the ordered set. If N is even, the median is not exceptional.. It is given by the two values that lie in the middle. If X is a numeric attribute, the median is considered by rule and handled as the mean of the values that lie in the very middle [33–39].

Example 2.5 *Median.* By using the data of Example 2.4 we can see that the observations are already ordered but even then the median is not unique. This means that the median may be any value that lie in the very middle between values 4 and 4.5. According to this example, it is within the sixth and seventh values in the list. It is a rule to assign the average of the two middle values as the median, which gives us: $\frac{4+4.5}{2} = \frac{8.5}{2} = 4.25$. From this, the median is 4.25.

In another scenario, let us say that only the first 11 values are in the list. With an odd number of values, the median is the middle most value. In the list, it is the sixth value with a value of 4.

The median is computationally difficult to calculate when there is a large number of observations. For numeric attributes, Boolean value can be calculated easily. Data may be grouped in intervals based on their x_i data values and the frequency (number of data values) of each interval may be known. For example, the individuals whose EDSS score is between 2 and 8.5. The interval that contains the median frequency is the *median interval*. In this case, we can estimate the median of the whole dataset (the median personal credit) by interpolation by using the following formula:

$$\text{median} = \text{L}_1 + \left(\frac{N/2 - (\sum \text{freq})_1}{\text{freq}_{\text{medium}}}\right)\text{width} \tag{2.3}$$

where L_1 is the lower boundary of the median interval, N is the number of values in the whole dataset, $(\sum \text{freq})_1$ is the frequencies' sum that belongs to the intervals that are lower than the median interval, $\text{freq}_{\text{median}}$ is the median interval's frequency and width i refers to the width of the median interval.

Median is the value that falls in the middle when the values are ranked from largest to smallest for ungrouped data. Median is the value that splits the array into two.

Equation 2.4(a) and (b) is applied depending on whether the number of elements in an array is even or odd.

$$x_{\text{median}} = x_{\left(\frac{N}{2}\right)} + x_{\left(\frac{N+1}{2}\right)} N \text{ is even} \tag{2.4a}$$

$$x_{\text{median}} = x_{\left(\frac{N+1}{2}\right)} N \text{ is odd} \tag{2.4b}$$

The *mode* is another measure of central tendency.

Definition 2.4.1.5. The mode is the value that occurs most frequently in the set. For this reason, it can be identified for qualitative and quantitative attributes. The greatest frequency may correspond to a number of different values, which brings about more than one mode. Datasets with one, two and three modes are called *unimodal, bimodal* and *trimodal*, respectively. Overall, a dataset that has two or more modes is *multimodal.* At the other extreme of the spectrum, if each data value is seen once only, we can say that there is no mode at all [7, 33–39].

Example 2.6 *Mode.* The data from Example 2.5 are bimodal. The two modes are 3.5, 4 and 5.

The following empirical relation is obtained for unimodal numeric data that are skewed moderately or happen to be asymmetrical [40].

$$\text{mean} - \text{mode} \approx 3 \times (\text{mean} - \text{median}) \tag{2.5}$$

This suggests that the mode for unimodal frequency curves, moderately skewed, can be estimated easily if we know the mean and median value.

We can also use the midrange for the assessment of the central tendency of a numeric dataset. *Midrange* is known to be the average of the largest and smallest values in the set. This measurement is convenient and easy kind when the SQL aggregate functions, max() and min(), are used [34].

Example 2.7: *Midrange.* The midrange of the data provided in Example 2.4 is $\frac{2.5+6.5}{2} = 4.5$

In a unimodal frequency curve with perfect *symmetric* data distribution, the mean, median and mode are all at the same center value (Figure 2.15(a)). Data in most applications in real life are not symmetric. Instead, they may be *positively skewed* (the mode occurs at a value smaller than the median (see Figure 2.15(b)), or *negatively skewed* (the mode occurs at a value greater than the median (see Figure 2.15(c)) [35–36].

2.5.2 Spread of data

The spread or dispersion of numeric data is an important measure to be assessed in statistics. The most common measures of dispersion are range, quantiles, quartiles,

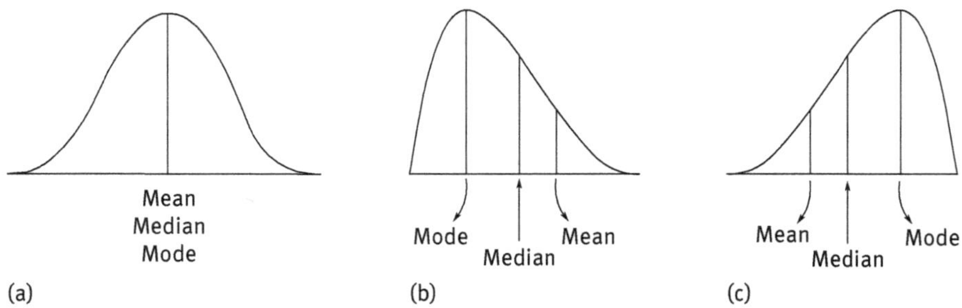

Figure 2.15: Mean, median and mode of symmetric versus positively and negatively skewed data: (a) symmetric data, (b) positively skewed data and (c) negatively skewed data.

percentiles and the interquartile range. Box plot is also another useful tool to identify outliers. Variance and standard deviation are the other measures that show the spreading of a data distribution.

Let $x : x_1, x_2, \ldots, x_N$ be a set of observations for numeric attribute. The *range* is the difference between the largest (max()) and smallest (min()) values.

$$\text{range} : \max() - \min() \tag{2.6}$$

Quantiles are points that are taken at regular intervals of a data spread or distribution. It is found by dividing the interval into equal size sequential sets.

The quartiles indicate a distribution's center, spread and shape. The *first quartile*, as denoted by Q_1, is the 25th percentile. It cuts off the lowest 25% of the data. The *third quartile*, as denoted by Q_3, is the 75th percentile. It cuts off the highest 25% or the lowest 75% of the data. The second quartile is known to be the 50th percentile (Q_2, *Median*). It is also the median, which yields the center of the data distribution [7, 33–39].

In order to find the middle half of the data, the distance between the first and third quartiles is seen as a simple measure of spread, which presents the range that is covered. *Interquartile range* (*IQR*) is the distance defined as follows:

$$\text{IQR} = Q_3 - Q_1 \tag{2.7}$$

Boxplot is a common technique used for the visualization of a distribution (see Figure 2.16). The ends of the box are at the quartiles. The box length is the interquartile range. A line within the box marks the median. Two lines are called *whiskers*, and these are outside the box extending to the minimum and maximum observations.

Example 2.8: Let us calculate the mean, median, interquartile range and min() and max() values of the numerical array $X = [1, 2, -3, 4, 5, -6, 2, -3, 5]$. Let us also draw its boxplot.

The mean calculation of the numerical array $X = [1, 2, -3, 4, 5, -6, 2, -3, 5]$ is based on eq. (2.1):

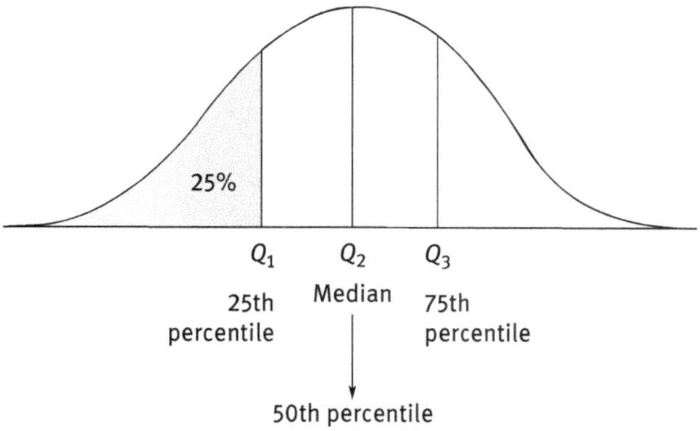

Figure 2.16: A plot of the data distribution for some attribute *X*. The quantiles that are plotted are quartiles. The three quartiles split the distribution into four consecutive subsets (each having an equal size). The second quartile corresponds to the median.

$$\bar{X} = \frac{\sum_{i=1}^{N} x_i}{N} = \frac{x_1 + x_2 + \ldots + x_N}{N}$$

$$\bar{X} = \frac{1 + 2 + (-3) + 4 + 5 + (-6) + 2 + (-3) + 5}{9} = 0.7$$

Median calculation is based on eq. (2.4(a))

$$x_{\text{median}} = x_{\left(\frac{N+1}{2}\right)}$$

$$x_{\text{median}} = x_{\left(\frac{9+1}{2}\right)} = x_{(5)} = 5$$

Range calculation is based on eq. (2.6).

Let us rank the numerical array $X = [1, 2, -3, 4, 5, -6, 2, -3, 5]$ from small to large values:

$$X = [-6, -3, -3, 1, 2, 2, 4, 5, 5]$$

interquartile range: $(\text{IQR}) = \max() - \min() = 5 - (-6) = 11$
max() value: 5
min() value: –6

Since boxplot is a plot of the distribution, so you may give a simple figure (see Figure 2.17) to show the boxplot for a simple distribution.

It is possible to make use of boxplots while making comparisons of several sets of compatible data. A method to calculate about boxplots is by making the computation in O(*nlogn*) time.

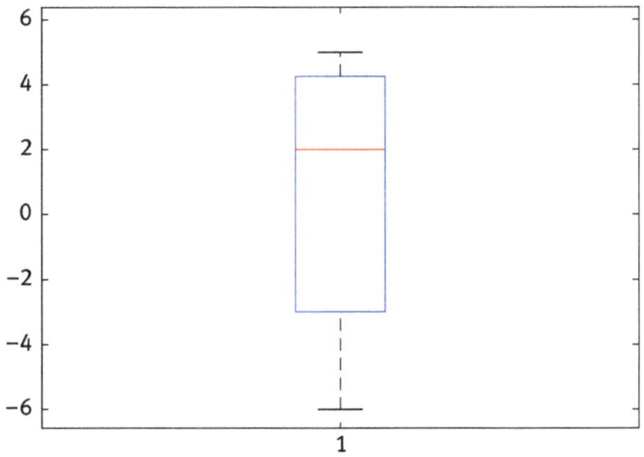

Figure 2.17: Boxplot of X vector.

- The ends of the box are usually at the quartiles, so that the box length is the interquartile range.
- A line marks the median within the box.
- Two lines are called *whiskers*. These lines are outside the box extending to the smallest (*minimum*) and largest (*maximum*) observations.

Example 2.9 *Boxplot.* Figure 2.18 presents the boxplots for the EDSS score and lesion counts that belong to the MR images of the patients within a given time interval and are given in Table 2.10. The initial value for EDSS is seen as 0 and for the third quartile the value for this is 5.

The figure presents the boxplot for the same data with the maximum whisker length specified as 5.0 times the interquartile range. Data points beyond the whiskers are represented using +.

2.5.3 Measures of data dispersion

These measures in a data distribution are spread out.

Let $x_1, x_2, \ldots, x_N$ be the values of a numeric attribute X, and the *variance* of N observations is given by the following equation:

$$\sigma^2 = \frac{1}{N}\sum_{i=1}^{N}(x_i - \bar{x})^2 = \left(\frac{1}{N}\sum_{i=1}^{N}x_i^2\right) - \overline{(x)}^2 \tag{2.8}$$

Here, $\bar{x}$ is the mean value of the observations. The *standard deviation* $\sigma$of the observations is the square root of the variance σ^2. A low standard deviation is a proof that the data observations are very close to the mean, and, a high standard deviation shows that the data are spread out over a large range of values [40–42].

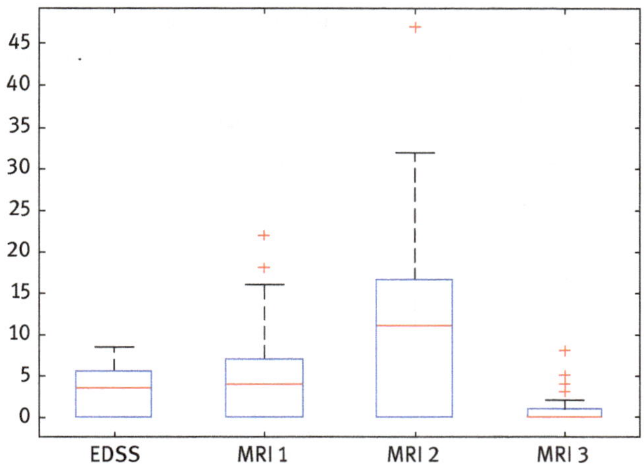

Figure 2.18: Boxplot depiction of EDSS and MRI lesion diameters of MS dataset based on a certain time interval.

Example 2.10 By using eq. (2.1) we found in Example 2.4 that the mean is $\bar{x} = 4.45$. In order to determine the variance and standard deviation of the data from that example, we set $N = 10$ and use eq. (2.8).

$$\sigma^2 = \frac{1}{10}\left(2.5^2 + 3.5^2 + 3.5^2 + 4^2 + 4^2 + 4.5^2 + 5^2 + 5^2 + 6^2 + 6.5^2\right) - 4.45^2 \cong 0.1$$

$$\sigma = \sqrt{0.1} \cong 0.316$$

The basic properties of the standard deviation are as follows:

It measures how the values spread around the mean.

$\sigma = 0$ when there is no dispersion, which means that all observations share the same value.

For this reason, the standard deviation is considered to be a sound indicator of the dataset dispersion [43].

2.5.4 Graphic displays

The graphic displays of basic statistical descriptions include histograms, scatter plots, quantile plot and quantile–quantile plot.

Histograms are extensively used in all areas where statistics is used. It displays the distribution of a given attribute X. If X is nominal, a vertical bar is drawn for each known value of this attribute X. The height of the bar shows the frequency (how many times it occurs) of the X. The graph is also known as *bar chart*. Histograms are preferred options when X is numeric. The subranges are defined as *buckets* or *bins*, which are separate sections for data distribution for X attribute. The width of the bar corresponds to the range of a bucket. The buckets have equal width.

Scatter plot is another effective graphical method to determine a possible relationship trend between two attributes that are numeric in nature. In order to form a scatter plot, each pair of values is handled as a pair of coordinates, and plotted as points in the plane. Figure 2.19 illustrates a scatter plot for the set of data in Table 2.22.

The scatter plot is a suitable method to see clusters of points and outliers for bivariate data. It is also useful to visualize some correlation.

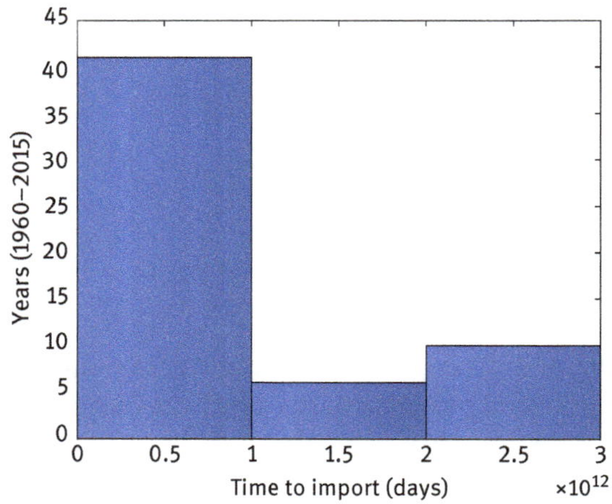

Figure 2.19: Histogram for Table 2.11.

Definition 2.5.4.1. Two attributes, X, and Y, are said to be correlated (more information in Chapter 3) if the variability of one attribute has some influence on the variability of the other.

Correlations can be positive, negative or null. In this case, the two attributes are said uncorrelated. Figure 2.15 provides some examples of positive and negative correlations. If the pattern of the plotted points is from lower left to upper right, this is an indication that values of X go up when the values of Y go up as well. is from lower left to upper right, this is an indication that values of X go up when the values of Y go up as well. As it is known, this kind of pattern suggests a positive correlation (Figure 2.21 (a)). If the pattern of plotted points is from upper left to lower right, with an increase in

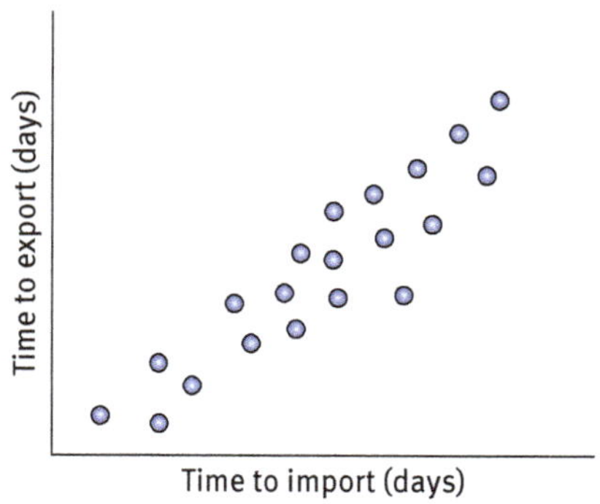

Figure 2.20: Scatter plot for Table 2.10: time to import–time to export for Italy.

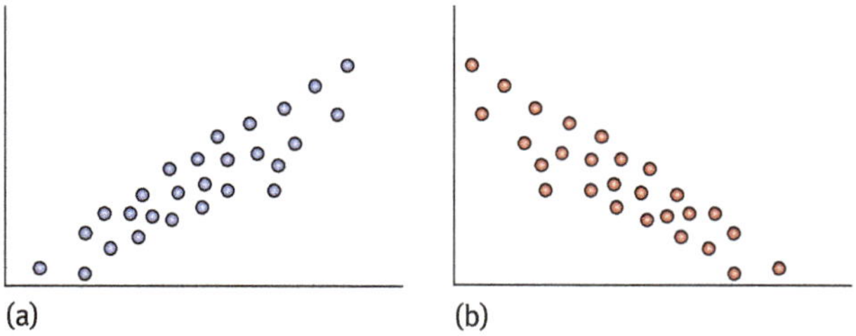

Figure 2.21: Scatter plot can show (a) positive or (b) negative correlations between attributes.

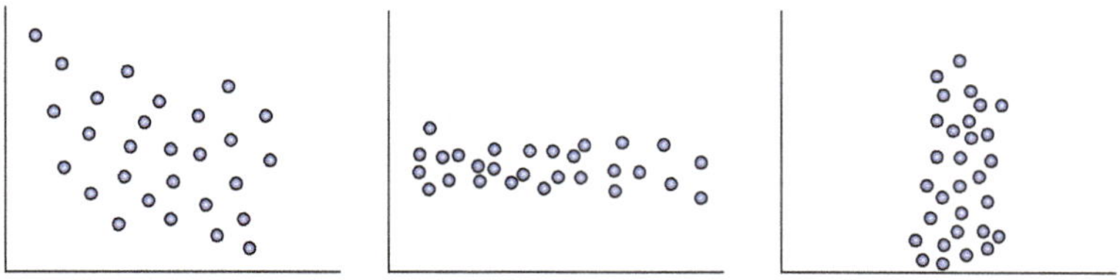

Figure 2.22: Three cases where there is no observed correlation between the two plotted attributes in each of the datasets.

values of X the values of Y go down. A negative correlation is seen in Figure 2.21 (b). Figure 2.22 Figure 2.24 indicates some cases in which no correlation relationship between the two attributes exists in the datasets given [42, 43].

Quantile plot: A *quantile plot* is an effective and straightforward way to know about a univariate data distribution. It displays all the data for the given attribute. It also plots quantile information. Let x_i, for $i = 1$ to N, be the data arranged in an increasing order, so we see x_1 as the smallest observation and x_N is the largest for numeric attribute X. Each observation, x_i, is coupled with a percentage, f_i. This shows that on average $f_i \times 100\%$ of the data are lower than the value, x_i. The word "on average" is used here since a value with exactly a fraction, f_i, of the data below x_i may not be found here. As explained earlier, the 0.25 percentile corresponds to quartile Q_1, the 0.50 percentile is the median and the 0.75 percentile is Q_3.

Let

$$f_i = \frac{i - 0.5}{N} \tag{2.9}$$

The numbers go up in equal steps of $1/N$, varying from $1/2N$ (as slightly above 0) to $1 - 1/2N$ (as slightly below 1). x_i is plotted against f_i on a quantile plot. This enables us to compare different distributions based on their quantiles. For instance, the quantile plots of sales data for two different time periods, their Q_1, median, Q_3 and other f_i values can be compared [44].

Example 2.11 Quantile plot. Figure 2.23 indicates a quantile for the time to import (days) and Time to exports (days) data for Table 2.8.

Quantile–quantile plot, also known as *q–q plot,* plots the quantiles of one univariate distribution against the corresponding quantiles of another. It is a powerful visualization tool because it enables the user to view if there is a shift going from one distribution to the other.

Let us assume that we have two sets of observations for the attribute or variable *interest* (*income* and *deduction*) that are obtained from two dissimilar branch locations. $x_1, \ldots, x_N$ is the data from the first branch, and $y_1, \ldots, y_M$ is the data from the second one. Each dataset is arranged in an increasing order. If $M = N$, then this is the case when the number of points in each set is the same. In this case, y_i is plotted against x_i where y_i and x_i are both $(i - 0.5)/N$quantiles of their datasets, respectively. If MtN (to exemplify, the second branch has fewer observations compared to the first), there can be only M points on the q–q plot. At this point, y_i is the $(i - 0.5) = M$ quantile of y [45].

Example 2.12 Table 2.15 is based on economy dataset. The histogram , scatter plot and q–q plot are derived from the time to import (days) and time to exports (days) that belong to Italy dataset (economy data).

Histograms are extensively used. Yet, they might not prove to be as effective as the quantile plot, q–q plot and boxplot methods while one is making comparisons in the groups of univariate observations.

In Table 2.24, Italy dataset (economy data) sets an example of time to import (days) and time to exports (days), normalized with max–min normalization.

Based on the data derived from Italy economy, q–q plot was drawn for the time to import (days) and time to export (days) values as shown in Figure 2.23 for Table 2.24.

Table 2.24: Italy dataset (*economy* data) presents an example of time to import (days) and time to exports (days).

Time to import (days)	Time to exports (days)
0	0
0	12,602,581,252
0	13,677,328,059
10.82	15,404,876,386
10.59	17,741,843,854
12.7	20,268,351,005
14.31	23,114,028,519
15.28	26,378,552,578
12.9116	30,442,035,460
11.7475	35,450,634,467
10.9975	42,550,883,916
8.885	51,776,353,504
...	...
...	...
...	...

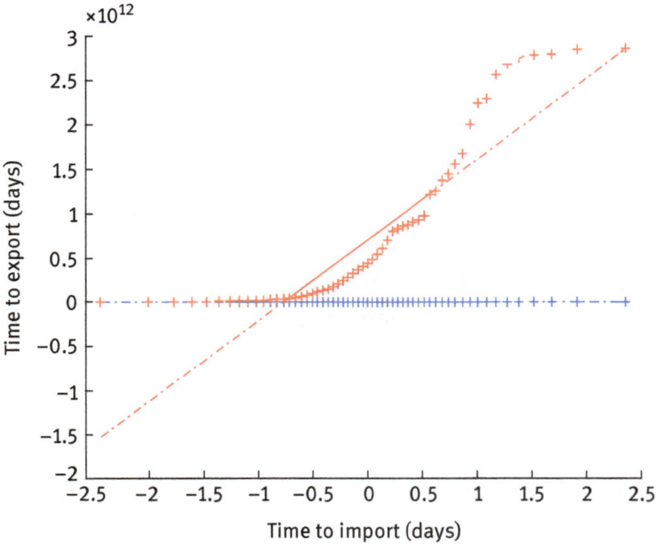

Figure 2.23: A q–q plot for time to import (days)–time to export (days) data from Table 2.24.

Figure 2.23 shows q–q plot graph of the attributes pertaining to time to import (days)–time to export (days) attributes as listed in Table 2.8. Based on Figure 2.23, while $x = 0$ for time to import (days) attribute, $y = 0$ between the period of 1960–2015. When $x = 0$, in $y = 0.5$, an increase has been observed in time to export (days). While time to import (days) attribute is seen within this interval, time to export (days) value has remained constant at 0.

2.6 Data matrix versus dissimilarity matrix

When the objects in question are one-dimensional, it means they can be described by a single attribute. The aforementioned section has covered objects that were described by single attribute. In this section, we cover *multiple* attributes. While dealing with multiple attributes, a change in notation is required. If you have n objects (e.g., MR images) described by p attributes, they are also referred to as *measurements* or *features*, such as age, income or marital status. The objects are $x_1 = (x_{11}, x_{12}, \ldots, x_{1p})$, $x_2 = (x_{21}, x_{22}, \ldots, x_{2p})$, and this goes on like this. Here, x_{ij} is the value for object x_i of the i_{th} attribute. For simple notation, x_i as object can be shown an s_i. The objects may be datasets in a relational dataset, which is also referred to as *feature vectors* or *data samples* [46, 47].

Data matrix is a kind of *object-by-attribute structure* that stores n data objects either in a relational table format or in $n \times p$ matrix (n objects $\times$ p attributes). An example for data matrix is as follows:

$$\begin{bmatrix} x_{11} & \dots & x_{1f} & \dots & x_{1p} \\ \dots & \dots & \dots & \dots & \dots \\ x_{i1} & \dots & x_{if} & \dots & x_{ip} \\ \dots & \dots & \dots & \dots & \dots \\ x_{n1} & \dots & x_{nf} & \dots & x_{np} \end{bmatrix} \tag{2.10}$$

In this matrix, each row relates to an object. f is used as the index through p attributes.

Dissimilarity matrix is a kind of *object-by-object structure* that stores a group of proximities available for all pairs of n objects, often represented by an $n \times n$ table as shown:

$$\begin{bmatrix} 0 & r & r & r & r \\ d(2,1) & 0 & r & r & r \\ d(3,1) & d(3,2) & 0 & r & r \\ \vdots & \vdots & \vdots & 0 & r \\ d(n,1) & d(n,2) & \dots & \dots & 0 \end{bmatrix} \tag{2.11}$$

where $d(i,j)$ refers to *dissimilarity between objects* i and j. $d(i,j)$ is generally a non-negative number close to 0. Objects i and j are similar to each other. $d(i,i) = 0$ shows the difference between an object and itself which is 0. In addition, $d(i,j) = d(j,i)$. The matrix is symmetric and for convenience, the $d(j,i)$ entries are not shown [46].

Measures of similarity are usually described as a function of measures of dissimilarity.

For nominal data, for instance,

$$\text{sim}(i,j) = 1 - d(i,j), \tag{2.12}$$

Here, *sim*(i, j) denotes the similarity between the objects i and j. A data matrix consists of two entities: rows for objects and columns for attributes. This is why we often call a data matrix a two-mode matrix. The dissimilarity matrix includes one kind of entity or dissimilarities; when there is one kind of entity, it is called a one-mode matrix. Most clustering and nearest neighbor algorithms also operate on dissimilarity matrix. Before such algorithms are applied, data in the form of a data matrix can be converted into a dissimilarity matrix [47].

Example 2.13 These matrices have eigenvalues 0, 0, 0, 0. They have two eigenvectors, one from each block. However, their block sizes do not match and they are not similar.

$$U = \begin{bmatrix} 0 & 1.3 & 0 & 0 \\ 0 & 0 & 0 & 0 \\ 0 & 0 & 0 & 1.3 \\ 0 & 0 & 0 & 0 \end{bmatrix}, \ I = \begin{bmatrix} 0 & 1.3 & 0 & 0 \\ 0 & 0 & 1.3 & 0 \\ 0 & 0 & 0 & 0 \\ 0 & 0 & 0 & 0 \end{bmatrix}$$

For matrix X, if $UX = XI$, then X is not invertible and so U is not similar to I.

Let $X = (x_{ij})$. Then

$$UX = \begin{bmatrix} x_{21} & x_{22} & x_{23} & x_{24} \\ 0 & 0 & 0 & 0 \\ x_{41} & x_{42} & x_{43} & x_{44} \\ 0 & 0 & 0 & 0 \end{bmatrix} \text{ and } XI = \begin{bmatrix} 0 & x_{11} & x_{12} & 0 \\ 0 & x_{21} & x_{22} & 0 \\ 0 & x_{31} & x_{32} & 0 \\ 0 & x_{41} & x_{42} & 0 \end{bmatrix}$$

If $UX = XI$, then $x_{11} = x_{22} = 0$, $x_{21} = 0$, $x_{31} = x_{42} = 0$ and $x_{41} = 0$. Thus, the first column of X is all zeros and X is not invertible.

If U were similar to I, *then* here would be an invertible matrix X that fulfills $I = X^{-1}UX$, and so XI = UI. Thus, it proves that there can be no such invertible matrix X. Hence, U is not similar to I.

References

[1] Ramakrishnan R, Gehrke J. Database Management Systems, New York, NY, USA: McGraw-Hill Companies, 2003.

[2] Marz N, Warren J. Big data: principles and best practices of scalable realtime data systems. Shelter Island. New York, NY, USA: Manning Publications, 2015.

[3] Kleppmann M. Designing data-intensive applications: The big ideas behind reliable, scalable, and maintainable systems. USA:O'Reilly Media, Inc., Publication, 2017.

[4] Mayer-Schönberger V, Cukier K. Big data: A revolution that will transform how we live, work, and think. London: John Murray Publishers, 2014.

[5] NYSE launched Intercontinental Exchange, with an electronic trading platform that brought more transparency and accessibility to the OTC energy markets, 2000.

[6] Wolcott HF. Transforming qualitative data: Description, analysis, and interpretation. California, USA: Sage Publications, Inc. 1994.

[7] Han J, Kamber M, Pei J, Data mining Concepts and Techniques. USA: The Morgan Kaufmann Series in Data Management Systems, Elsevier, 2012.

[8] Witten IH, Frank E, Hall Ma, Pal CJ. Data mining: Practical machine learning tools and techniques. USA: Morgan Kaufmann Series in Data Management Systems, Elsevier, 2016.

[9] Fayyad UM, Wierse A, Grinstein GG. Information visualization in data mining and knowledge discovery. USA: Morgan Kaufmann Publishers. 2002.

[10] Gupta GK. Introduction to data mining with case studies. Delhi: PHI Learning Pvt. Ltd., 2014.

[11] Kelleher JD, Mac NB , D'Arcy A. Fundamentals of machine learning for predictive data analytics: algorithms, worked examples, and case studies. USA: MIT Press, 2015.

[12] Murray, TJ, Saunders C, Holland NJ. Multiple sclerosis: A guide for the newly diagnosed. New York, NY, USA: Demos Medical Publishing, 2012.

[13] Lafferty J, McCallum A, Pereira F. C. Conditional random fields: Probabilistic models for segmenting and labeling sequence data, 2001.

[14] Harrington P. Machine learning in action. USA: Manning Publications, 2012.
[15] Karaca Y, Onur O, Karabudak R. Linear modeling of multiple sclerosis and its subgroups. *Turkish Journal of Neurology*, 2015, 21(1), 7–12.
[16] Rubin DB. Statistical disclosure limitation. *Journal of official Statistics*, 1993, 9(2), 461–468.
[17] Little RJ. Statistical analysis of masked data. *Journal of Official statistics-Stockholm*, 1993, 9(2), 407–426.
[18] Barbosa D, Mendelzon A, Keenleyside J, Lyons K. ToXgene: a template-based data generator for XML. In *Proceedings of the 2002 ACM SIGMOD international conference on Management of data*, 2002, 616.
[19] Stephens JM, Poess M. MUDD: A multi-dimensional data generator. *In ACM SIGSOFT Software Engineering Notes*. 2004, 29(1), 104–109.
[20] Van den Bulcke T, Van Leemput K, Naudts B, et al. SynTReN: a generator of synthetic gene expression data for design and analysis of structure learning algorithms. BMC bioinformatics, 2006, 7(1), 43.
[21] Sharing knowledge will be crucial to end extreme poverty and boost shared prosperity around the World. (Accessed http://www.worldbank.org).
[22] Karaca Y, Zhang YD, Cattani C, Ayan U. The differential diagnosis of multiple sclerosis using convex combination of infinite Kernels. *CNS & Neurological Disorders-Drug Targets (Formerly Current Drug Targets-CNS & Neurological Disorders* 2017, 16(1), 36–43.
[23] Gilroy J. Basic Neurology. USA: The McGraw-Hill Companies, 2000.
[24] Poser CM, Paty DW, Scheinberg L et al. New diagnostic criteria for multiple sclerosis: Guidelines for research protocols. *Annals of Neurology*, 1983, 13(3),227–231.
[25] Karabudak R. Temel ve Klinik Nöroimmünoloji. Turkey: Akademisyen Press, 2013.
[26] Kurtzke JF. Rating neurologic impairment in multiple sclerosis an expanded disability status scale (EDSS). *Neurology*, 1983, 33(11), 1444–1452.
[27] Jellinger KA. Multiple Sclerosis. A guide for the newly diagnosed. *European Journal of Neurology*, 2002, 9(4), 429–440.
[28] Weschler D. The Measurement of Adult Intelligence.: Baltimore (MD), USA: The Williams & Wilkins Company, 1939.
[29] Weschler D. WAIS-R Manual: Wechsler Adult Intelligence Scale-Revised. New York: The Psychological Corporation, 1981.
[30] Burgess A, Flint J, Adshead H. Factor structure of the Wechsler Adult Intelligence Scale–revised (WAIS–R): A clinical sample. British journal of clinical psychology, 1992, 31(3), 336–338.
[31] Wechsler D. The measurement of adult intelligence. Baltimore: Williams and Wilkins Co, 1944.
[32] Ashendorf L. An Exploratory Study of the Use of the Wechsler Digit–Symbol Incidental Learning Procedure with the WAIS-IV. Applied Neuropsychology: Adult, 2012, 19(4), 272–278.
[33] Friedman J, Hastie T, Tibshirani, R. The Elements of Statistical Learning: Data Mining, Inference, and Prediction. New York, NY, USA: Springer series in statistics, 2009.
[34] Nisbet R, Elder J, Miner G. Handbook of statistical analysis and data mining applications. Canada: Academic Press publications, Elsevier, 2009.
[35] Aggarwal CC. Data mining: the textbook. New York, USA: Springer, 2015.
[36] Zaki MJ, Wagner M Jr, Wagner M. Data mining and analysis: Fundamental concepts and algorithms. Cambridge University Press, 2014.
[37] Benesty J, Chen J, Huang Y, Cohen I. Pearson correlation coefficient. Berlin Heidelberg: In Noise reduction in speech processing, Springer, 37-39, 2009.
[38] Linoff GS., Michael JAB. Data mining techniques: for marketing, sales, and customer relationship management. Indiana: Wiley Publishing Inc., Publication, 2011.
[39] Muller, AC, Guido S, Introduction to machine learning with Python: a guide for data scientists. USA: O'Reilly Media, Inc, Publication, 2017.

[40] Ott RL, Longnecker MT. An introduction to statistical methods and data analysis. USA: Nelson Education. 2015.
[41] Larose DT, Chantal DL. Discovering knowledge in data: an introduction to data mining. USA: John Wiley & Sons, Inc., Publication, 2014.
[42] Tan PN, Steinbach M, Kumar V. Data mining cluster analysis: basic concepts and algorithms. Introduction to data mining, London: Pearson, 2013.
[43] Berson A, Smith SJ. Data warehousing, data mining, and OLAP. New York: McGraw-Hill, Inc., Publication, 2013
[44] Provost F, Fawcett T. Data Science For Business: What You Need To Know About Data Mining and Data-Analytic Thinking. USA O'reilly Media, Inc., Publication, 2013.
[45] Raudenbush SW, Anthony SB. Hierarchical linear models: Applications and data analysis methods. USA: Sage Publications, 2002.
[46] Giudici P. Applied data mining: statistical methods for business and industry. Great Britain: John Wiley & Sons Inc., Publication, 2005.
[47] John LZQ. The elements of statistical learning: data mining, inference, and prediction. *Journal of the Royal Statistical Society: Series A (Statistics in Society)*, 2010, 173(3),693-694.

3 Data preprocessing and model evaluation

This chapter covers the following items:
- Data preprocessing, data cleaning, data integration, data reduction and data transformation
- Attribute subset selection: normalization
- Classification of data
- Model evaluation and selection

Today, owing to their large sizes and heterogeneous sources, real-world datasets are prone to noisy, inconsistent data and missing data. For high-quality mining, it is vital that data are of high quality. Several data preprocessing techniques exist so as to enhance the quality of data, which result in mining. For example, data cleaning is applied for the removal of noise and correction of data inconsistency. Data integration unites data that come from varying sources, making them integrated into coherent store like data dataset. Normalization is an example of data transformation, which is applied to improve the efficiency and accuracy of mining algorithms that include distance measurements. Such techniques can be applied together. For example, transformation procedures are applied in data cleaning so as to correct the wrong data. For this, all entries in a field are transformed to a common format. In Chapter 2, we have gone through different types of attributes. These can also prove to be helpful to identify the wrong values and outliers. Such identification is of help for steps pertaining to data cleaning and integration. The following parts shed light on data cleaning, data integration, data reduction as well as data transformation [1–5].

In this chapter, fundamental concepts of data preprocessing are introduced (see Section 3.2). The data preprocessing methods are categorized as follows: data cleaning (Section 3.2.1), data integration (Section 3.2.3), data reduction (Section 3.5) and data transformation (Section 3.6).

This chapter presents an overall picture of data preprocessing. Section 3.1 discusses many elements that define data quality. This provides a reason behind data preprocessing. In Section 3.2, the most important tasks involved in data preprocessing are discussed.

3.1 Data quality

It is considered that if data can fulfill the requirements of intended use, then it is of quality. This also refers to validity of data, enabling it to be used for intended purposes. Certain factors indicate *data quality*; these being *accurate, consistent, complete, reliable* and *interpretable* [6]. Accuracy refers to yielding correct attribute values. Consistency refers to the process of keeping information uniform as it

https://doi.org/10.1515/9783110496369-003

progresses along a network and/or between various applications. Consistent data do not contain discrepancies. Completeness refers to not having any lacking attribute values. Reliability refers to yielding the same results when different applications or practices are performed. Interpretability refers to how easy the data are understood. Following these brief explanations, let us, at this point, provide a real-world example: let's say you are asked to analyze patients' data but you have discovered that some information has not been recorded. In addition, you may notice some data having inconsistencies or reported errors for some patient characteristics. All these indicate that the data you are asked to analyze through data mining techniques are not complete, which means they do not have attribute values or certain attributes that may be of interest to you. Sometimes data may be unavailable for the access of the researcher. It can also be said that the data are noisy or inaccurate, which refers to the fact that the data have errors or may have values that deviate from the expected. Data containing discrepancies are said to be inconsistent. In theory, everything about data may seem clear-cut and organized; however, as you may see in the future or you have already experienced, real world is not as smooth as the theory suggests. For this reason, we should have the required tools in our toolbox so that we can be equipped with sorting out problems related to inconsistent, inaccurate or disorganized data [7, 8].

3.2 Data preprocessing: Major steps involved

This section explains the major steps involved in data preprocessing and relevant methods for each. Major steps under discussion for data preprocessing are *data cleaning, integration, data reduction* and *data transformation* as shown in Figure 3.1.

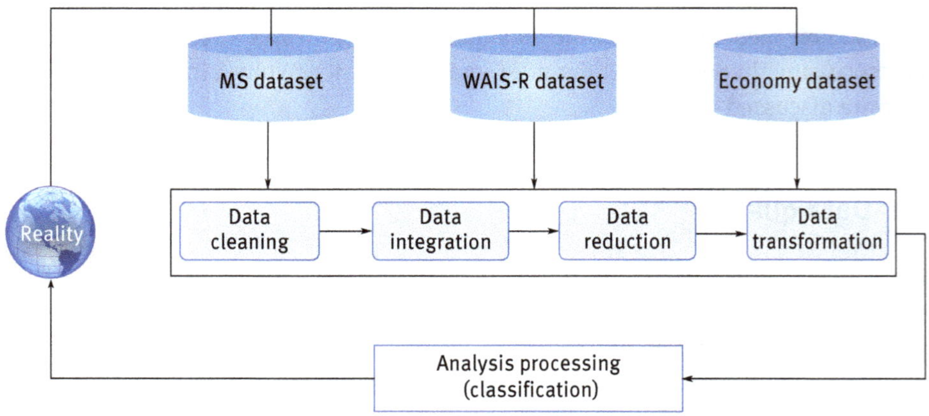

Figure 3.1: Stages of data preprocessing and data analysis processing.

3.2.1 Data cleaning and methods

Routines involved in data cleaning literally clean the data. In order to do that, noisy data are smoothed, missing values are filled in, outliers are either identified or removed, finally inconsistencies, if there are any, are solved. If a researcher thinks that the data are dirty, he/she knows that it is not trustworthy. In addition, another disadvantage that dirty data can cause is that it may confuse the mining procedure, which in the end results in unreliable outputs. As for the basic methods used for data cleaning, we can talk about certain steps for dealing with missing values . Reminding ourselves of the real-life example, you may need to analyze the data of multiple sclerosis (MS) patients in a clinic. You see that some magnetic resonance (MR) image processing data for some patients are missing. How can one proceed covering those missing values for this attribute (MR data)? To sort out this problem, you need to follow certain methods. The first of which is ignoring these ordered list of elements in the dataset. This is usually followed if the class label is not present. It is considered that this method is not very effective if the dataset does not contain several attributes that have missing values . This might mean adopting a pragmatic attitude, using what works and what is available. Another method involved is missing the value manually [9]. This method also has drawbacks: one of which is that it takes time and it may not be applicable if large dataset with numerous values that are missing is at stake. The third method for data cleaning is using a global constant to cover the missing value. In this method, all missing attribute values are replaced with the same constant by a label *unknown* This method is not complicated. It is simple, but it is with foolproof since when missing values are replaced by the label *unknown*, then the mining program may by mistake consider that they form an interesting concept with a value in common that is *unknown*. Another method we can speak of at this point is using a measure of central tendency for the attribute (e.g., the mean or median) to cover the missing value. Central tendency indicates the middle value of a data distribution. The mean can be used for normal (symmetric) data distributions, whereas median is a better choice to be employed by skewed data distribution. The fifth method is using the attribute mean or median for all samples that belong to the same class. For example, the distribution of data in the gross domestic product (GDP) per capita for the economy of U.N.I.S. (USA, New Zealand, Italy, Sweden) countries is symmetrical with a mean value of 29,936 (according to Table 2.8). This value is valued to replace the missing value for income. The last method to be used for data cleaning is to use the most likely value to fill in the missing value. This can be found out by regression, an inference-based tool using Bayesian type of formalism, or decision tree induction. This method is considered to be a popular strategy using most of the information from the present data so that the researcher can predict the values that are missing. Overall, in a good design pertaining to dataset, data entry helps to find the number of missing values, thereby minimizing errors initially while one is trying to do his/her best to clean the data after having it seized [10–12].

3.2.2 Data cleaning as a process

Missing values, inconsistencies and noise lead to inaccurate data. Data cleaning is an important and big task; hence, it is a process. The first step for the data cleaning process is *discrepancy detection*. Discrepancies can emerge due to several factors such as designed data entry forms that are poorly designed including many optional fields, human error in data entry, intentional errors (respondents being unwilling to give information about themselves) and data decay (outdated email address). Discrepancies may also be brought about by inconsistent use of codes and inconsistent data representations. Other sources of discrepancies involve system errors and errors in instrumentation devices recording data. Errors may occur when the data are inadequately used for intentions aside from the ones originally aimed at. For discrepancy detection, one may use any knowledge he/she has already had as to the properties of the data. This kind of knowledge is known as *metadata* ("data about data"). Knowledge on attributes can be of help at this stage. Ask yourself the questions: "what are the data type and domain of each attribute?" and "what are the acceptable values for each attribute?" to understand data trends and identify the anomalies. Finding the mean, median and mode values are useful, which will help you find out if the data are skewed or symmetric, what the range of values is or if all values belong to the expected range or not [13].

Field overloading [14] is another source of error that occurs when developers crowd new attribute definitions into unused (bit) portions of attributes that are already defined. The data should be examined for unique rules, consecutive rules and null rules. A *unique rule* suggests that all values of the given attribute have to differ from all the other values for that attribute. A *consecutive rule* suggests that there cannot be any missing values between the minimum and maximum values for the attribute, with all values being unique (as given in Table 2.12, the ID of the individuals). A *null rule* postulates the use of special characters, punctuation symbols such as blanks or other strings that show the null condition (in cases where a value for a given attribute is not available) and the way these values are to be handled. The null rule should show us to be able to indicate the procedure of recording the null condition. A number of different commercial tools exist for helping the discrepancy detection step. Relying on fuzzy matching and parsing techniques while cleaning data from various sources, *data scrubbing tools* utilize simple domain knowledge (knowledge of postal addresses and spell-checking) to find errors and make the required corrections in the data. *Data auditing tools* can also identify discrepancies by making the analysis of data to find out relationships and rules, thus data are detected violating these conditions. These are variants for data mining tools. For instance, statistical analysis can be utilized for the purpose of finding correlations or clustering for the identification of the outliers. Basic statistical data descriptions presented in Section 2.4 can also be used. In addition to these mentioned earlier, it is also possible to correct some of the data inconsistencies manually by using external references. Errors at the data entry can be corrected by carrying out a paper trace. Most errors, nevertheless, require the preference of *data transformations*. Once discrepancies are found, transformations are used for definition and application for the purpose of correction.

There exist some commercial tools that can be of help for the data transformation step. *Data migration tools* can ensure simple transformations by replacing the string *gender* with *sex*. *Extraction/transformation/loading tools* [15] render users to identify changes through a graphical user interface. Such tools assist only a limited set of transforms, though. For this reason, it is also possible to prefer writing custom scripts for this particular step of the data cleaning process.

There are novel approaches employed for data cleaning, and emphasis is placed for an increased sort of interactivity. Users progressively build a sequence of transformations by creating and debugging discrete transformations on a spreadsheet-like interface. The transformations can be displayed in a graphic format or through coming up with examples. Results are displayed straightaway on the records that are visible on the screen. The user can prefer undoing the transformations as well for the erasing of additional errors. The tool automatically conducts checking of discrepancies in the background on the latest transformed view of the data. Users can in this way develop and also refine transformations gradually as the discrepancies are detected. This procedure can ensure a data cleaning process to be more effective and efficient.

Developing declarative languages for the arrangement of data transformation operator is acknowledged as another approach for augmented interactivity in data cleaning. Such efforts are oriented toward the definition of powerful extensions to Structured Query Language (SQL) and algorithms that allow users to state data cleaning specifications more competently.

If one discovers further regarding the data, it is important that he/she keeps updating the metadata to reveal this knowledge, which will certainly be helpful for speeding up data cleaning for the future versions of the same data store.

3.2.3 Data integration and methods

Data integration is required for the purpose of data mining. This involves the incorporation of multiple datasets, data cubes or flat files into a coherent data store. The structure of data and semantic heterogeneity pose challenges in data integration process. In some cases, we come across a situation where some attributes represent a given concept that may have different name(s) in different datasets. Such a situation would lead to inconsistency and redundancy for sure. For example, the attribute for patient names may be referred to as *individual (ID)* in one data store and *individual (ID)* in another one. The same naming inconsistency may also occur in attribute values. For instance, the same first name could be recorded as "Donald" in one dataset, "Michael" in another one and "D." in the third. If you have a large amount of redundant data in your cases, the knowledge discovery process may be slowed down or may lead to confusion. Besides data cleaning, it is important to take steps to assist in avoiding redundancies throughout data integration [16]. Generally, data integration and data cleaning are conducted as a preprocessing step while one is preparing data for a data dataset. Further data cleaning can be carried out to find out and eliminate redundancies that could have been a result of data integration.

Entity identification problem is an issue related to data integration. For this reason, one should be particularly attentive to the *structure* of the data while matching attributes from one dataset to another one. *Redundancy* is another noteworthy issue in data integration. If you can derive an attribute from another attribute or set of attributes, such an attribute is known to be redundant. Some redundancies can be found out through correlation analysis. Based on the available data, this analysis can quantify how effectively one attribute infers the other attribute. For nominal data, the x^2 (*chi-square*) test is administered. As for numeric attributes, the *correlation coefficient* and *covariance* are often opted for. Thus, it becomes possible to access how the value of one attribute differs from that of the other.

3.2.3.1 x^2 correlation test for nominal data

A correlation relationship between two attributes, A and B, can be found out by x^2 (*chi-square*) test for nominal data as mentioned earlier. Let's say A has c distinct values, which are $a_1, a_2, \ldots, a_r$. B has r distinct values, which are $b_1, b_2, \ldots, b_r$. The described datasets A and B can be presented as a *contingency table*, with c values of A constituting the columns and r values of B constituting the rows. (A_i, B_j) shows the joint event with attribute A assuming the value a_i and attribute B taking on the value b_j, where $(A = a_i, B = b_j)$. Each possible (A_i, B_j) joint event has its own cell (or slot) in the table [16]. The x^2 value (also referred to as the *Pearson x^2 statistic*) is calculated in the following manner:

$$x^2 = \sum\nolimits_{i=1}^{c} \sum\nolimits_{j=1}^{r} \frac{(o_{ij} - e_{ij})^2}{e_{ij}} \tag{3.1}$$

where o_{ij} is the observed frequency (i.e., actual count of the joint event (A_i, B_j)) and e_{ij} is the expected regularity of (A_i, B_j), which is calculated as follows:

$$e_{ij} = \frac{\text{count}(A = a_i) \times \text{count}(B = b_j)}{n} \tag{3.2}$$

where n is the frequency observed for the dataset, count$(A = a_i)$ is the number of datasets that have the value a_i for A and count$(B = b_j)$ is the number of datasets with the value b_j for B. The sum in eq. (3.2) is calculated for all $r \times c$ cells. The cells that have the most contribution to the x^2 value are the ones and regarding these the actual count highly differs from the one expected [17].

The hypothesis that A and B are *independent* is verified by the x^2 statistic. This indicates that no correlation exists between them. The test is based on a significance level, which shows $(r-1) \times (c-1)$ degrees of freedom. Example 3.1 presents the use of this statistic. If it is possible to reject the hypothesis, then it is said that A and B are statistically correlated.

Example 3.1 Correlation analyses of nominal attributes with x^2 being used.

$$e_{11} = \frac{\text{count(male)} \times \text{count(healthy)}}{n} = \frac{204 \times 200}{400} = 102$$

Consider that the sum of the expected frequencies in any row must be equal to the total frequency observed for that row. The sum of the expected frequencies in the columns must be equal to the total frequency observed for the column (Table 3.1).

Table 3.1: 2 × 2 contingency table for Table 2.19.

Class	Gender		Total
	Male	Female	
Healthy	100 (104)	100 (96)	200
Patient	104 (100)	96 (100)	200
Total	204	196	400

Note: Are gender and psychological_analysis correlated?

For x^2 when eq. (3.1) is used,

$$x^2 = \frac{(100-104)^2}{104} + \frac{(104-100)^2}{100} + \frac{(100-96)^2}{96} + \frac{(96-100)^2}{100}$$
$$= 0.15 + 0.16 + 0.16 + 0.16 = 0.63$$

For this 2×2 table, the degrees of freedom are $(2-1)(2-1) = 1$. For 1 degree of freedom, x^2 value needed to discard the hypothesis at the 0.001 significance level is 10.828, which is obtained from the table of upper percentage points of the x^2 distribution. Course books provide this distribution. As the value calculated is above this, the hypothesis that *health status* and *gender* are independent is rejected. As such, it can be concluded that both attributes have (strong) correlation for the particular group of people.

3.2.3.2 x^2 correlation test for numeric data

As mentioned earlier, the correlation between two attributes, A and B, can be evaluated by calculating the *correlation coefficient* (also referred to as *Pearson's product moment coefficient* [18], named after its inventor, Karl Pearson) for numeric attributes. The formula for this calculation is

$$r_{A,B} = \frac{\sum_{i=1}^{n}(a_i - \bar{A})(b_i - \bar{B})}{n\sigma_A\sigma_B} = \frac{\sum_{i=1}^{n}(a_i b_i) - (n\bar{A}\bar{B})}{n\sigma_A\sigma_B} \tag{3.3}$$

where n is the number of datasets, a_i and b_i are the particular values of A and B in i, $\bar{A}$ and $\bar{B}$ are the corresponding mean values of A and B, σ_A and σ_B are the corresponding standard deviations of A and B. $\sum(a_i b_i)$ is the sum of the AB cross-product (i.e., for each dataset, the value of A is multiplied by the value of B in that dataset). Note that $-1 \le r_{A,B} \le +1$. If $r_{A,B}$ is higher than 0, A and B have positive correlation, which means that as the values of A go up so do the values of B. The higher the value is, the stronger the correlation will be. Thus, a higher value can also suggest that A (or B) may be deleted as a redundancy.

In a case when the result value equals to 0, A and B are independent and they have no correlation that exists. In cases where the resulting value is lower than 0, A and B have negative correlation between them, and the values of one attribute go up as the values of the other attribute go down, which means that each attribute dejects the other. It is also possible to use scatter plots if one wishes to view the correlations between attributes.

3.2.3.3 Covariance of numeric data

Correlation and covariance are two similar measures employed in probability theory and statistics – the evaluation of what extent two attributes alter together. Let's say A and B are two numeric attributes. In a set of n observations $\{(a_1, b_1), \ldots, (a_n, b_n)\}$, the mean values of A and B, in that order, are also regarded as the *expected values* on A and B based on formulas

$$E(A) = \bar{A} = \frac{\sum_{i=1}^{n} a_i}{n}$$

and

$$E(B) = \bar{B} = \frac{\sum_{i=1}^{n} b_i}{n}$$

between A and B, and the *covariance* is characterized as

$$\mathrm{Cov}(A, B) = E((A - \bar{A})(B - \bar{B})) = \frac{\sum_{i=1}^{n} b_i}{n} \tag{3.4}$$

By comparing eq. (3.3) for $r_{A,B}$ (correlation coefficient) with eq. (3.4) for covariance, it can be seen that

$$r_{A,B} = \frac{\text{Cov}(A,B)}{\sigma_{A,B}} \tag{3.5}$$

where σ_A and σ_B are the standard deviations of A and B in the relevant order. It is also possible to show that

$$\text{Cov}(A,B) = E(A \cdot B) - \bar{A}\bar{B} \tag{3.6}$$

This equation may render the calculations more simplified.

Let's say that A and B, as two attributes, are likely to change together. We mean that if A is larger than $\bar{A}$ (the expected value of A), B is likely to be larger than $\bar{B}$ (the expected value of B). For this reason, the covariance between A and B is said to be *positive*. Alternatively, if one attribute is prone to be higher than its expected value when the other attribute is below its expected value, in this case the A and B covariance is said to be *negative*.

In a case where A and B are independent (they are not correlated), we can say that $E(A \cdot B) = E(A) \cdot E(B)$.

In such a case, the covariance is $Cov(A,B) = E(A \cdot B) - \bar{A}\bar{B} = E(A) \cdot E(B) - \bar{A}\bar{B} = 0$. Yet, the converse does not hold either [12–14].

Some random variable (attribute) pairs may have 0 covariance but they are not independent. They are said to be independent only under some additional assumptions.

Example 3.2 (Covariance analysis of numeric attributes). Considering Table 3.2 for the economy, data analysis, *tax payments and GDP per capita* attributes with the disorder definition will be analyzed for four countries' economy on which U.N.I.S. economy data were administered.

Table 3.2: Sample result for *tax payments and GDP per capita* as per Table 2.8.

Tax payments	GDP per capita
1	0.4712
1	0.4321
2	0.4212
3	0.3783
4	0.3534

$$E(\text{tax payments}) = \frac{1+1+2+3+4}{5} = \frac{11}{5} = 2.2$$

and

$$E(\text{GDP per capita}) = \frac{0.4712 + 0.4321 + 0.4212 + 0.3783 + 0.3534}{5} = \frac{2.0562}{5} = 0.411$$

Thus using eq. (3.6), the following is calculated:

$$\begin{aligned}&\text{Cov}(\text{tax payments, GDP per capita})\\&\quad = \frac{1\times 0.4712 + 1\times 0.4321 + 2\times 0.4212 + 3\times 0.3783 + 4\times 0.3534}{5} - 2.2\times 0.411\\&\quad = 0.858 - 0.90 = -0.042\end{aligned}$$

As a result, given the negative covariance.

Example 3.3 (Covariance analysis of numeric attributes). Considering Table 3.3 for the psychiatric analysis, *three-dimensional ability* and *numbering* attributes with the disorder definition will be analyzed for five individuals on whom WAIS-R test was administered.

Table 3.3: Sample result for *three-dimensional ability* and *numbering* as per Table 2.19.

Individual (ID)	Three-dimensional ability	Numbering
1	9	11
2	17	13
3	1	2
4	8	13
5	10	11

$$E(\text{three-dimensional ability}) = \frac{9 + 17 + 1 + 8 + 10}{5} = \frac{45}{5} = 9$$

and

$$E(\text{numbering}) = \frac{11 + 13 + 2 + 13 + 11}{5} = \frac{50}{5} = 10$$

Thus using eq. (3.6), the following is calculated:

$$\begin{aligned}&E(\text{three-dimensional ability, numbering})\\&\quad = \frac{9\times 11 + 17\times 13 + 1\times 2 + 8\times 13 + 10\times 11}{5} - 9\times 10\\&\quad = 107.2 - 90 = 17.2\end{aligned}$$

As a result, given the positive covariance it can be said that *three-dimensional ability* and *numbering* are the attributes concerned with the diagnosis of mental disorder.

3.3 Data value conflict detection and resolution

Among data integration procedures, the *detection and resolution of data value conflicts* is also worthy of being mentioned. For instance, in real-world instances, attribute values that originate from different sources may vary from one another. This is because of the differences in scaling, representation or even encoding. For example, a *weight* attribute may be recorded in metric units in a system, and it may be recorded in another kind of unit in a different system. Education-level codification or naming or university grading systems can also differ from one country to another or even from one institution to another. For this reason, an exact course-to-grade transformation rules between two universities could be highly difficult, which makes information exchange difficult as a result. The discrepancy detection issue is described further in Section 3.2.2 [15].

In brief, a meticulous combination can be of help for the reduction and avoiding of inconsistencies and redundancies in the resulting dataset, which will in return improve both the speed and the accuracy of the following data mining process.

3.4 Data smoothing and methods

Before discussing the data smoothing issue, let us focus on *noisy data*. *Noise* is defined as a random error or variance in a variable measured. Chapter 2 provided some ways as to how some basic statistical description techniques (boxplots and scatter plots) along with methods of data visualization can be utilized so as to detect outliers that might signify noise. Speaking of a numeric attribute such as *age*, how would it be possible to smooth out the data for the purpose of removing the noise? The answer to such question leads us to analyze data smoothing techniques as provided below.

Binning methods are used for smoothing a sorted data value by referring to its "neighborhood," which refers to the values around it. The arranged values are distributed into a number of buckets or, in other words, *bins*. Binning methods perform *local* smoothing because the neighborhood of values is consulted. Figure 3.2 presents some of the relevant binning techniques. The data for *age* are first assorted, and then partitioned into *equal-frequency* bins of size 3. Each bin includes three values. In *smoothing by bin means*, one replaces each value in a bin by mean value of the bin. The mean of the values 15, 17 and 19 in bin 1 is 17. Thus, each original value in this bin is substituted by 17 as the value. Correspondingly, *leveling by bin medians*

Partition into (equal-frequency) bins:
Bin 1: 15, 17, 19
Bin 2: 22, 23,24
Bin 3: 30, 30,33

Smoothing by bin means:
Bin 1: 17, 17, 17
Bin 2: 23, 23, 23
Bin 3: 31, 31, 31

Smoothing by bin boundaries:
Bin 1: 15, 15, 19
Bin 2: 22, 22, 24
Bin 3: 30, 30, 33

Figure 3.2: Binning methods for data smoothing.

can be used. Each bin value is substituted by the bin median. In *smoothing by bin boundaries*, the lowest and highest values in a given bin are recognized as the *bin boundaries* [16].

Arranged data for *age* (WAIS-R dataset, see Table 2.19): 15, 17, 19, 22, 23, 24, 30, 30, 33.

Each bin value is substituted by the boundary value that closes. The greater the width is, the bigger the effect of smoothing is. Bins may have *equal width* and the interval range of values in each bin is constant. It is also known that binning is used as a discretization technique (to be discussed in Section 3.6).

There are several data smoothing methods used for data discretization which is known to be a form of data reduction and data transformation. For instance, the binning techniques diminish the number of distinct values for each attribute. These act as a procedure of data reduction for logic-based data mining methods such as decision tree induction. With this, value comparisons on arranged data are made repetitively. As a form of data discretization, we can give concept hierarchies as an example also being employed for data smoothing.

Another technique used for data smoothing is *regression*, which accords data values to a function. *Linear regression* includes spotting the "best" line to fit two attributes in order that one attribute can be utilized for the prediction of other attribute. Extrapolating linear regression in *multiple linear regression*, there are more than two attributes, where data are being fit into a surface that is multidimensional. Regression is described in Section 3.5.

In *outlier analysis*, outliers can be identified by clustering, which suggests that you organize the similar values into groups or "clusters." Naturally, values falling outside of the group of clusters can be regarded as outliers (Figure 3.3).

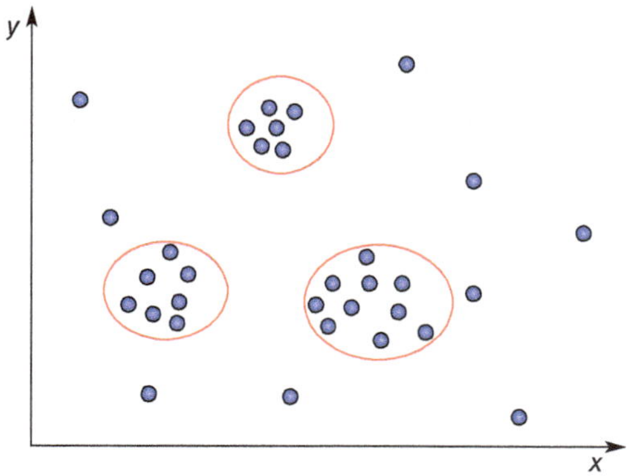

Figure 3.3: Clustering of the subjects according to MS subgroups along with EDSS (Expanded Disability Status Scale) scores. The data lying outside the circle are those outside the cluster.

3.5 Data reduction

Data reduction attains a lessened representation of the dataset smaller in volume. Nevertheless, either the same or almost the same analytical results are yielded. *Data reduction* techniques can be implemented to obtain a reduced representation of the dataset that is smaller in volume, but one that keeps original data's integrity. *Dimensionality reduction*, *numerosity reduction* and *data compression* are among the data reduction strategies [16]. In *dimensionality reduction*, data encoding schemes are applied in order that one can obtain a "compressed" representation of the original data. This process is a reducing process of the number of random attributes or variables at stake. Data compression techniques are *wavelet transforms* and *principal components analysis*, *attribute subset selection* (removing irrelevant attributes) and *attribute construction* (a small set of more useful attributes is obtained from the original set). *Numerosity reduction* techniques replace the original data volume through alternative and smaller forms of data representation. Through this strategy, alternative smaller representations replace the data. And during this process, parametric models are used. Some examples of parametric models are *regression* or *log-linear models*, or nonparametric models such as *histograms*, *clusters*, *sampling* or *data aggregation* can also be employed. It is also preferred to use a distance-based mining algorithm for analyses using nearest neighbor classifiers and neural networks [15, 16]. These methods yield more accurate results if the data to be analyzed have been *normalized*, or in other words, mounted to a smaller range of [0.0, 1.0]. *Data compression* transformations are implemented to get a reduced or "compressed" original data representation. If it is possible to *reconstruct* the original data from the compressed data without being subjected to information loss, then we can call such a data reduction *lossless*. However, if only an estimate of the original data can be reconstructed, the data reduction would be named *lossy*. Several lossless algorithms exist for string compression; yet they permit only limited data manipulation. Numerosity reduction and dimensionality reduction techniques can also be considered as forms of data compression. Some other ways also exist for methods of data reduction organization. The calculation time that allocated for data reduction should not be greater than or "erase" the time saved by mining on a reduced dataset size [17].

3.6 Data transformation

Concept hierarchy generation and *discretization* can also be useful. Raw data values for attributes are substituted by ranges or higher conceptual levels [16, 17]. For instance, raw values for Expanded Disability Status Scale *(EDSS) score* may be substituted by higher level concepts, such as relapsing remitting multiple sclerosis (*RRMS*), secondary progressive multiple sclerosis (*SPMS*) and primary progressive multiple sclerosis (*PPMS*). Concept hierarchy generation and discretization are prevailing tools for data mining since they ensure data mining at multiple abstraction levels. Data discretization, normalization and concept hierarchy generation are forms of *data transformation*.

To recap, as we have mentioned earlier, data in the real world tend to be incomplete, inconsistent and noisy. For this reason, data preprocessing techniques help us to render us to improve data quality, which in return helps us improve the efficiency and accuracy of the mining process as the next level. Since quality decisions are to be reliant on quality data, finding out data anomalies, correcting them promptly and decreasing the data to be analyzed can contribute to significant benefits for us in decision making.

3.7 Attribute subset selection

Datasets as to analysis may include a high number of attributes and some of which could be redundant or irrelevant for the mining procedure. Eliminating relevant attributes or keeping the irrelevant ones may have detrimental effects because this leads to misunderstanding for the mining algorithm used. This could bring about revealed patterns of poor quality. Moreover, a further volume with redundant or irrelevant attributes may cause a slowdown in the mining process.

Attribute subset selection [15, 16, 19] is recognized as feature subset selection in machine learning and it decreases the dataset size by removing the redundant or irrelevant attributes. Here the aim is to find a minimum set of attributes so that the resulting probability distribution of the data classes will be close to the original distribution derived by using all attributes as much as possible. Mining on a reduced set of attributes offers an extra benefit since it curtails the number of attributes that appear in the patterns discovered, which makes the patterns easier to comprehend. For *n* attributes, there are $2n$ possible subsets when one wants to find a good subset of the original attributes. A comprehensive search for the optimal subset of attributes can be costly as *n* and the number of data classes increases. For this very reason, heuristic methods exploring a reduced search space are generally used for the selection of attribute subset. These methods are generally known to be *greedy* since they always make what seems to be the best choice when looking for through attribute space. This strategy lends itself to make an optimal choice that is local for an optimal solution that is global. These greedy methods are known to be effective in real life and practice. In addition, it may be close to an estimation regarding an optimal solution.

Generally speaking, the "best" (and "worst") attributes are identified by employing tests regarding statistical significance. This suggests that the attributes are independent of each other. Many other evaluation measures for attributes can be used. The *information gain* measure can be used in building decision trees for the purpose of classification. (The details of the information gain measure are provided in Chapter 6.)

There are basic heuristic methods of attribute subset selection which encompass certain techniques as described in Figure 3.4.

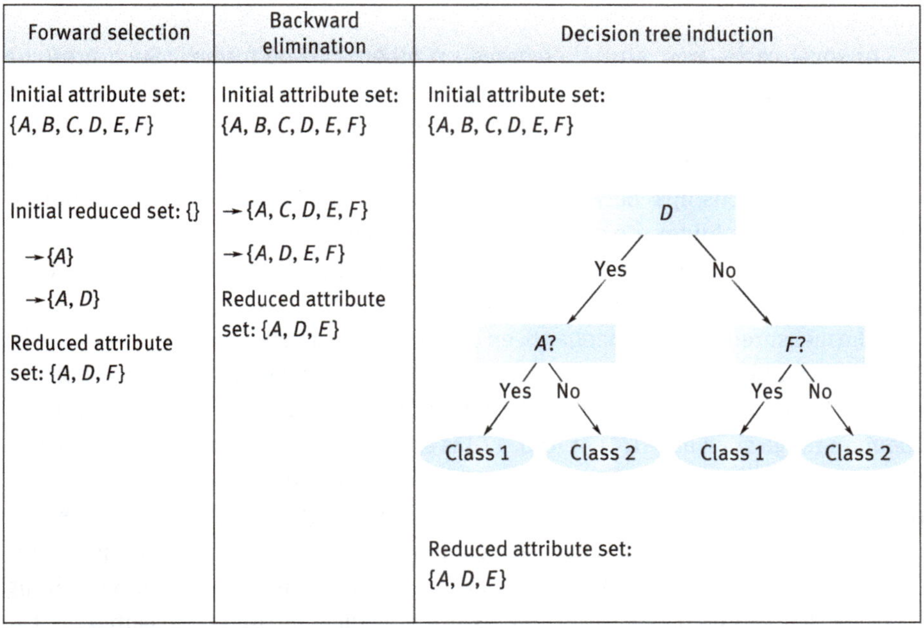

Figure 3.4: Heusristic or greedy methods for attribute subset selection.

1. **Stepwise forward selection**: The procedure begins with an empty set of attributes, which make up the reduced set. Among the original attributes, the best are determined and included in the reduced set. At each next step or iteration, the best of the original attributes that remain are added into the set.
2. **Stepwise backward elimination**: The procedure begins with the complete set of attributes. The worst attribute remaining in the set is eliminated in each step.
3. **Combination of forward selection and backward elimination**: It is possible to merge the methods such as the stepwise backward elimination and forward selection. In this way, the best attribute is selected by the procedure and the worst among the remaining attributes is discarded accordingly.
4. **Decision tree induction**: Decision tree algorithms such as ID3, C4.5 and CART were originally intended for the purpose of classification. Decision tree induction forms a structure like a flowchart. Here, each internal node or nonleaf shows a test on an attribute and each branch corresponds to a test outcome and each external node or leaf represents a class prediction. The algorithm chooses the best attribute to divide the data into separate classes. A tree is formed based on the given data when decision tree induction is implemented for attribute subset selection. All attributes not seen in the tree are said to be unrelated. The set of attributes that appear in the tree constitutes the reduced subset of attributes.

Stopping criteria for the methods might change. A threshold on the measure may be used by the procedure so as to determine when the attribute selection process is to be stopped.

In some cases, new attributes based on others can be formed. Such *attribute construction* can be of help to increase accuracy and comprehension of structure in data. For instance, attribute marital status based on the attributes married, single and divorced may be added. Attribute construction can spot the missing information about the relationships between data attributes through combining attributes, which can be helpful for the discovery of knowledge.

Data normalization in which the attribute data are scaled to be able to fall within a smaller range could be –1.0 to 1.0, or 0.0 to 1.0.

The measurement unit that has been used can have an effect on the data analysis. The data should be normalized or standardized to avoid dependence on the choice of measurement units. In this process, transforming the data to fall within a smaller or common range such as [–1, 1] or [0.0, 1.0] is at question. If the data are normalized, this will give all attributes an equal weight. Normalization is a procedure which proves to be useful for classifying algorithms that involve neural networks or distance measurements (e.g., the nearest neighbor classification and clustering). If one uses the neural network backpropagation algorithm for classification mining (Chapter 10), normalizing the input values for each attribute measured in the training dataset will make the learning phase faster. In distance-based methods, normalization is also helpful in preventing attributes that have initially large ranges (e.g., *GDP per capita*) from outweighing attributes that have initially smaller ranges (e.g., binary attributes). This could help in cases when no prior knowledge of the data is available [16].

There exist several ways of data normalization. Let us assume that min_A and max_A are the minimum and maximum values of an attribute, A. Min–max normalization plots a value, v_i, of A to v_i' in the range $[new\_min_A, new\_max_A]$, hence, one calculates:

$$v_i' = \frac{v_i - min_A}{max_A - min_A} \cdot (new\_max_A - new\_min_A) + new\_min_A \quad (3.7)$$

The relationships among the original data values are persevered by min–max normalization. It may face an "out-of-bounds" error if a future input case for normalization lies outside the original data range for A [16, 17].

Example 3.4 (Min–max normalization). Let us assume that the minimum and maximum values for the attribute *deposits* are \$24,000 and \$96,000. *GDP per capita* can be mapped to the range [0.0, 1.0]. By min–max normalization, a value of \$51,200 for *deposit* is changed into

$$\frac{51,200 - 24,000}{96,000 - 24,000}(1.0 - 0) + 0 = 0.377$$

As for *z-score normalization* (*zero-mean normalization*), the values of an attribute A are normalized relying on the mean (average) and standard deviation of A. A value v_i of A is normalized to v_i' by calculating

$$v_i' = \frac{v_i - \bar{A}}{\sigma_A} \quad (3.8)$$

In this denotation, $\bar{A}$ and σ_A are the standard deviation and mean of attribute A, respectively. The mean and standard deviation have been discussed in Section 2.4.

$\bar{A} = \frac{1}{n}(v_1 + v_2 + \cdots + v_n)$ and σ_A are computed as the square root of the variance of A (see eq. (2.8)). This normalization method proves to be of use when the actual minimum and maximum of attribute A are not known or in cases when there are outliers that dominate the min–max normalization.

Example 3.5 (z-Score normalization). Let us suppose that the mean and standard deviation of the values for the attribute *income* are 190 and mean (μ) of 150 and a standard deviation (σ) of 25, respectively. With z-score normalization, $190 - 150/25 = 1.6$.

A variation of this z-score normalization takes the place of the standard deviation of eq. (3.9) by the *mean absolute deviation* of A [16, 17]. The *mean absolute deviation* of A, shown by s_A, is

$$s_A = \frac{1}{n}\left(|v_1 - \bar{A}| + |v_2 - \bar{A}| + \cdots + |v_n - \bar{A}|\right) \tag{3.9}$$

Accordingly, z-score normalization that uses the mean absolute deviation is

$$v_i' = \frac{v_i - \bar{A}}{s_A} \tag{3.10}$$

The mean absolute deviation, s_A, is stronger to outliers than the standard deviation, σ_A. When you calculate the mean absolute deviation, you do not square the deviations from the mean (i.e., $|x_i - \bar{x}|$). For this very reason, the effect of outliers is reduced to some extent.

Another method for data normalization is *normalization by decimal scaling*. The normalization takes place by moving the decimal point of values of attribute A. Moreover, the number of decimal points that is moved is based on the maximum absolute value of A. A value, v_i, of A is normalized to v_i' by doing the following calculation [16]:

$$v_i' = \frac{v_i}{10^j} \tag{3.11}$$

Here, j is the smallest integer with $\max(|v_i'| < 1)$.

Example 3.6 (Decimal scaling). The recorded values of *Sweden Economy* (Table 2.8) vary from 102 to 192. The maximum absolute value pertaining to *Sweden* is 192. In order to normalize by decimal scaling, each value is divided by 1,000 (i.e., $j = 3$) and by doing so 102 normalizes to 0.102, and 192 normalizes to 0.192.

It is possible to see cases when normalization can alter the original data slightly, when using z-score normalization or decimal scaling, particularly. At the same time, it is required to save the normalization parameters (e.g., the mean and standard deviation if one uses z-score normalization). If this is done, then future data can be normalized uniformly.

3.8 Classification of data

This section provides definition of classification, general approach to classification along with decision tree induction, attribute selection measures and Bayesian classification methods.

3.8.1 Definition of classification

A specialist in his/her area can make use of the parameters provided in Table 2.12 (in Section 2) in order to make the diagnosis whether an individual has MS or is healthy. In the example given above, the data analysis task is in fact a task of *classification*. In classification, a model or *classifier* is created in order to estimate *class labels*, or in other words *categorical labels*, such as "RRMS," "SPMS," "PPMS" or "healthy" for the medical application data; or "USA," "New Zealand," "Italy" or "Sweden" for the economy data. Discrete values can represent these categories. The ordering among values has no meaning. In some types of data analysis, the task may be an example of *numeric prediction*. If that is so, the model constructed makes a prediction of a *continuous-valued function*, or *ordered value* which is unlike the class label. This is a *predictor* model. *Regression analysis* is a statistical methodology frequently used for numeric predictions. The two terms are generally used with the same meaning but there are also other methods for numeric predictions. Regarding the types of *prediction problems*, classification and numeric prediction are the two most important types.

Let us start with the *general approach to classification. Data classification* is a process with two steps that include a learning step. In the training phase or learning step, a classification model is formed. Classification step is used for predicting class labels for the given data. Figure 3.5 details such processes.

A classifier is built to describe a predetermined set of data classes (concepts) in the first step. This is called the *learning step* or in some contexts we also call it training phase. In this step, a classification algorithm builds the classifier through the analysis or learning from a *training set*. This training set constitutes datasets and their associated class labels. A dataset, $\boldsymbol{X}$, is denoted by an n-dimensional *attribute vector*, $\boldsymbol{X} = (x_1, x_2, ..., x_k)$ which shows n measurements on the dataset from n dataset attributes in the following order: $A_1, A_2, ..., A_n$. (Each attribute represents a "feature" of $\boldsymbol{X}$.) Thus, literature regarding the pattern recognition makes use of *feature vector* term instead of the term *attribute vector*. Each dataset, $\boldsymbol{X}$, belongs to a class that is predetermined. It is in fact determined by another dataset attribute referred to as the *class label attribute*. The class label attribute has a discrete value; and it is unordered, being *categorical* (or nominal) (since each value functions as a category or class). The separate datasets constitute the training set, known as *training datasets*. The sampling is done randomly from the dataset under discussion. For classification, datasets are also known as *instances, data points, objects, samples or examples. Training sample* is the term used commonly in the literature regarding the machine learning [16–20].

This step is also called *supervised learning* since the class label of each training dataset is given. For *unsupervised learning* or *clustering*, we can say that unsupervised learning has class label of each training dataset that is not known. The number or set of classes that are going to be learned might not be known in advance either.

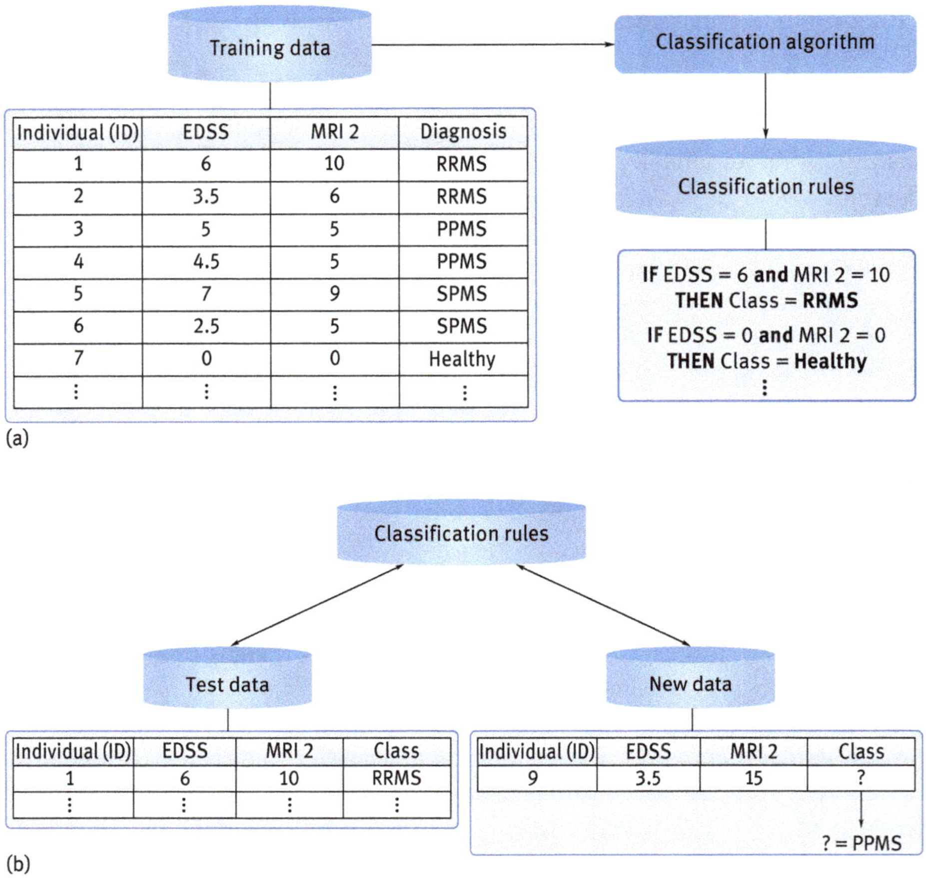

Figure 3.5: The data classification process. (a) *Learning*: A classification algorithm analyzes the training data. The class label attribute is *class*. Classifier or the learned model is shown in the form of classification rules. (b) *Classification*: It is important that test data are utilized for the estimation of the accuracy regarding the classification rules. If the accuracy is deemed to be acceptable, it is possible to apply the rules to the new dataset classification.

The first step regarding the classification process is also regarded as the learning of a mapping or function, $y = f(\boldsymbol{X})$. This is capable of making a prediction of the connected or associated class label y of a given dataset $\boldsymbol{X}$. It is important in this context that function or mapping separates the data classes. Such mapping is characteristically characterized in classification rules, decision trees or mathematical formula form. In our example, the mapping is depicted as classification rules that can identify medical data applications as RRMS, SPMS, PPMS or healthy (Figure 3.5(a)) as subgroups of MS. The rules have the function of classifying future datasets and eliciting deeper understanding as to the data content. A compressed data representation is also provided under such light.

In the second step (Figure 3.5(b)), we can see that the model is used for the purposes of classification including estimation of the predictive accuracy of the classifier. If we wanted to use the training set for the accuracy measurement of the classifier, this estimate would possibly be optimistic because the classifier tends to *overfit* the data (e.g., during learning it may include some specific anomalies of the training data that are not present in the general dataset). Hence, we use a *test set* which constitutes *test datasets* and their associated class labels. They are independent of the training datasets, which means that they were not used for the construction of the classifier.

Accuracy is an important feature in estimation. *Accuracy* of a classifier is the percentage of test set datasets correctly classified by the classifier. A comparison is made between the associated class label of each test dataset and the class prediction of the learned classifier regarding that dataset. Section 3.9 provides a number of methods that can be employed for the estimation of the classifier accuracy. If the classifier accuracy is deemed acceptable, then that classifier can be employed for the classification of the future datasets. And for these datasets, we do not know the class label. In literature with regard to machine learning, the data that are not known are referred to as either "unknown" or "previously unseen" data [16–18]. For example, classification rules that have been learned as in Figure 3.5(a) based on the data analysis of previous medical data applications can be utilized so as to approve or reject new or future medical data applicants.

3.9 Model evaluation and selection

Even though we have built a classification model, there may still be many questions that require more elucidation. For instance, suppose that you are using data from previous sales in order to construct a classifier to do the prediction of customer purchasing behavior. You aim at having a prediction of how correctly the classifier can estimate the buying behavior of prospective customers. In other words, this refers to the prospective customer data on which the classifier has not been trained. It is possible that you tried different methods so that you could build more than one classifier and have the intention of doing comparison regarding their accuracy [21]. At this point, we may ask what accuracy is and how we can estimate it. Other relevant questions that follow are some measures that indicate the accuracy of a classifier is more appropriate than the other ones. Furthermore, we also may know how we can obtain a *reliable* accuracy estimate. You will find the answers to such questions in this section of the book.

As common techniques for assessing accuracy, we see holdout and random subsampling (Section 3.9.1.1) and cross-validation (Section 3.9.1.2) based on randomly sampled partitions of the given data. What about the case when there is more than one classifier and aim at choosing the "best" one? This is referred to as *model selection* (choosing one classifier over the other one).

Following this introductory part, you may still be confused about certain issues. MS subgroups are classified based on the existence of lesion entity from the previously taken MR images and EDSS score. Then, you can ensure the MR image and EDSS score data to be tested for the purposes of classification.

It is possible that one might have attempted using different methods so as to build more than one classifier and compare the accuracy thereof. One may pose the question as to the definition and the way of estimating accuracy [22, 23]. You can further question if some measures of a classifier's accuracy are more appropriate than others or about the ways of getting a reliable estimation of accuracy.

You can find the answers and explanations for such questions in this section of the book. Section 3.9.1 presents various metrics for the evaluation of predictive accuracy pertaining to a classifier. Two common techniques are used for the assessment: holdout and random subsampling (Section 3.9.1.1), and cross-validation (Section 3.9.1.2). These are based on randomly sampled partitions of the data. Let's say you have more than one classifier and you are in a position to choose the "best" one. This is related to what is known as *model selection*; in other words, choosing one classifier instead of another one.

3.9.1 Metrics for evaluating classifier performance

In this section, measures for the assessment of how "accurate" a classifier are shown in the context of predicting the class label of datasets. Table 3.4 presents a summary of the classifier evaluation measures which encompass accuracy, sensitivity, or recall, specificity, precision, F_1, and F_β. Accuracy refers to the prediction abilities of a classifier. It is a better option to measure the accuracy of a classifier on a test set that includes datasets with class labels (datasets that were not actually used for the training of the model) [16, 21–23]. In machine learning literature, *positive samples* refer to datasets with main class of interest, whereas *negative samples* refer to all the other datasets. With two classes, the positive datasets may be *patient* = *yes*, whereas the negative datasets are *healthy* = *no*. Assume that our classifier on a test set of labeled datasets is used. For denotation, P is the number of positive samples and N is the number of negative samples. For each of them, the class label prediction of the classifier is compared with the known class label of the dataset.

Below you can find some of the "building block" terms that are used.

The *confusion matrix* provided in Table 3.5 summarizes these building block terms. The confusion matrix is a beneficial tool used for the analysis of how well the classifier can identify datasets belonging to different classes. *TP* and *TN* show when the classifier does things right, whereas *FP* and *FN* show when the classifier gets things wrong such as doing mislabeling.

A confusion matrix for the two classes *patient* = *yes* (positive) and *healthy* = *no* (*negative*) is provided in Table 3.6.

Table 3.4: Evaluation measures.

Measure	Formula
Accuracy, recognition rate	$\frac{TP+TN}{P+N}$
Error rate, misclassification rate	$\frac{FP+FN}{P+N}$
Sensitivity, true positive rate, recall	$\frac{TP}{P}$
Specificity, true negative rate	$\frac{TN}{N}$
Precision	$\frac{TP}{TP+FP}$
F, F_1, F-score, harmonic mean of precision and recall	$\frac{2\times \text{precision}\times \text{recall}}{\text{precision}+\text{recall}}$
F_β, where β is a non-negative real number	$\frac{(1+\beta^2)\times \text{precision}\times \text{recall}}{\beta^2\times \text{precision}+\text{recall}}$

Note: It should be noted that some measures are known by more than one name. *TP, TN, FP, P* and *N* show the number of true positive, true negative, false positive, positive and negative samples, respectively (see text).

Table 3.5: Confusion matrix that is presented with total for positive and negative datasets.

Actual Class	Predicted Class	
	yes	***no***
yes	*TP*	*FN*
no	*FP*	*TN*
Total	*P'*	*N'*

Table 3.6: Confusion matrix for the classes *patient, yes* and *healthy, no.*

Class	Patient	Healthy	Total	Recognition (%)
Patient	192	8	200	96
Healthy	4	196	200	98
Total	196	204	400	97

Note: An entry in row *i* and column *j* presents the number of datasets of class *i* labeled by the classifier as class *j*. In an ideal situation, the nondiagonal entries should have a value that equals to zero or a value that is close to zero.

With m classes, in which $m \geq 2$, a *confusion matrix* will be a table with at least size $m \times m$. An entry, $CM_{i,j}$ in the first m rows and m columns presents the number of datasets of class *i* labeled by the classifier as class *j*. If one wants a classifier to have decent accuracy, most of the datasets are preferably to be represented along the confusion matrix diagonal. That will be from entry $CM_{1,1}$ to entry $C_{m,m}$, and the rest of the entries will be either zero or a value close to zero. *FP* and *FN* are close to zero. That is the ideal situation in fact.

For providing the totals, it is possible for the table to have additional rows or columns. *P* and *N* are shown in the confusion matrix given in Table 3.5. In addition, P' is the number of datasets labeled as positive ($TP + FP$), whereas N' denotes the number of datasets labeled as negative ($TN + FN$). $TP + TN + FP + TN$, or $P + N$, or $P' + N'$ yields the total number of datasets. You may notice here that despite confusion matrix being shown for a binary classification problem, it is also possible to draw such matrices for multiple classes in a similar manner.

At this point, we can handle the evaluation measures with an onset with accuracy. On a given test, the *accuracy* of a classifier set is the percentage of test set datasets that the classifier correctly classifies, which can be shown as

$$\text{Accuracy} = \frac{TP + TN}{P + N} \tag{3.12}$$

Another common term used for this is the classifier's *recognition rate*, which shows how well the classifier recognizes the datasets of the various classes. The confusion matrix example for two classes "*patient* = *yes* (positive) and *healthy* = *no* (negative)" has been presented in Table 3.6. Recognition rates per class and overall as well as the totals are shown. With an overall look, it is not hard to understand if the corresponding classifier is causing a confusion in two classes. We see that it has mislabeled 600 *no* datasets as *yes*. When the class distribution is relatively balanced, accuracy seems to be the most effective.

Other terms that one can consider are *misclassification rate* or *error rate* of a classifier, *M*, which is basically 1 – *accuracy* (*M*), where M is the accuracy. The relevant calculation can be done using this formula:

$$\text{Error rate} = \frac{FP + FN}{P + N} \tag{3.13}$$

If one wants to use the training set rather than the test set for estimating the error rate of a model, we call this quantity *resubstitution error*, which is an error approximation optimistic of the true error rate since the testing of the model on a sample (not yet seen) has not been done. Accordingly, you get an optimistic kind of a corresponding accuracy rate.

As for the *class imbalance problem*, the main class of interest is uncommon, which means the dataset distribution reveals a minority positive class and a substantial majority of the negative class. For instance, in disease identification applications, the class of interest (or positive class) is *patient*, which occurs less often than the negative *healthy* class. To give an example from medicine, we can consider medical data with a rare class such as *RRMS*. Let's assume that the researcher trained a classifier for the purpose of classification of medical datasets, where we see *RRMS* as the class label attribute along with the possible class values *yes* and *no*. An accuracy rate of 98% is likely to make the classifier appear accurate. Yet, what would we say for a 2% of the training datasets having actually *RRMS (patient)*? Apparently 98% would not be an accuracy rate that is acceptable (the classifier could accurately be labeling only the datasets that are not *RRMS [patient])*. As a result of this, all the MS dataset in Table 2.12 would be misclassified. Rather than such a result, the researcher may require some other measures that have an access to how well the classifier is able to recognize the positive datasets/samples (*RRMS* = *yes*) and how well it can recognize the negative datasets/samples (*RRMS* = *no*).

For this purpose, a researcher can use the measures of *sensitivity* and *specificity*, respectively. *True positive* (*recognition*) *rate* (with the proportion of positive datasets identified in the approved manner) is another term sensitivity is also referred to and *true negative rate* is used for the specificity (with the proportion of negative datasets identified correctly). The definitions of these two measures are

$$\text{Sensitivity} = \frac{TP}{P} \tag{3.14}$$

$$\text{Specifity} = \frac{TN}{N} \tag{3.15}$$

We see accuracy as a function of both sensitivity and specificity:

$$\text{Accuracy} = \text{sensitivity}\frac{P}{(P+N)} + \text{specifity}\frac{N}{(P+N)} \tag{3.16}$$

Recall and *precision* measures are other terms used extensively for the purposes of classification. *Precision* is regarded as a measure of *exactness* (e.g., what percentage of datasets that are labeled positive are actually positive) while *recall* is a measure of *completeness* (e.g., what percentage of positive datasets are labeled as positive). These measures may be calculated through the following formulae:

$$\text{Precision} = \frac{TP}{TP+FP} \tag{3.17}$$

$$\text{Recall} = \frac{TP}{TP+FN} = \text{precision} = \frac{TP}{P} \tag{3.18}$$

For a class C, 1.0 is a perfect precision score of means with each dataset having the classifier being labeled as belonging to class C in fact belongs to class C. Yet, this does not give us any information regarding the number of class C datasets mislabeled by the classifier. A perfect recall score of 1.0 for C suggests that every item from class C was labeled accordingly. However, it does not provide us with the information concerning how many other datasets were labeled incorrectly as the ones falling into or belonging to class C. It is likely that there exists an inverse relationship between precision and recall so you can increase one of them at the expense of reducing the other one. For instance, it is possible that our medical classifier can achieve a high precision through the labeling of all MS datasets that present a certain way as the *MS subgroups* disorder; however, this might achieve a low recall should it mislabel many other instances of MS datasets on Table 2.12. It is a typical practice to use precision and recall scores together with each other with precision values being compared for a fixed value of recall or vice versa. To illustrate this point, the researcher might compare precision values at 0.75 as a recall value.

Alternative means for using recall and precision is to merge them into a single measure. This approach concerns the F measure, referred to as the F_1-score or F-score) and the F_β measure is defined as

$$F = \frac{2 \times \text{precision} \times \text{recall}}{\text{precision} + \text{recall}} \tag{3.19}$$

$$F_\beta = \frac{(1+\beta^2) \times \text{precision} \times \text{recall}}{\beta^2 \times \text{precision} + \text{recall}} \tag{3.20}$$

According to these formulae, β is a nonnegative real number. F measure is the *harmonic mean* of precision, with equal weight to precision and recall. F_β measure is a weighted measure of recall and precision. F_2 (this weights recall twice as much as precision) and $F_{0.5}$ (this weights precision twice as much as recall) are two of the F_β measures generally used.

Another question to pose at this point could be if there are *other cases in which accuracy may perhaps not be appropriate*. It is a common assumption that in classification problems all datasets are classifiable in a unique way, which means that each training dataset can only belong to one single class. However, large datasets have an extensive variety of data. For this reason, it would not be always wise to suppose that all of the datasets can be classified in a unique fashion. Instead of such an assumption, it is more likely to suppose that each dataset has the chance of belonging to more than one single class. Given this situation, we can pose the question of how to measure the accuracy of classifiers on large datasets. In this case, the accuracy measure will not be appropriate since it does not take into consideration the likelihood of datasets to belong to more than one class. In these instances, it will be beneficial to return a probability class distribution instead of returning a class label [24]. Then, it will be possible that accuracy measures use a *second guess* heuristic. In this way, a class prediction is deemed as correct if it agrees with the first or second most probable class. This approach will not take into account the nonunique classification of datasets to some extent, and it is not a complete solution either. Besides using the accuracy-based measures, it is also possible to compare the classifiers in regard to the additional aspects as provided. *Speed* is one way. It refers to the computational costs related during production and use of the given classifier. *Robustness* is known as the classifier's ability while carrying out correct predictions with noisy data or data with missing values. Robustness is usually assessed by means of a sequence of synthetic datasets that signify the increasing degrees related to noise as well as the missing values. Another measure is *scalability*, which is known as the capability of constructing the classifier efficiently with bulky amounts of data. It is usually assessed with a sequence of datasets that have increasing size. Finally, we can mention *interpretability*. This is regarded as the level of understanding that the predictor of the classifier provides. It is known that interpretability is subjective. For this very reason, its assessment would be harder to make. It is easy to do the interpretation of classification rules or the decision trees. However, it is possible that their interpretability may be lessened as they become more complex and complicated (Figure 3.6) [55].

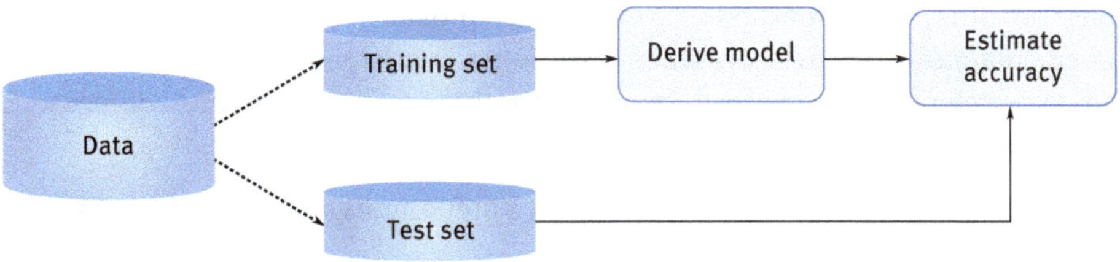

Figure 3.6: Estimating accuracy with the holdout method.

So far, some evaluation measures have been presented. Among the measures, what works best is the accuracy measure when the data classes are moderately distributed in an even way. Other measures such as precision, sensitivity/ recall, specificity, F and F_β are more appropriate for the class imbalance problem with the main class of interest being unusual.

In the following sections, we focus on the fact of getting classifier accuracy estimates in a reliable manner.

3.9.1.1 Holdout method and random subsampling

One of the ways to obtain reliable classifier accuracy estimation is the *holdout* method by which the given data are divided randomly into two independent sets: *training set* and *test set*. In general, two-thirds of the data are assigned to the training set and the remaining one-third to the test set. The training set is used for the purpose of deriving the model. The accuracy of the model is subsequently

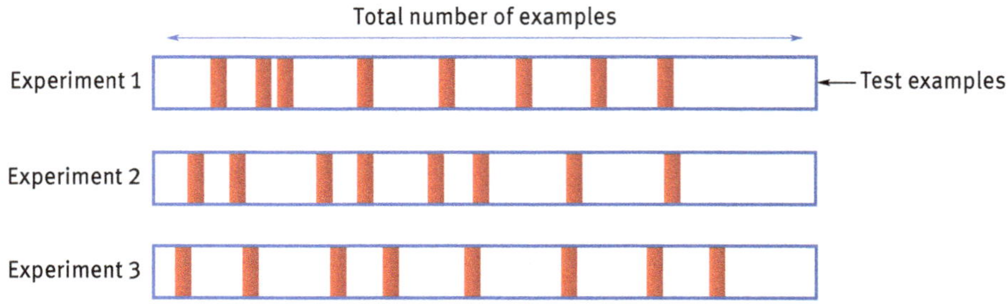

Figure 3.7: Random subsampling carried out k data splits of the dataset.

estimated with the test set (Figure 3.7). In this case, pessimistic estimate is observed because only a segment of the initial data is used for deriving the model. Another method is the *random subsampling* which is a variation of the holdout method. In random subsampling methods, one repeats the holdout at repeated times of iteration and takes the general accuracy prediction as average of accuracies as gained from each of the iteration [25].

Based on holdout method experiment 1, experiment 2 and experiment 3 in Figure 3.7, the same dataset is formed by being split into k number of proportions (in different proportions). Now, let us provide the procedural steps for the calculation of the test accuracy rates regarding experiment 1, experiment 2 and experiment 3 as obtained by splitting the same dataset into different k proportions.

Step (1) By splitting the dataset into different k proportions (test data), experiment 1, experiment 2 and experiment 3 are formed.

The following are the steps about how one may choose the experiment regarding the best result as to the classification accuracy rate (in the following steps) for experiment 1, experiment 2 and experiment 3 (best validation).

Step (2) The data to be allocated for the test data should be smaller than the data of the training.

Step (3) The classification accuracy rates are obtained from experiment 1, experiment 2 and experiment 3, and k-split with the highest classification accuracy rate among the results is preferred.

Please find below how holdout method is applied for the MS dataset (see Table 2.12).

Example 3.7 MS dataset has been formed by being split into k number of proportions (in different proportions) based on holdout method. Now, let us provide the procedural steps for the calculation of the test accuracy rates regarding experiment 1, experiment 2 and experiment 3 as obtained by splitting the same dataset into different k proportions based on k-NN (k-nearest neighbor) algorithm.

Step (1) By splitting the MS dataset into different k proportions (test data), experiment 1 and experiment 2 are formed.

The following are the steps about how one may choose the experiment (experiment 1 and experiment 2) regarding the best result as to the classification accuracy rate with k-NN algorithm (best validation).

Step (2) For experiment 1, MS dataset has been split as follows: one-third of the dataset as test dataset and two-thirds of the dataset as the training dataset. For experiment 2, the MS dataset has been split as follows: one-fourth of the dataset as test dataset and three-fourths as the training dataset.

Step (3) The classification accuracy rate of the MS dataset obtained based on Figure 3.8 according to k-NN algorithm has been calculated in two different experiments (pertaining to Step (2)). For experiments 1 and 2, results have been obtained as 68.31% and 60.5%, respectively. As can be seen, result of experiment 1 is higher than that of experiment 2. The best classification accuracy rate according to experiment 1 is yielded with a split of MS dataset having two-thirds allocated as training data and one-third as test data.

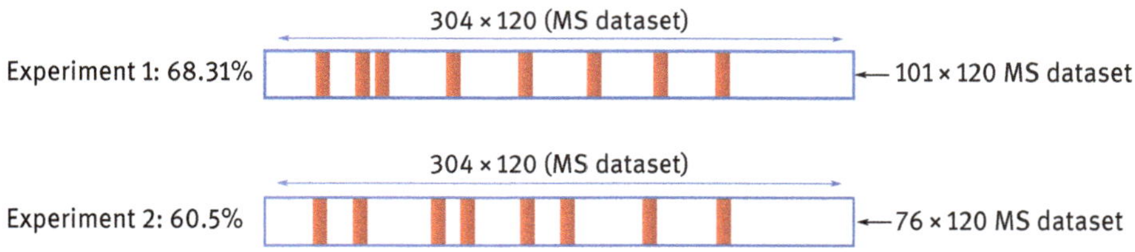

Figure 3.8: The classification of MS dataset based on holdout method with k-NN algorithm.

This book is categorized as follows: Chapter 6 describes decision tree algorithms (ID3, C4.5 and CART), Chapter 7 is about naïve Bayes algorithm, Chapter 8 describes SVM algorithms (linear and nonlinear), Chapter 9 describes k-NN algorithm and Chapter 10 describes ANN algorithms (FFBP and LVQ). The accuracy rate in the classification of data has been received through the application of holdout method. In the chapters specified, MS dataset based on holdout method (see Table 2.12), economy U.N.I.S. dataset (see Table 2.8) and WAIS-R dataset (see Table 2.19) have been split as follows: two-thirds (66.66 %) for the training procedure and one-third for the test procedure.

3.9.1.2 Cross-validation method

The initial data are randomly divided into k mutually exclusive subsets or "folds" in *k-fold cross-validation*. The denotation is $D_1, D_2, \ldots, D_k$ each having a fairly accurate equal size. The training and testing procedures are performed k times. As for the iteration process, i, partition D_i is reserved as the test set with the remaining partitions being used together for the training of the model. Let us give further details on this process: in the first iteration, subsets $D_2, \ldots, D_k$ serve as the training set together so that one can obtain a first model (tested on D_1). One does the training of the second iteration on subsets $D_1, D_3, \ldots, D_k$ with the test being done on D_2, and the sequence goes on and on. In this method, we see that each sample is used the same number of times for the training and only once for the testing procedure as opposed to the methods of holdout and random subsampling. The accuracy estimation is the overall number of correct classifications from the k iterations as they are divided by the total number of datasets in the initial data for the classification process. Stratified 10-fold cross-validation is suggested to be employed for the estimation of accuracy in general despite the fact that computation power allows the use of more folds because of its comparatively low bias as well as its variance. In *stratified cross-validation*, the folds are stratified in order that the class distribution of datasets in each fold will be generally the same as the one in the initial data. *Leave-one-out* is considered to be a special case of k-fold cross-validation in which k set is arranged in accordance with the number of initial datasets. Only one single sample is left out at a time for the test set. This is how the leave-one-out is termed [26, 27].

The procedural steps for cross-validation method are as follows:

Step (1) *k-Fold* data are split from the dataset for the test procedure.

Step (2) The dataset is split into *k-fold* equal number of subsets. While the $k-1$ number of dataset is used for the dataset training procedure, one is excluded from the training procedure so that it can be used for the purposes of testing.

Step (3) The application procedure in Step (2) is iterated k times. In the meantime, the accuracy rates of each application are calculated. By doing the calculation of the accuracy rate average, test accuracy rate is obtained.

Now, at this point, let us try to understand the application steps of cross-validation method in line with the procedural steps specified earlier.

Example 3.8 Let us get the test accuracy rate of WAIS-R dataset (Table 2.19) by splitting it according to 10-fold cross-validation method through k-NN algorithm.

Step (1) WAIS-R dataset (400×21) is split for the 10-fold data test procedure.

Step (2) The dataset is split into *k-fold* equal number of subsets. While the $10 \times (40 \times 21) -1$ number of dataset is used for the dataset training procedure, one is excluded from the training procedure so that it can be used for the purposes of testing.

Step (3) The application procedure in Step (2) is iterated $k = 10$ times. In the meantime, the accuracy rates of each application are calculated. The cells marked in dark color in Figure 3.9 are calculated. By calculating the accuracy rate average based on Figure 3.9 ((50% + 51% +42% + … + 75%)/10), test accuracy rate is obtained. According to k-NN algorithm, the test accuracy rate is obtained as 70.01%.

$k = 1$. ***step*** (50% test accuracy rate)

40 × 21	40 × 21	40 × 21	40 × 21	40 × 21
40 × 21	40 × 21	40 × 21	40 × 21	40 × 21

$k = 2$. ***step*** (51% test accuracy rate)

40 × 21	**40 × 21**	40 × 21	40 × 21	40 × 21
40 × 21	40 × 21	40 × 21	40 × 21	40 × 21

$k = 3$. ***step*** (42% test accuracy rate)

40 × 21	40 × 21	**40 × 21**	40 × 21	40 × 21
40 × 21	40 × 21	40 × 21	40 × 21	40 × 21

⋮

$k = 10$. ***step*** (75% test accuracy rate)

40 × 21	40 × 21	40 × 21	40 × 21	40 × 21
40 × 21	40 × 21	40 × 21	40 × 21	**40 × 21**

Figure 3.9: WAIS-R dataset 10-fold cross-validation.

As shown in Figure 3.9, $k = 1, 2, \ldots, 10$ steps, 40×21 matrix training and test data in each cell are different from one another.

References

[1] Larose DT, Larose CD. Discovering knowledge in data: an introduction to data mining. USA: John Wiley & Sons, Inc., Publication, 2014.

[2] Ott RL, Longnecker MT. An introduction to statistical methods and data analysis. USA: Nelson Education. 2015.

[3] Linoff GS, Berry MJ. Data mining techniques: for marketing, sales, and customer relationship management. USA: Wiley Publishing, Inc., 2011.

[4] Tufféry S. Data mining and statistics for decision making. UK: John Wiley & Sons, Ltd., Publication, 2011.

[5] Kantardzic M. Data mining: concepts, models, methods, and algorithms. USA: John Wiley & Sons, Ltd., Publication, 2011.

[6] Giudici P. Applied data mining: statistical methods for business and industry. England: John Wiley & Sons, Ltd., Publication, 2005.

[7] Myatt GJ. Making sense of data: a practical guide to exploratory data analysis and data mining USA: John Wiley & Sons, Ltd., Publication, 2007.

[8] Harrington P. Machine learning in action. USA: Manning Publications, 2012.

[9] Mourya SK, Gupta S. Data Mining and Data Warehousing. Oxford, United Kingdom: Alpha Science International Ltd., 2012.

[10] North M. Data mining for the masses. Global Text Project, 2012.
[11] Nisbet R, Elder J, Miner G. Handbook of statistical analysis and data mining applications. Canada: Academic Press publications, Elsevier, 2009.
[12] Pujari AK. Data mining techniques. United Kingdom: Universities Press, Sangam Books Ltd., 2001.
[13] Borgelt C, Steinbrecher M, Kruse R R. Graphical models: representations for learning, reasoning and data mining. USA: John Wiley & Sons, Ltd., Publication, 2009.
[14] Brown ML, John FK. Data mining and the impact of missing data. *Industrial Management & Data Systems*, 2003, 103(8), 611–621.
[15] Gupta GK. Introduction to data mining with case studies. Delhi: PHI Learning Pvt. Ltd., 2014.
[16] Han J, Kamber M, Pei J, Data mining Concepts and Techniques. USA: The Morgan Kaufmann Series in Data Management Systems, Elsevier, 2012.
[17] Raudenbush SW, Anthony SB. Hierarchical linear models: Applications and data analysis methods. USA: Sage Publications, 2002.
[18] John Lu ZQ. The elements of statistical learning: Data mining, inference, and prediction. *Journal of the Royal Statistical Society: Series A (Statistics in Society)*, 2010, 173(3), 693–694.
[19] LaValle S, Lesser E, Shockley R, Hopkins MS, Kruschwitz N. Big data, analytics and the path from insights to value. *MIT Sloan Management Review*, 2011, 52(2), 21–31.
[20] Witten IH, Frank E, Hall MA, Pal CJ. Data Mining: Practical machine learning tools and techniques. USA: Morgan Kaufmann Series in Data Management Systems, Elsevier, 2016.
[21] Arlot S, Alain C. A survey of cross-validation procedures for model selection. *Statistics Surveys*, 2010, 4, 40–79.
[22] Picard RR, Cook DR. Cross-validation of regression models. *Journal of the American Statistical Association*, 1984, 79(387), 575–583.
[23] Whitney AW. A direct method of nonparametric measurement selection. *IEEE Transactions on Computers*, 1971, 100 (9), 1100–1103.
[24] Kohavi R. A study of cross-validation and bootstrap for accuracy estimation and model selection. *Appears in the International Joint Conference on Articial Intelligence (IJCAI)*, 1995, 14(2), 1137–1145.
[25] Gutierrez-Osuna R. Pattern analysis for machine olfaction: A review, *IEEE Sensors Journal*, 2002, 2 (3), 189–202.
[26] Golub GH, Health M, Wahba G. Generalized cross-validation as a method for choosing a good ridge parameter. Technometrics, 1979, 21(2), 215–223.
[27] Arlot S, Matthieu L. Choice of V for V-fold cross-validation in least-squares density estimation. The Journal of Machine Learning Research, 2016, 17(208), 7256–7305.

4 Algorithms

This chapter covers the following items:
- Defining algorithm, pseudocode
- Image on coordinate systems
- Statistical application through algorithms image on coordinate systems

In this chapter, we look at what an algorithm is and what it isn't. We also look at an example of a common algorithm shown as both a numbered list and a flowchart, and then we briefly analyze what it does. This section includes some fundamental concepts you have to know before moving onto learning how to apply the algorithms with our available data. The answers to the following questions will be provided in this chapter:
- What is an algorithm?
- What is a flowchart?
- How can we do the calculations of mean, median, range, variation and standard deviation of an image data through algorithm?

What is an algorithm? You can consider an *algorithm* as a recipe that describes the precise steps required for the computer either to figure out the solution of a problem or attain a goal. As we know a food recipe specifies the ingredients required and an algorithm is similar to that. In computing terms, the ingredients are called inputs and a recipe corresponds to a *procedure*. The computer looks at your procedure and follows it exactly, and you eventually see the results and club these *outputs*. A programming algorithm describes the way of handling something and your computer will follow accordingly every time. Once you transform your algorithm into a language, it understands the essence.

It is important to consider that a programming algorithm is not a computer code. The language used is plain and simple English. Just like Aristotelian story principles, it has a start, a middle, and an end. One possibly labels the first step as "start" and the last one as "end". It includes only what you have to conduct the task and not unclear or ambiguous items.

It always comes to a solution that this solution is often the most efficient one you may imagine. Numbering the steps is often a good idea but there is no obligation to do it. Some people prefer using identification and write in *pseudocodes* instead of numbering process. Pseudocode is a semiprogramming language used to describe the steps in an algorithm. Some others just use a flowchart which is discussed in the following sections.

The fundamental concepts about algorithm are elaborated further as follows (Section 4.1): Section 4.1.1 describes the flowcharts and meanings, Section 4.1.2

https://doi.org/10.1515/9783110496369-004

covers the fundamental concepts of programming (token concept, expression concept, notification and identification concepts), Section 4.1.3. describes expressions and repetition statements and Section 4.2 elaborates as to how to calculate the mean, median, mode, range, variance and standard deviation of an image data through algorithm.

4.1 What is an algorithm?

Algorithm is the general name in the field of mathematics and computer sciences given to a systematic method of any kind of numerical calculation for the solution of a problem. In other words, an algorithm is a route drawn in order to solve a given problem or achieve a goal. It is often used in programming, and the language of all the programming relies on algorithm. At the same time, algorithm is a step-by-step provision of commands or expressions that do the basic chores or a pattern that will solve a single problem. It is important to pay attention to the ordering of these steps. There are two applicable approaches while solving a problem: algorithmic approach and heuristic approach. The most suitable among the possible methods for solution are chosen in algorithmic method. There are two applicable methods to identify the algorithm: textual plain expression and flowchart. Algorithms can be run by computers [1–3].

These days "algorithm" is uttered nearly by everyone, but the most exact definition remains a mystery. In this chapter, we get into the depths of this mystery. In addition to history, we also address the methodological aspects of algorithmic work as well as different programming techniques. We investigate the Babylonian–Sumerian method of extracting a root, one of the first documented examples of mathematical algorithms, and names like "Euklid" or "al-Khwarizmi" [4]. The word algorithm can be traced back to the ninth century, so it has a long history. Persian mathematician, scientist and astronomer named Abdullah Muhammad bin Musa al-Khwarizmi, known as "the father of Algebra," was responsible for the creation of the term "algorithm" in an indirect way. One of his books was translated into Latin in the twelfth century, and his name was converted into Latin as "Algorithm." Nevertheless, this was not the beginning of algorithms. We can wonder more about them [5].

4.1.1 What is the flowchart of an algorithm?

A flowchart is a pictorial or graphical representation of an algorithm by means of different shapes, arrows or symbols so that one can show a program or a process. We can understand a program with ease owing to algorithms. The main purpose of a flowchart is to analyze different processes [6]. Shapes in the flowchart change depending on the algorithm used and they may have different meanings accordingly. The shapes and meanings to be used in the algorithms to be handled in the following sections are provided in Table 4.1.

Table 4.1: Shapes and meanings used in the flowchart of algorithms.

Shape	Description
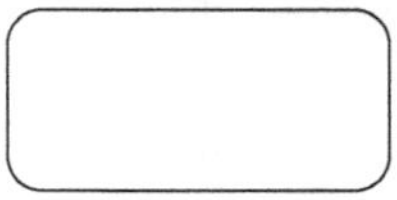	**Start, End:** Start and end procedures are standard in an algorithm. For this reason, each flow chart starts with the shape of "start" and ends with the shape "end". The procedures in algorithm are conducted between start and end.
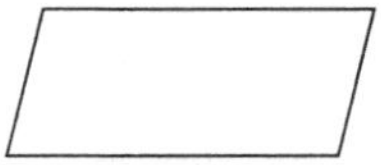	**Data input:** The shape which represents the data input encoded by the user.
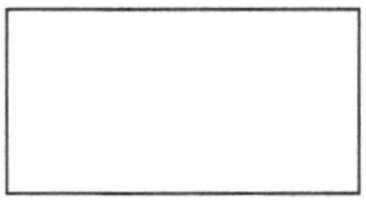	**Procedure:** This is used for the denotation of the procedures to be conducted during the execution of the procedures during the program. The procedures like formula and assigning the results of the procedures are written as they are. When the procedural flow reaches this step, the procedures written in this box are executed.
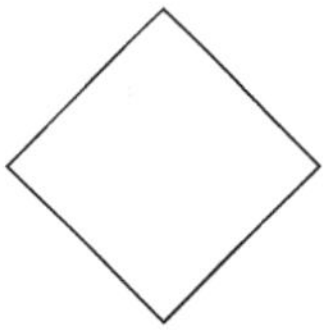	**Decision:** It is the shape that represents the decision and comparison procedures. It is also called condition or branching shape.It corresponds to the following in the programming language: ***if*** (condition) is true ... ***else*** ...
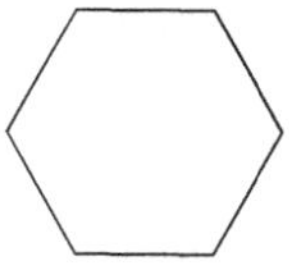	**Repetition Statement:** Different procedures are repeatedin different places. This repetitive procedure is called loop or iteration. The initial value, end value and increase value are assigned in the repetition statement shape. It corresponds to the following in the programming language: ***for*** (; ;) ***for each*** ( ) ***do-while*** (condition) ***while*** (condition)

4.1.2 Fundamental concepts of programming

It is important to know what token, variable and expression concepts mean in order to better understand the algorithms to be covered in the following sections. Let us analyze these fundamental concepts [4–8]:

Token concept: The smallest unit that bears a meaning in a programming language is called token. Tokens are the plainest elements that cannot be further split into more meaningful units in the algorithm.

We can categorize atoms as follows:

1. **Keywords**: They are the words that bear meaning for algorithm and are not allowed to be used as variables. For example, *for, if,* etc.

2. **Variables**: These are the tokens we can give name as we like provided that we obey the certain rules that have been specified previously. For example, *x*, *n*, *a*, *b*, *c*, *d*, *i*, etc.
3. **Operators**: These are the atoms that conduct certain procedures that have been previously specified: +, −, *, /, <, > are operators.
4. **Constants**: They are the tokens that are directly exposed to procedure and they do not contain variable data. In $c = k + l$, the numbers in the *k* and *l* variables are added and assigned to **c**. In fact, in the expression $c = k + 10$, the constant 10 is added to *a*.
5. **Brackets or punctuation marks:** Atoms used for bracket and termination are

   {

   ;

   ,

   }

Expression: The combination of variable, operator and constant is called expression.

The following examples explain who we can show the variables and procedure operator within algorithm:

Example 4.1 Let us define *a*, *b*, *c* variables and + procedure operator within the algorithm.

The expression is made up of *a+b//a*, *b*, *c* variable and + procedure operator.

Example 4.2 Let us define *d* variable, 2 constant and – procedure operator within algorithm.

The expression is made up of *d* – 2//*d* variable, 2 constant and – procedure operator.

Left side value: The expressions that depict object (variable) are called left side values. Since the assignment operator is on the left (=: equation), these expressions are called left side values.

Example 4.3 How can *k* variable value with 5 within the algorithm be defined as the left side value?

In order to define the value of *k* variable, *a* variable like **c** (integer, float, etc.) should be defined within the algorithm.

```
{
   k = c;
}
```

Right side value: Expressions that do not show object are called *right side value.* Since they are positioned on the right of the assigning operator (=: equation), they are named as right side values.

Example 4.4 Let us take k and l as variables (integer, double, etc.). Can we define $k + l$ expression within the algorithm as right side value?

In other words, by adding values of k and l variables, how can we make the interpretation by writing them as the right side value (which means on the right of the equation)?

```
{
  m = k + l;
}
```

The result of the summing of $k + l$ values within the algorithm given above is assigned to a variable like m, which means it shows that $k + l$ expression is a right side value.

4.1.2.1 Notification and identification concepts

Giving information about the attributes of the objects prior to their use is called *variable definition.*

In the following sections, the giving of information on attributes before using objects in algorithms has been defined outside the blocks[7].

What do we mean by the block concept? The sites in between the set brackets are called *blocks.*

There are two blocks in Figure 4.1; the second block is within the first block.

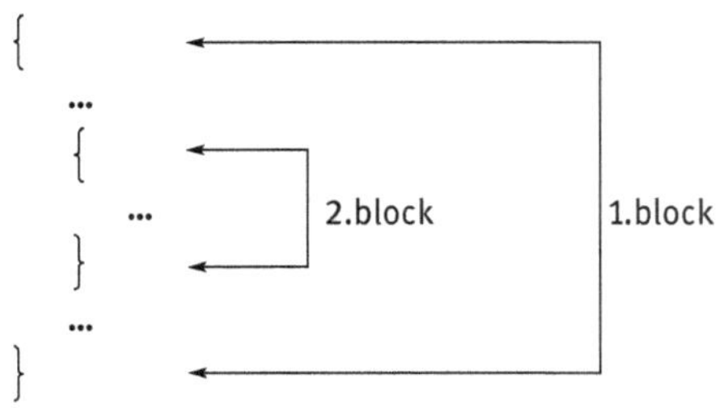

Figure 4.1: Block concept in notification and definition procedure.

Example 4.5 How do we define the priority about procedures in blocks in which variables are defined in the algorithm?

There are two blocks in the following algorithm. Different procedures will be done in two different blocks. The priority for the procedure is given to the block that is defined in the innermost part of the algorithm. The procedure of the block that is in the innermost section is completed and it goes on with the outermost block from inside to the external side.

```
…
{
    a * b; // a, b variable, * operator, multiplication procedure operator
    …
    {
        a = a * 2; // a variable, 2 constant value, * operator, multiplication procedure
        operator
    }
    …
}
…
```

a = ***a*** * **2** procedure within the algorithm calculates before the ***a*** * ***b*** operation in the first block.

Constant concept: Data are loaded as constant directly or within the objects. Constants are the data that are not in the form of object (integer, float, etc.).

Example 4.6 How can we transfer the summation operation of values in the algorithm pertaining to *a* and *b* variables to *a* variable like **c**?

```
{
    c = a + b;
}
```

In the algorithm, the value obtained from the summation operation within ***a*** and ***b*** variables is transferred to *c* variable.

4.1.3 Expressions and repetition statements

Expressions are a group of token that is made up of expression and terminator (as split from one another with semicolon)[7].

Example 4.7 ***i*** = ***a*** * ***b*** – ***d*** is an expression because (***a*** * ***b*** – ***d***) expression is not split with semicolon (;) or, in other words, with the terminator.

Example 4.8 ***i*** = ***a*** * ***b*** – ***d*** is an expression because (***a*** * ***b*** – ***d***) expression is split with semicolon (;) or, in other words, with the terminator.

In the following sections, expressions within the algorithms have been used for three purposes:

1. *Simple expressions* are formed by bringing a terminator (;) to the end of an expression within the algorithm.

For example, let us do the application of calculation for $a = b + 3$ operation in the algorithm.

```
{
    a = b + 3; // single operation is carried out.
}
```

2. *Compound expressions* constitute the summation of more than one expression within a block.

For example, let us do the application of calculation for $a = b + k - c$ and b++ operation as compound expressions:

```
...
{
   a = b + k - c;
   b++;
}
```

Here, operations a and b++ are done at the same time within the operational block, so a compound expression is at stake.

3. *Control expressions* are the ones used for controlling the program flow.

```
if(expression)
{
   Expressions 1
   Expressions 2
}
else
{
   Expressions 3
   Expressions 4
}
```

The compiler calculates the numerical value of an expression within the if parenthesis. It interprets this numerical value computed logically as follows: *True* (*if* it is a value other than zero) or *False* (*if* it is 0). If the result of the expression is *True*, the part up to else keyword is handled as operation. If the result of the expression is *False*, the part following the else keyword is taken under operation (or executed) (see Figure 4.2).

Let us now see how *if* expressions are applied to the algorithm through Example 4.9.

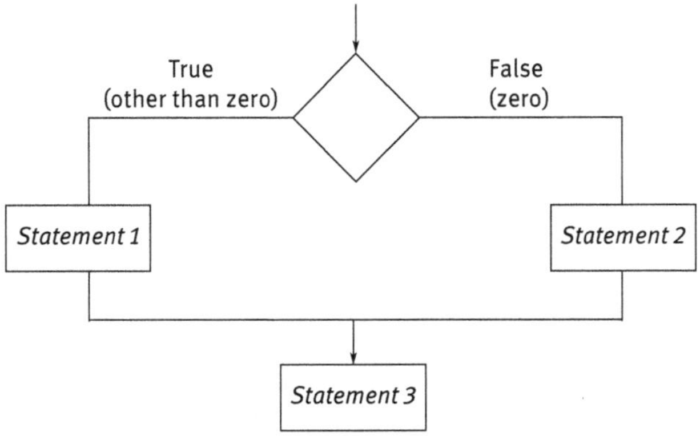

Figure 4.2: The flowchart of *if expressions*.

Example 4.9 Let us show if the value 5 assigned to variable *x* by an algorithm is an even or an odd number.

(1) $x = 5$; // the value 5 is assigned to the *x* variable
(2) ...
(3) ***if***($x \% 2 == 1$)
(4) *Statement 1*
(5) ***else***
(6) *Statement 2*
(7) *Statement 3*
(8) ...

Steps (1–3) The numerical value of the expression within *if* parenthesis is calculated.

Is the mode of the *x* variable 1 according to 2 or not?
if $x = 5$ has a mode that is 1 according to 2, in the algorithm it is operated with

Statement 1.
Result: Correct, which means $x = 5$ has a mode 1 according to 2 and it is operated with *Statement 1* in the algorithm.

Steps (3–4) Since the statement is *True*, the part up to the *else* keyword (*Statement 1*) is executed.

Steps (5–8) By skipping the *else* part, *Statements 2* and *3* are executed.

Repetition statements enable the execution of a part of the algorithm in an iterative manner. It is an indispensable component of algorithms. Let us illustrate at this point *for, while, do-while* and *foreach* repetition statements:

1. ***for*** repetition statement:
It is used to ensure a certain number of iteration in the *for* repetition statement.

Example 4.10 Let us apply the algorithm that makes the numbers from 1 to 100 count with *for* repetition statement:

```
...
for (i = 1; i < 100; i = i + 1)
...
```

The variable *i* of *for* loop makes us count one by one starting from 1 up until 100 (1, 2, 3, 4,..., 100).

2. ***while*** repetition statement:
In such repetition statements, the condition is generally given based on the variable.

Example 4.11 Let's apply an algorithm that makes numbers from 1 to 100 count through *while* repetition statement.

```
...
i = 1;
while( i < 100 )
  i++;
...
```

In Example 4.11, the *i* variable started with 1 ($i = 1$).

i++ operation is increased by 1 until the *i* variable reaches the value 100. As a result, by an incremental increase of 1, the while loop initiated with 1 reaches 100 and along this process it enables a counting from 1 to 100 (1, 2, 3, 4, ..., 100).

3. *foreach* repetition statement:
foreach is a repetition statement that proceeds over the array elements.

The general use of this repetition statement is as follows:

```
foreach(variable)
{
Statement 1;
}
```

Example 4.12 Let us obtain all the numbers in a number set through *foreach* repetition statement:

```
A = {1, 2, 3, 4, 5, 6, 7, 8, 9, 10};
```

```
foreach (i in A)
{
  ...
}
```

foreach repetition statement in Example 4.12 enables the obtaining of each element separately within the number set of *A* array 1, 2, 3, 4, 5, 6, 7, 8, 9, 10 from the *i* variable value.

4.2 Image on coordinate systems

A color is assigned to each point in the image in order to create a two-dimensional (2D) image. A point in 2D can be identified by a pair of numerical coordinates. It is also possible to indicate the colors numerically. For this reason, some time should be allocated to study the *coordinate systems* – a way of assigning numerical coordinates to geometric points. Each point corresponds to a pair of numbers in two dimensions. As the name suggests, each point corresponds to a triple of numbers in three dimensions. Numbers are associated with points, and *color models* -- a way of assigning numerical coordinates to geometric points. Each point corresponds to a pair of numbers in two dimensions and to a triple of numbers in three dimensions, and numbers are associated with colors accordingly [7–9].

4.2.1 Pixel coordinates

A digital image constitutes rows and columns of pixels. A pixel in such an image can be identified by stating which column and which row include it. A pixel can be identified by a pair of integers in terms of coordinates, yielding the column number and the row number. For instance, the pixel with coordinates (4,8) lies in column number 4 and row number 8. Columns are usually numbered from left to right, and they start with zero. Including the ones to be handled in this book, most graphic systems have the number rows from top to bottom, and they start from zero [10–14].

In Figure 4.3, you can see the 5×9 pixel grid, as depicted with row and column numbers. Rows are numbered from top to bottom.

Consider that the pixel identified by a pair of coordinates (x, y) is dependent on the choice of a coordinate system.

Row and column numbers specify a pixel, not a point. A pixel includes many points. It mathematically contains an infinite number of points. Computer graphics is not essentially oriented to color pixels and the goal is in fact to create and manipulate images. In ideal cases, an image should be defined by identifying a color for each point, and not just for each pixel. If it is assumed that there exists a true and ideal

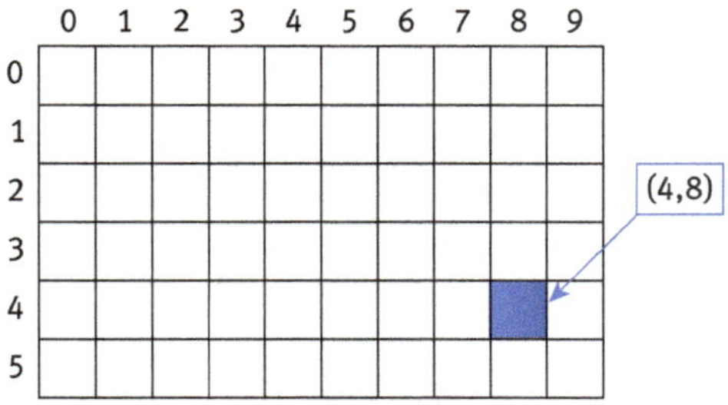

Figure 4.3: Row and column numbers on 5 × 9 pixel grid.

image to be displayed, then any image we display by coloring pixels will be an approximation.

4.2.2 Color models

Here what we discuss is the most basic and essential foundations regarding computer graphics: one of which is the coordinate systems and the other one is the color which is actually a complex and surprising subject matter. All of the modern monitors take digital input for the "value" of a pixel and makes conversion of this into an intensity level. When they are off, real monitors have nonzero intensity since the screen reflects some light. For the current purposes, we can regard this as "black" and monitor fully on as "white." We assume a numeric description of pixel color varying from zero to one. Black is zero, white is 1 and gray halfway between black and white (BW) is 0.5 [11–13].

The colors on a computer screen are generated as combinations of red, green and blue light. Different colors are made by varying the intensity of each type of light. It is possible to specify a color by three numbers, with the intensity of red, green and blue in the color. Intensity can be stated as a number in the range zero, for minimum intensity, to one, for maximum intensity. We call this method of specifying the color the *RGB color* model where RGB means red/green/blue. For instance, the number triple (1, 0, 0) signifies the color obtained by setting red to full intensity, whereas green and blue to half the intensity in the RGB color model [13–17]. The red, green and blue values for a color are termed as *color components*.

Figure 4.4(a) shows the RGB quantized form and Figure 4.4(b) shows $X_{\mathrm{R}}, X_{\mathrm{G}}, X_{\mathrm{B}}$, which are the weights applied to RGB colors to be able to figure out X. Figure 4.4(c) shows $X_{\mathrm{B}}, X_{\mathrm{W}}$ which are the weights applied to BW colors to be able to figure out X.

Actual RGB levels are often given in quantized form. Each component is indicated with an integer. One byte each is the most common size for these integers. Therefore, each of the three RGB components is an integer between 0 and 255. The three integers together take three bytes, corresponding to 24 bits. Hence, a system with "24 bit color" has 256 possible levels for each of the primary color.

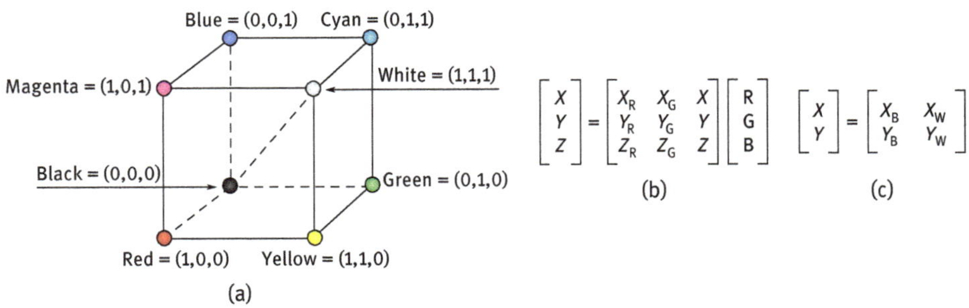

Figure 4.4: RGB–BW model. From Russ et al. [14].

4.3 Statistical application with algorithms: Image on coordinate systems

This section explains as to how to calculate mean, median, range, variance and standard deviation through statistical formulae as defined in Chapter 2 (Section 2.5) through the algorithms on the image coordinate systems.

Up to this point, we have mainly dealt with quantitative data. Let us assume that we have an image instead and we would like to compute some statistic indicators such as mean, median, range, variance and standard deviation. This section explains how to compute these values on an image. An *image, digital image* or *still image* is known as the binary representation of visual information, which may be exemplified as pictures, drawings, logos, graphs or individual video frames. In this study, we have worked on the .jpeg image format. JPEG (pronounced JAY-peg) is a graphic image file.

An image is a 2D plot measured in terms of pixels. In particular, an image of $K \times L$ pixels is a collection of pixels with screen (image) coordinates:

$$\left(x_i\ ,\ y_j\right),\ i=1,\ldots,K\ ;\ j=1,\ldots,L$$

In order to do the calculation of statistical operations (mean, median, range, variance and standard deviation) for any image data, we can apply the general algorithm provided in Figure 4.5.

Our aim regarding the algorithm in Figure 4.5 is to obtain the pixel value that corresponds to x- and y-coordinate axes. The pixel values obtained are calculated through algorithm using statistical operations.

Figure 4.5 shows the steps of procedures pertaining to the general algorithm *for an image* data to be applied to do calculations regarding statistical methods.

For an image data in x- and y-coordinate system

Iteration 1:

For $i=1; j=1,\ldots,L,\ \left(x_i, y_j\right)=\left(x_1, y_1, \ldots, y_L\right)$

```
foreach i = 1 in K// image in x coordinate
{
  foreach j = 1 in L// image in y coordinate
  {
    // Statement (Mean, Median, Range, Variance, Standard Deviation)
  }
}
```

Figure 4.5: An algorithm for an image data (general algorithm).

Step (1) When $i = 1$, $(x_i, y_j) = (x_1, y_1, \ldots, y_L)$

Step (2) With the loop to be formed for $j = 1, 2, \ldots, L$, $(x_i, y_j) = (x_1, y_1 \ldots, , y_L)$, the pixel value that will correspond to x- and y-coordinates are obtained.

In the first iteration, image in x- and y-coordinate axes, the pixel values in all columns of the y-axis corresponding to the first row of the x-axis are obtained.

Iteration 2: For $i = 2; j = 1, \ldots, L$, $(x_i, y_j) = (x_2, y_1, \ldots, y_L)$,

Step (1) When $i = 2$, $(x_i, y_j) = (x_1, y_1, \ldots, y_L)$

Step (2) When $j = 1, 2, \ldots, L$, with the loop to be formed for $(x_i, y_j) = (x_1, y_1, \ldots, y_L)$, the pixel value that will correspond to x- and y-coordinates are obtained.

In the second iteration, image in x- and y-coordinate axes, the pixel values in all the columns of the y-axis corresponding to the second row of the x-axis are obtained.

...

...

...

Iteration *K*: For $i = K; j = 1, \ldots, L$, $(x_i, y_j) = (x_K, y_1, \ldots, y_L)$

Step (1) When $i = K$, $(x_i, y_j) = (x_1, y_1, \ldots, y_L)$

Step (2) When $j = 1, 2, \ldots, L$, with the loop to be formed for $(x_i, y_j) = (x_1, y_1, \ldots, y_L)$, the pixel value that will correspond to x- and y-coordinates are obtained.

In the K iteration, image in x- and y-coordinate axes, the pixel values in all columns of the y-axis corresponding to the Kth row of the x-axis are obtained.

In images on coordinate, the general algorithm for an image data in Figure 4.5 can be applied as to the statistical methods (mean, median, range, variance and standard deviation). Section 4.3.1 explains in depth how the original magnetic resonance imaging (MRI) of a relapsing remitting multiple sclerosis (RRMS) patient can be calculated through algorithms with statistical methods on coordinates of RGB.

4.3.1 Statistical application with algorithms: BW image on coordinate systems

BW image is an image in which each pixel can either be black or white and nothing else. Grayscale image has gray values that vary from 0 to 255 where 0 = black and 255 = white, while the BW image has only 0 and 1 values where 0 = black and 1 = white.

Figure 4.6(a) and (b) shows a BW image from MRI of an RRMS patient.

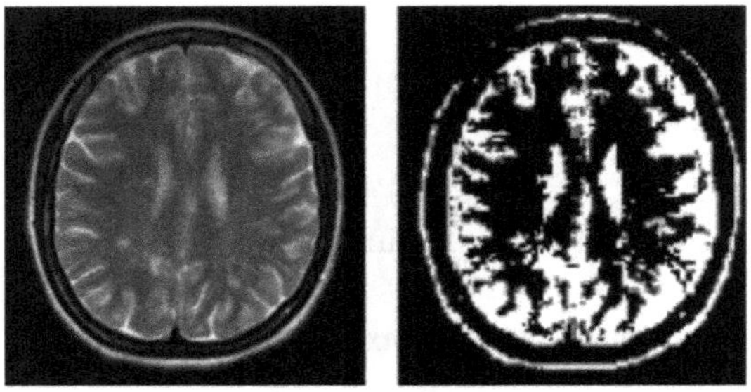

(a) (b)

Figure 4.6: BW application of original MRI belonging to RRMS (obtained from Hacettepe University): (a) original RRMS MRI and (b) BW application of MRI belonging to an RRMS patient.

Images used in Examples 4.13–4.17 belong to the RRMS patient of MS. The MRI images pertain to BW. For BW MRI images that have different colors and dimensions, mean, median, range, variance and standard deviation have been calculated.

How the original MRI of an RRMS patient can be calculated through algorithms with statistical methods in coordinates of BW is provided in the following steps:

Step 1 First of all, the pixel values that correspond to *x*- and *y*-coordinate axes in the image used are obtained.

Step 2 Statistical methods are used along with the algorithm through the pixel values obtained (Figure 4.5).

Now, let us apply the algorithm for an image data for the calculation of BW image of an RRMS patient (see Figure 4.6(b)) using statistical methods such as mean XE, median, range, variance and standard deviation in Examples 4.13–4.17.

Example 4.13 The application of *BW* image in Figure 4.7 for the *Mean* function algorithm has been elaborated step by step. The mean function algorithm has steps as elicited in Figure 4.7. In Figure 4.8, you can see how BW is applied to the MRI image of a patient suffering from RRMS.

(1) ***foreach*** $i = 1$ **in** K//image in x coordinate value, K is 200
(2) {
(3) ***foreach*** $j = 1$ **in** L//image in y coordinate value, L is 212
(4) {
(5) //**Mean**
(6) $M_{BW_i} = \sum_{i=1}^{K} \sum_{j=1}^{L} \frac{B(x_i, y_j) + w(x_i y_j)}{L}$
(7) }
(8) }

Figure 4.7: BW image **Mean()** function algorithm.

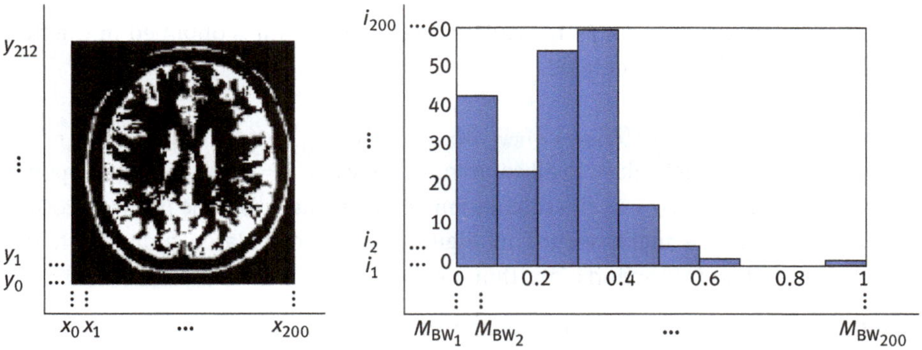

(a) (b)

Figure 4.8: The depiction of *Mean()* calculation regarding BW MRI image of an RRMS patient through histogram. (a) BW MRI image for pixel values and (b) BW MRI image of histogram for *Mean()*.

The histogram graphs are provided, owing to the application of Steps (1–8) for the *Mean()* function algorithm (see Figure 4.7) for the BW image in Figure 4.8.

Figure 4.8(a) shows in BW MRI image, x-axis $x = (x_1, x_2, ..., x_{200})$ row and y-axis $y = (y_1, y_2, ..., y_{212})$ column as the corresponding pixel values.

In the BW MRI image of histogram (Figure 4.8(b)), x-axis represents the results as obtained by calculating the mean value in Figure 4.8(a) (for each pixel values in x- and y-axes) and y-axis represents the iteration numbers in Figure 4.8(a) (iteration $i = 1, 2, ..., 200$) (mean values for pixel values in x- and y-axes).

In order to calculate the mean value in BW image (200 × 212 dimension), the following iteration steps are applied ((i): 1–200) along with the steps of *Mean()* function algorithm (Figure 4.7) (which means while passing from Figure 4.8(a) BW MRI image for pixel value to Figure 4.8(b) BW MRI image of histogram for *Mean()*).

For BW MRI image,

Iteration (i) = 1
Step (1) When ($i = 1$), with the loop to be formed for $(x_i, y_j) = (x_1, y_1, ..., y_{212})$, the pixel values that correspond to x- and y-coordinate axes are obtained.

Step (2) When ($i = 1$), for $j = 1, ..., 212$, with the pixel values obtained in the first iteration, the mean values are calculated.

...

...

...

Iteration (i) = 200
Step (1) When ($i = 200$), for $(x_i, y_j) = (x_{200}, y_1, ..., y_{212})$, the pixel values that correspond to x- and y-coordinate axes are obtained.

Step (2) When ($i = 200$), for $j = 1, ..., 212$, with the pixel values obtained in the first iteration, mean value is calculated.

The mean (M) values above ($M_{BW_i} = M_{BW_1}, M_{BW_2}, ..., M_{BW_{200}}$): for the iteration of ($i = 1, 2, ..., 200$) and for the pixel values in all columns in the y-axis (j=1, 2, ..., 212) corresponding to x-coordinate axis ($i = 1$), the calculation of mean values is done until ($i = 200$).

In Table 4.2, the mean values in Figure 4.8 are calculated along with the application of *Mean()* function algorithm for ($i = 1, ..., 200$).

Example 4.14: Application of *BW* image in Figure 4.9 for the *Median* function algorithm has been provided step by step. The steps of median function algorithm are shown in Figure 4.9. In Figure 4.10, BW is applied to the MRI image of a patient who is afflicted with RRMS.

The histogram graphs are provided owing to the application of Steps (1–8) for the *Median()* function algorithm (see Figure 4.9) for the BW image in Figure 4.10.

Figure 4.10(a) shows in BW MRI image, x-axis$x = (x_1, x_2, ..., x_{200})$ row and y-axis $y = (y_1, y_2, ..., y_{212})$ column as the corresponding pixel values.

In BW MRI image of histogram (Figure 4.10(b)), x-axis represents the results as obtained by calculating the median value in Figure 4.10(a) (for each pixel values in x- and y-axes) and y-axis represents the iteration numbers in Figure 4.10(a) (iteration $i = 1, 2, ..., 200$) (median values for pixel values in x- and y-axes).

In order to calculate the median value in BW image (200×212 dimension), the following iteration steps are applied **((*i*):** 1–200) along with the steps of *Median()* function algorithm (Figure 4.9) (where medians pass from Figure 4.10(a) BW MRI image for pixel value to Figure 4.10(b) BW MRI image of histogram for *Median()*).

For BW MRI image,

Iteration (i) = 1
Step (1) When ($i = 1$), with the loop to be formed for $(x_i, y_j) = (x_1, y_2, ..., y_{212})$, the pixel values that correspond to x- and y-coordinate axes are obtained.

Table 4.2: The detailed illustration of black and white MRI image *Mean()* calculation.

Iteration	Mean() Calculation for i^{th} iteration	Mean() Value for i^{th} iteration	BW MRI Image for pixel value-BW MRI Image of Histogram for Mean()
$i = 1$	$M_{BW_1} = \frac{[B(x_1, y_1) + W(x_1, y_1)] + \ldots + [B(x_1, y_{212}) + W(x_1, y_{212})]}{212}$	$M_{BW_1}\,[B(x_1, y_j) + W(x_1, y_j)] = [0.01]$	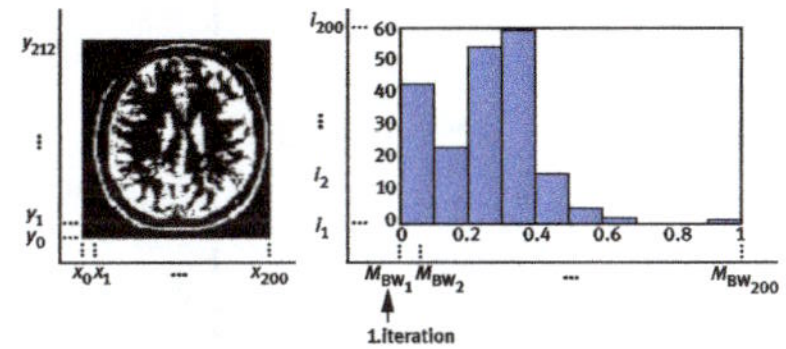
$i = 2$	$M_{BW_2} = \frac{[B(x_2, y_1) + W(x_2, y_2)] + \ldots + [B(x_{200}, y_{212}) + W(x_{200}, y_{212})]}{212}$	$M_{BW_2}\,[B(x_2, y_j) + W(x_2, y_j)] = [0]$	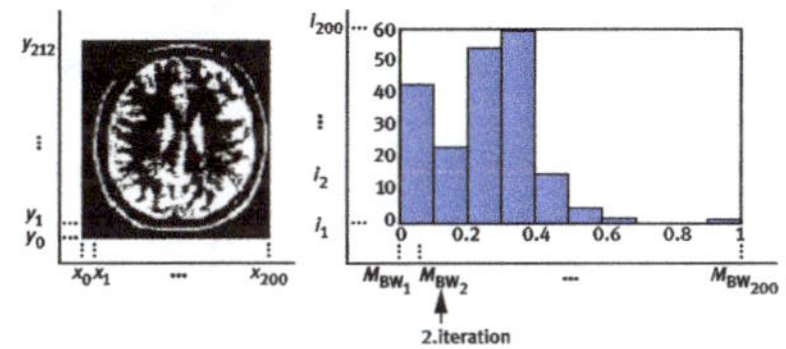
...	...	...	...
...	...	...	...
...	...	...	...
$i = 200$	$M_{BW_{200}} = \frac{[B(x_{200}, y_1) + W(x_{200}, y_1)] + \ldots + [B(x_{200}, y_{212}) + W(x_{200}, y_{212})]}{212}$	$M_{BW_{200}}\,[B(x_{200}, y_j) + W(x_{200}, y_j)] = [1]$	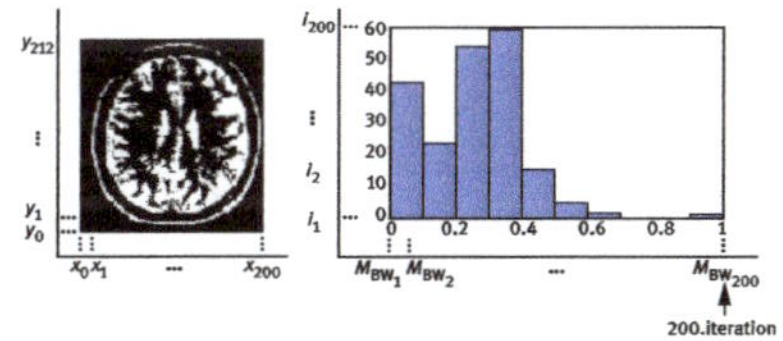

```
foreach i = 1 in K//image in x coordinate value K is 200
{
    foreach j = 1 in L//image in y coordinate value L is 212
    {
        //Median
        Mdn_BW_i = ( Σ_{i=1}^{K} Σ_{j=1}^{L} y_j(L/2) + y_j(L/2+1) ) / 2
    }
}
```

$$Mdn_{BW_i} = \frac{\sum_{i=1}^{K}\sum_{j=1}^{L} y_{j_{\left(\frac{L}{2}\right)}} + y_{j_{\left(\frac{L}{2}+1\right)}}}{2}$$

Figure 4.9: BW image ***Median()*** function algorithm.

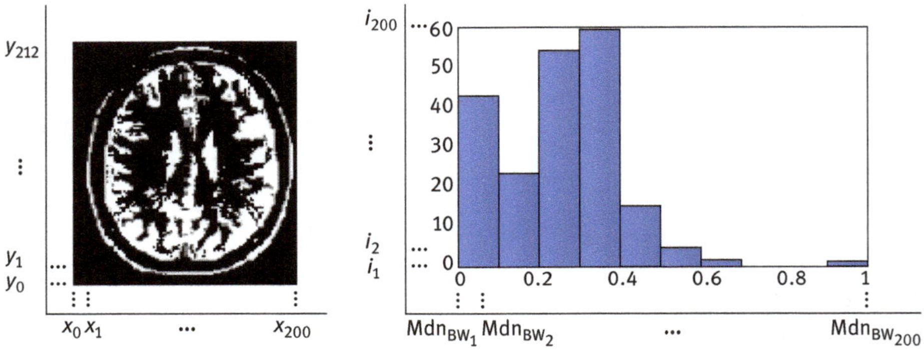

(a) (b)

Figure 4.10: The depiction of the *Median()* calculation regarding *BW MRI image* of an RRMS patient: (a) BW MRI image for pixel value and (b) BW MRI image for *Median()*.

Step (2) When (i = 1), for $j = 1, \ldots, 212$, with the pixel values obtained in the first iteration, the median values are calculated.

...

...

...

Iteration (i) = 200

Step (1) When ($i = 200$), for $(x_i, y_j) = (x_{200}, y_1, \ldots, y_{212})$, the pixel values that correspond to x- and y-coordinate axes are obtained.

Step (2): When (i = 200), for $j = 1, \ldots, 212$, with the pixel values obtained in the first iteration, median value is calculated.

The median (Mdn) values above ($\mathrm{Mdn}_{\mathrm{BW}_i} = \mathrm{Mdn}_{\mathrm{BW}_1}, \mathrm{Mdn}_{\mathrm{BW}_2}, \ldots, \mathrm{Mdn}_{\mathrm{BW}_{200}}$): for the iteration of (i = 1, 2, ..., 200) and for the pixel values in all columns in the y-axis (j = 1, 2, ..., 212) corresponding to x-coordinate axis (i = 1), the calculation of medianvalues is done until (i = 200).

In Table 4.3, the median values in Figure 4.10 are calculated along with the application of *Median()* function algorithm for ($i = 1,..., 200$).

Example 4.15: The application of *BW* image in Figure 4.11 for the *Range* function algorithm has been provided step by step. The steps of range function algorithm have been elicited in Figure 4.11. In Figure 4.12, you can see how BW is applied to the MRI image of a patient who is afflicted with RRMS.

The histogram graphs are provided owing to the application of Steps (1–8) for the *Range()* function algorithm (see Figure 4.11) for the BW image in Figure 4.12.

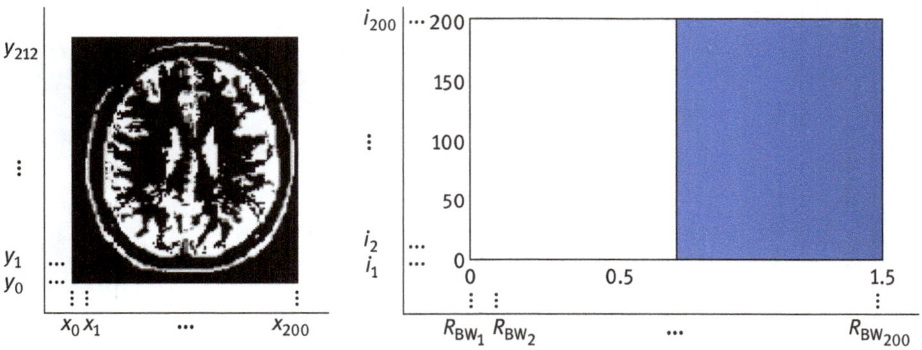

(a) (b)

Figure 4.12: The depiction of the *Range()* calculation regarding the BW MRI image of an RRMS patient: (a) BW MRI image for pixel value and (b) BW MRI image of histogram for *Range()*.

Figure 4.12(a) shows in BW MRI image, x-axis $x = (x_1, x_2, ..., x_{200})$ row and y-axis $y = (y_1, y_2, ..., y_{212})$ column as the corresponding pixel values.

In BW MRI image of histogram (Figure 4.12(b)), x-axis represents the results as obtained by calculating the range value in Figure 4.12(a) (for each pixel values in x- and y-axes) and y-axis represents the iteration numbers in Figure 4.12(a) (iteration $i = 1, 2, ..., 200$) (range values for pixel values in x- and y-axes).

In order to calculate the range value in BW image (200×212 dimension), the following iteration steps are applied ((i): 1–200) along with the steps of *Range()* function algorithm (Figure 4.11) (where ranges pass from Figure 4.12(a) BW MRI image for pixel value to Figure 4.12(b) BW MRI image of histogram for *Range()*).

For BW MRI image,

Iteration (i) = 1

Step (1) When ($i = 1$), with the loop to be formed for $(x_i, y_j) = (x_1, y_1, ..., y_{212})$, the pixel values that correspond to x- and y-coordinate axes are obtained.

Table 4.3: The detailed illustration of black and white MRI image *Median()* calculation.

Iteration	Median() Calculation for i^{th} iteration	Median() Value for i^{th} iteration	BW MRI Image for pixel value-BW MRI Image of Histogram for Median()
$i=1$	$Mdn_{BW_1}=[(B(x_1,y_1)+W(x_1,y_1))+\ldots +(B(x_1,y_{212})+W(x_1,y_{212}))]_{\frac{212}{2}} +[(B(x_1,y_1)+W(x_1,y_1))+\ldots +(B(x_1,y_{212})+W(x_1,y_{212}))]_{\frac{212}{2}+1}$	$Mdn_{BW_1}\ [B(x_1,y_j)\ +\ W(x_1,y_j)] = [0]$	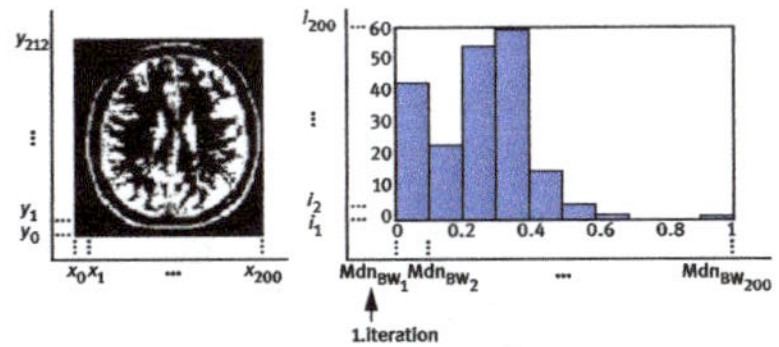
$i=2$	$Mdn_{BW_2}=[(B(x_2,y_1)+W(x_2,y_1))+\ldots +(B(x_2,y_{212})+W(x_2,y_{212}))]_{\frac{212}{2}} +[(B(x_2,y_1)+W(x_2,y_1))+\ldots +(B(x_2,y_{212})+W(x_2,y_{212}))]_{\frac{212}{2}+1}$	$Mdn_{BW_2}\ [B(x_2,y_j)\ +\ W(x_2,y_j)] = [0.2]$	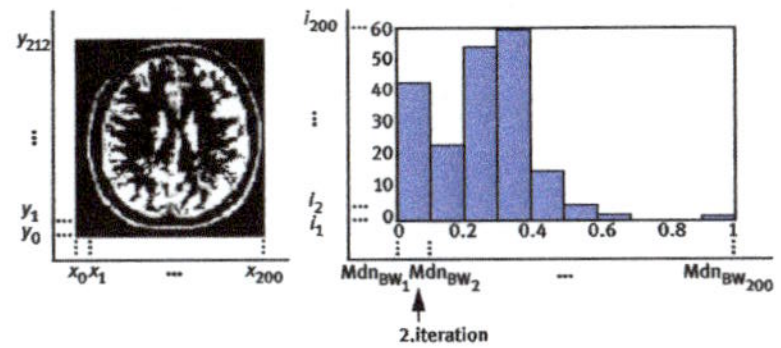
… … …	… … …	… … …	… … …
$i=200$	$Mdn_{BW_{200}}=[(B(x_{200},y_1)+W(x_{200},y_1))+\ldots +(B(x_{200},y_{212})+W(x_{200},y_{212}))]_{\frac{212}{2}} +[(B(x_{200},y_1)+W(x_{200},y_1))+\ldots +[(B(x_{200},y_{212})+W(x_{200},y_{212}))]_{\frac{212}{2}+1}$	$Mdn_{BW_{200}}\ [B(x_{200},y_j)\ +\ W(x_{200},y_j)] = [0]$	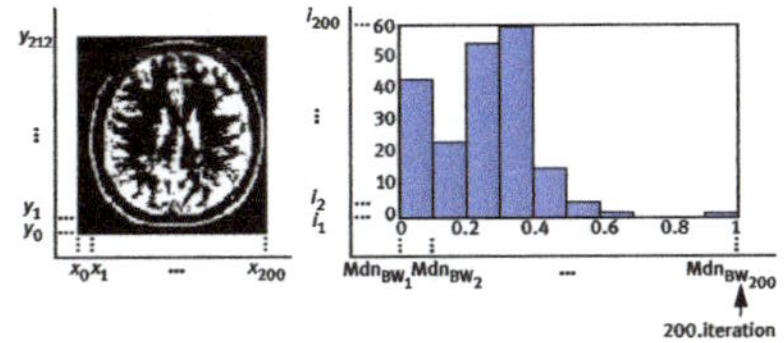

```
(1) foreach i = 1 in K//image in x coordinate value K is 200
(2) {
(3)   foreach j = 1 in L//image in y coordinate value L is 212
(4)   {
(5)     //Range
(6)     R_BW_i = max(x_i, y_j) − min(x_i, y_j)
(7)   }
(8) }
```

Figure 4.11: BW image *Range()* function algorithm.

Step (2) When (i = 1), for $j = 1, \ldots, 212$, with the pixel values obtained in the first iteration, the range values are calculated.

...

...

...

Iteration (*i*) = 200

Step (1) When ($i = 200$), for $(x_i, y_j) = (x_{200}, y_1, \ldots, y_{212})$, the pixel values that correspond to *x*- and *y*-coordinate axes are obtained.

Step (2) When (i = 200), for $j = 1, \ldots, 212$, with the pixel values obtained in the first iteration, range value is calculated.

The range (R) values above ($R_{BW_i} = R_{BW_1}, R_{BW_2}, \ldots, R_{BW_{200}}$): for the iteration of (i = 1, 2, ..., 200), for the pixel values in all the columns in the *y*-axis (j = 1,2, ..., 212) corresponding to *x*-coordinate axis (i = 1), the calculation of range values is done until (i = 200).

In Table 4.4, the range values in Figure 4.12 are calculated along with the application of *Range()* function algorithm for (i = 1,..., 200).

Example 4.16 The application for *BW* image in Figure 4.13 for the *Variance* function algorithm has been provided step by step. The steps of *Variance* function algorithm elicited in Figures 4.13 and 4.14 show how BW is applied to the MRI image of a patient who is afflicted with RRMS.

The histogram graphs are provided owing to the application of Steps (1–8) for the Variance() function algorithm (see Figure 4.13) for the BW image in Figure 4.14.

Figure 4.14(a) shows in BW MRI image, *x*-axis $x = (x_1, x_2, \ldots, x_{200})$ row and *y*-axis $y = (y_1, y_2, \ldots, y_{212})$ column as the corresponding pixel values.

In BW MRI image of histogram (Figure 4.14(b)), *x*-axis represents the results as obtained by calculating the variance value in Figure 4.14(a) (for each pixel values in *x*- and *y*-axes) and *y*-axis represents the iteration numbers in Figure 4.14(a) (iteration i = 1, 2, ..., 200) (variance values for pixel values in *x*- and *y*-axes).

Table 4.4: The detailed illustration of black and white MRI image *Range ()* calculation.

Iteration	Range() Calculation for i^{th} iteration	Range() Value for i^{th} iteration	BW MRI Image for pixel value-BW MRI Image of Histogram for Range()
$i=1$	$R_{BW_1}=[Max(B(x_1,y_1)+W(x_1,y_1))-Min(B(x_1,y_1)+W(x_1,y_1))]+\ldots+Max(B(x_1,y_{212})+W(x_1,y_{212}))-Min(B(x_1,y_{212})+W(x_1,y_{212}))]$	$R_{BW_1}\,[B(x_1,y_j)+W(x_1,y_j)]=[0]$	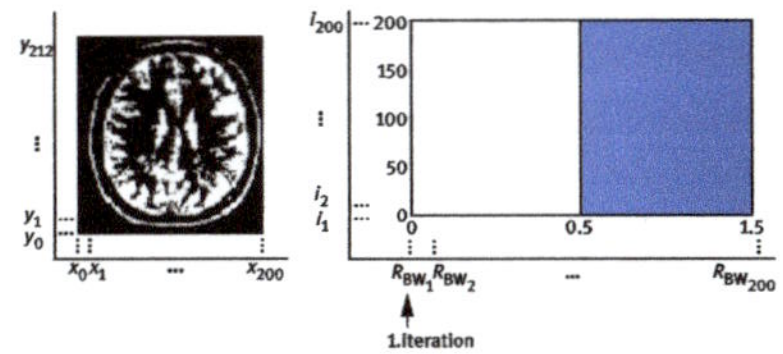
$i=2$	$R_{BW_2}=[Max(B(x_2,y_1)+W(x_2,y_1))-Min(B(x_2,y_1)+W(x_2,y_1))]+\ldots+[Max(B(x_2,y_{212})+W(x_2,y_{212}))-(Min(B(x_2,y_{212})+W(x_2,y_{212})])$	$R_{BW_2}\,[B(x_2,y_j)+W(x_2,y_j)]=[0]$	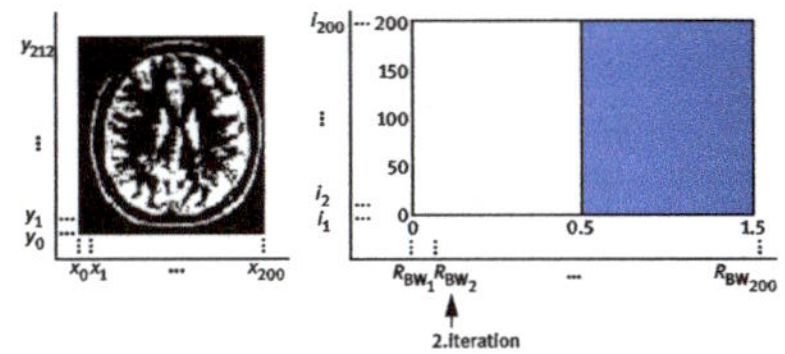
…	…	…	…
…	…	…	…
…	…	…	…
$i=200$	$R_{200}=[Max(B(x_{200},y_1)+W(x_{200},y_1))-Min(B(x_{200},y_1))]+\ldots+[Max(B(x_{200},y_{212})+W(x_{200},y_{212}))-Min(B(x_{200},y_{212})+W(x_{200},y_{212}))]$	$R_{BW_{200}}\,[B(x_{200},y_j)+W(x_{200},y_j)]=[0]$	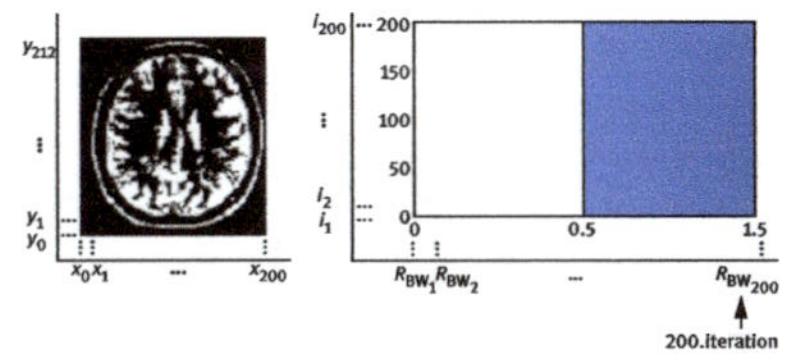

```
(1) foreach i = 1 in K//image in x coordinate value K is 200
(2) {
(3)   foreach j = 1 in L//image in y coordinate value L is 212
(4)   {
(5)     //Variance
(6)     σ²_BW_i = 1/(K×L) Σ_{i=1}^{K} Σ_{j=1}^{L} (x_i − x̄)² (y_j − ȳ)²
(7)   }
(8) }
```

Figure 4.13: *BW* image *Variance* function algorithm.

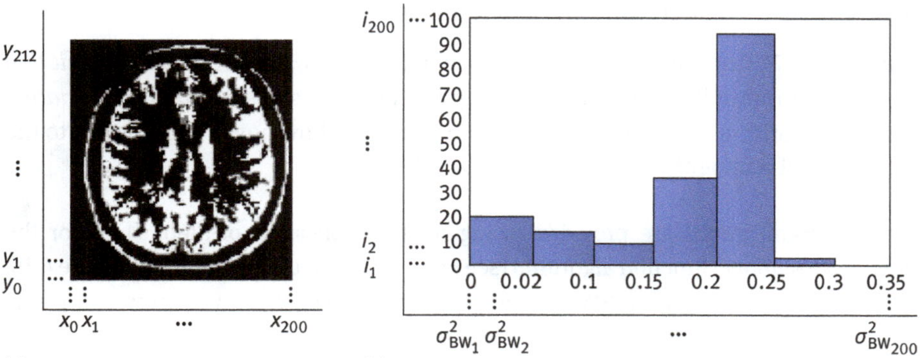

(a) (b)

Figure 4.14: The depiction of the *Variance()* calculation regarding the BW MRI image of an RRMS patient: (a) BW MRI image for pixel value and (b) BW MRI image of histogram for *Variance()*.

In order to calculate the variance value in BW image (200 × 212 dimension), the following iteration steps are applied ((*i*): 1–200) along with the steps of *Variance()* function algorithm (Figure 4.14) (where variances pass from Figure 4.14(a) BW MRI image for pixel value to Figure 4.14(b) BW MRI image of histogram for *Variance()*).

For BW MRI image,

Iteration (*i*) = 1

Step (1) When ($i = 1$), with the loop to be formed for $(x_i, y_j) = (x_1, y_1, \ldots, y_{212})$, the pixel values that correspond to *x*- and *y*-coordinate axes are obtained.

Step (2) When ($i = 1$), for $j = 1, \ldots, 212$, with the pixel values obtained in the first iteration, the variance values are calculated.

…

…

…

Iteration $(i) = 200$
Step (1) When $(i = 200)$, for $(x_i, y_j) = (x_{200}, y_1, \ldots, y_{212})$, the pixel values that correspond to x- and y-coordinate axes are obtained.

Step (2) When $(i = 200)$, for $j = 1, \ldots, 212$, with the pixel values obtained in the first iteration, variance value is calculated.

The variance(σ^2) values above $(\sigma^2_{\mathrm{BW}} = \sigma^2_{\mathrm{BW}_1}, \sigma^2_{\mathrm{BW}_2}, \ldots, \sigma^2_{\mathrm{BW}_{200}})$: for the iteration of $(i = 1, 2, \ldots, 200)$ and for pixel values in all columns in the y-axis $(j = 1, 2, \ldots, 212)$ corresponding to x-coordinate axis $(i = 1)$, the calculation of variance values is done until $(i = 200)$.

In Table 4.5, the variance values in Figure 4.13 are calculated along with the application of *Variance()* function algorithm for $(i = 1, \ldots, 200)$.

Example 4.17 The application of *BW* image in Figure 4.15 for the *standard deviation* function algorithm has been provided step by step. The steps of *standard deviation* function algorithm as elicited in Figures 4.15 and 4.16 show how BW is applied to the MRI image of a patient who is afflicted with RRMS.

The histogram graphs are provided owing to the application of Steps (1–9) for the *StandardDeviation()* function algorithm (see Figure 4.15) for the BW image in Figure 4.16.

Figure 4.16(a) shows in BW MRI image, x-axis $x = (x_1, x_2, \ldots, x_{200})$ row and y-axis $y = (y_1, y_2, \ldots, y_{212})$ column as the corresponding pixel values.

In BW MRI image of histogram (Figure 4.16(b)), x-axis represents the results as obtained by calculating the standard deviation value in Figure 4.16(a) (for each pixel values in x- and y-axes) and y-axis represents the iteration numbers in Figure 4.16(a) (iteration $i = 1, 2,, \ldots, 200$) (standard deviation values for pixel values in x- and y-axes).

In order to calculate the standard deviation value in BW image (200 × 212 dimension), the following iteration steps are applied ((i): 1–200) along with the algorithm steps of StandardDeviation() function algorithm (Figure 4.16) (where standard deviations pass from Figure 4.16(a) BW MRI image for pixel value to Figure 4.16(b) BW MRI image of histogram for *StandardDeviation()*.

For BW MRI image,

Iteration$(i) = 1$
Step (1) When $(i = 1)$, with the loop to be formed for $(x_i, y_j) = (x_1, y_1, \ldots, y_{212})$, the pixel values that correspond to x- and y-coordinate axes are obtained.

Step (2) When $(i = 1)$, for $j = 1, \ldots, 212$, with the pixel values obtained in the first iteration, the standard deviation values are calculated.

...

...

...

Table 4.5: The detailed illustration of black and white MRI image *Variance()* calculation.

Iteration	Variance() Calculation for i^{th} iteration	Variance() Value for i^{th} iteration	BW MRI Image for pixel value-BW MRI Image of Histogram for Variance()
$i=1$	$\sigma^2_{BW_1}=\left[\frac{1}{200\times 212}\sum_{j=1}^{212}\begin{matrix}(B(x_1-\bar{x})^2+(y_j-\bar{y})^2)+\\(W(x_1-\bar{x})^2+(y_j-\bar{y})^2)+\ldots\end{matrix}+\frac{1}{200\times 212}\sum_{j=1}^{212}\begin{matrix}(B(x_1-\bar{x})^2+(y_j-\bar{y})^2)+\\(W(x_1-\bar{x})^2+(y_j-\bar{y})^2)\end{matrix}\right]$	$\sigma^2_{BW_1}\,[B(x_1,y_j),\ W(x_1,y_j)]=[0.1]$	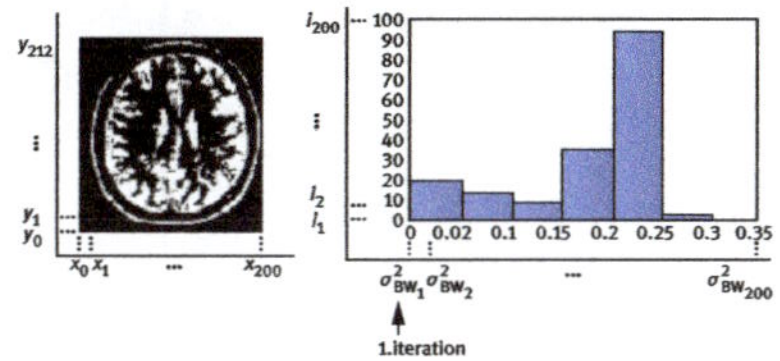
$i=2$	$\sigma^2_{BW_2}=\left[\frac{1}{200\times 212}\sum_{j=1}^{212}\begin{matrix}(B(x_2-\bar{x})^2+(y_j-\bar{y})^2)+\\(W(x_2-\bar{x})^2+(y_j-\bar{y})^2)+\ldots\end{matrix}+\frac{1}{200\times 212}\sum_{j=1}^{212}\begin{matrix}(B(x_2-\bar{x})^2+(y_j-\bar{y})^2)+\\(W(x_2-\bar{x})^2+(y_j-\bar{y})^2)\end{matrix}\right]$	$\sigma^2_{BW_2}\,[B(x_2,y_j),\ W(x_2,y_j)]=[0]$	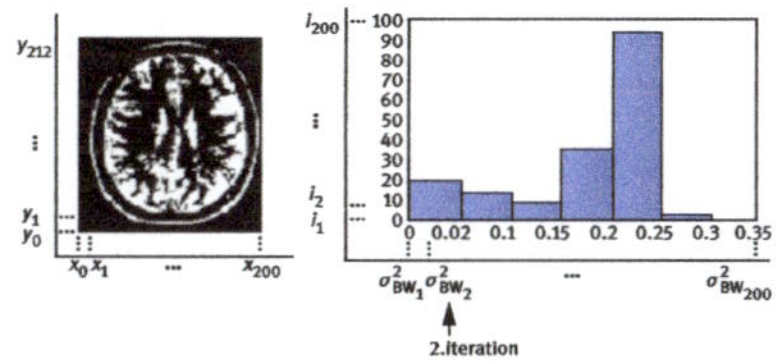
...	...	...	...
...	...	...	...
...	...	...	...
$i=200$	$\sigma^2_{BW_{200}}=\left[\frac{1}{200\times 212}\sum_{j=1}^{212}\begin{matrix}(B(x_{200}-\bar{x})^2+(y_j-\bar{y})^2)+\\(W(x_{200}-\bar{x})^2+(y_j-\bar{y})^2)+\ldots\end{matrix}+\frac{1}{200\times 212}\sum_{j=1}^{212}\begin{matrix}({}_{200}(\bar{x}^2-x)_j+(\bar{y}^2-y))+\\(W(x_{200}-x)^{-2}+(y-y_j)^{-2}\end{matrix}\right.$	$\sigma^2_{BW_{200}}\,[B(x_{200},y_j),\ W(x_{200},y_j)]=[0]$	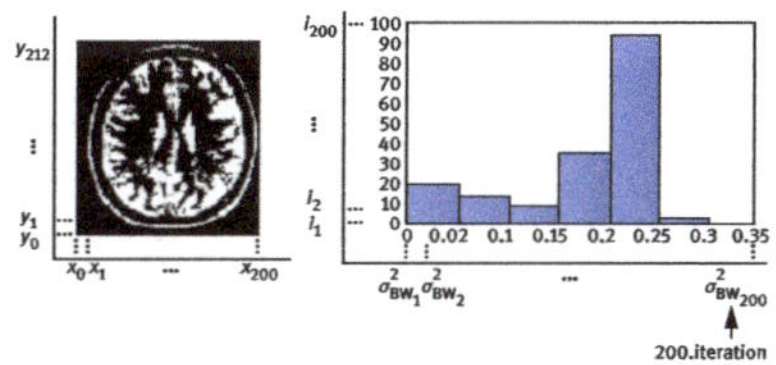

```
foreach i = 1 in K//image in x coordinate value, K is 200
{
    foreach j = 1 in L//image in y coordinate value, L is 212
    {
        //standard deviation
        σ² = 1/(K×L) Σ_{i=1}^{K} Σ_{j=1}^{L} (x_i − x̄)² (y_j − ȳ)²
        SD_BW_i = √σ²
    }
}
```

Figure 4.15: *BW* image *StandardDeviation* function algorithm.

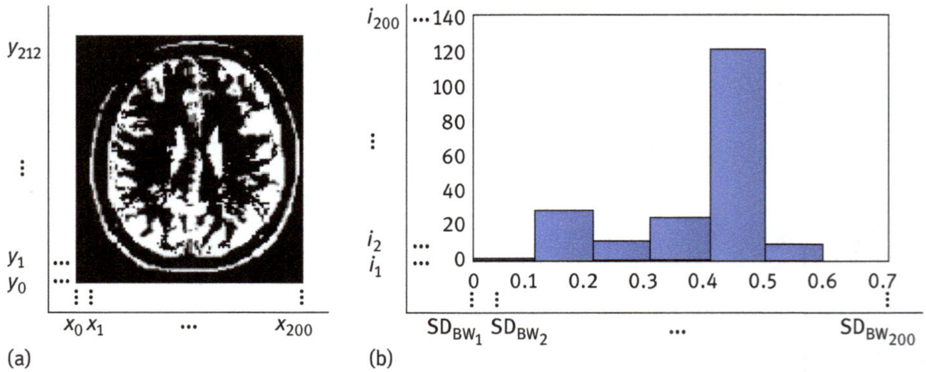

Figure 4.16: The depiction of *StandardDeviation()* calculation regarding BW MRI image of an RRMS patient. (a) BW MRI image for pixel value and (b) BW MRI image of histogram for *StandardDeviation()*.

Iteration (i) = 200

Step (1) When ($i = 200$), for $(x_i, y_j) = (x_{200}, y_1, \ldots, y_{212})$, the pixel values that correspond to x- and y-coordinate axes are obtained.

Step (2) When ($i = 200$), for $j = 1, \ldots, 212$, with the pixel values obtained in the first iteration, standard deviation value is calculated.

The standard deviation ($SD = \sigma$) values above ($SD_i = SD_{BW_1}, SD_{BW_2}, \ldots, SD_{BW_{200}}$): for the iteration of ($i = 1, 2, \ldots, 200$) and for the pixel values in all columns in the y-axis ($j = 1, 2, \ldots, 212$) corresponding to x-coordinate axis ($i = 1$), the calculation of standard deviation values is done until ($i = 200$).

In Table 4.6, the standard deviation values in Figure 4.16 are calculated along with the application of *StandardDeviation*() function algorithm for ($i = 1, \ldots, 200$).

Table 4.6: The detailed illustration of black and white MRI image *StandardDeviation()* calculation.

Iteration	StandardDeviation() Calculation for i^{th} iteration	StandardDeviation() Value for i^{th} iteration	BW MRI Image for pixel value-BW MRI Image of Histogram for StandardDeviation()
$i=1$	$SD_{BW_1}=[\sqrt{\frac{1}{200\times 212}\sum_{j=1}^{212}(B(x_1-\bar{x})^2+(y_j-\bar{y})^2)+(W(x_1-\bar{x})^2+(y_j-\bar{y})^2)+\ldots+\frac{1}{200\times 212}\sum_{j=1}^{212}(B(x_1-\bar{x})^2+(y_j-\bar{y})^2)+(W(x_1-\bar{x})^2+(y_j-\bar{y})^2)}]$	$SD_{BW_1}[B(x_1,y_j),W(x_1,y_j)]=[0.3]$	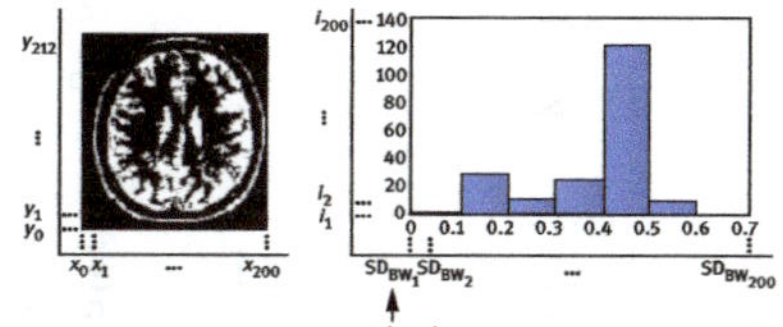
$i=2$	$SD_{BW_2}=[\sqrt{\frac{1}{200\times 212}\sum_{j=1}^{212}(B(x_2-\bar{x})^2+(y_j-\bar{y})^2)+(W(x_2-\bar{x})^2+(y_j-\bar{y})^2)+\ldots+\frac{1}{200\times 212}\sum_{j=1}^{212}(B(x_2-\bar{x})^2+(y_j-\bar{y})^2)+(W(x_2-\bar{x})^2+(y_j-\bar{y})^2)}]$	$SD_{BW_2}[B(x_2,y_j),W(x_2,y_j)]=[0]$	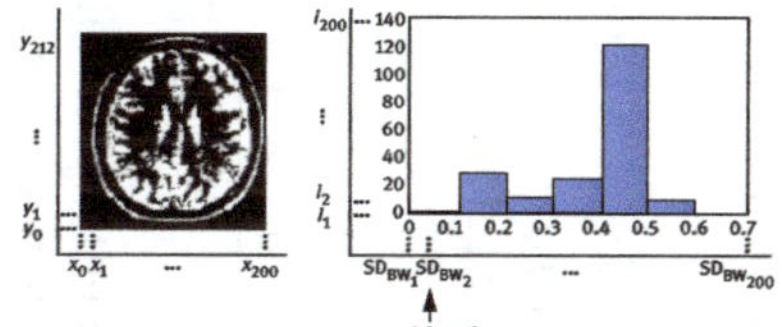
...	...	...	...
...	...	...	...
...	...	...	...
$i=200$	$SD_{BW_{200}}=[\sqrt{\frac{1}{200\times 212}\sum_{j=1}^{212}(B(x_{200}-\bar{x})^2+(y_j-\bar{y})^2)+(W(x_{200}-\bar{x})^2+(y_j-\bar{y})^2)+\ldots+\frac{1}{200\times 212}\sum_{j=1}^{212}(B(x_{200}-\bar{x})^2+(y_j-\bar{y})^2)+(W(x_{200}-\bar{x})^2+(y_j-\bar{y})^2)}]$	$SD_{BW_{200}}[B(x_{200},y_j),W(x_{200},y_j)]=[0]$	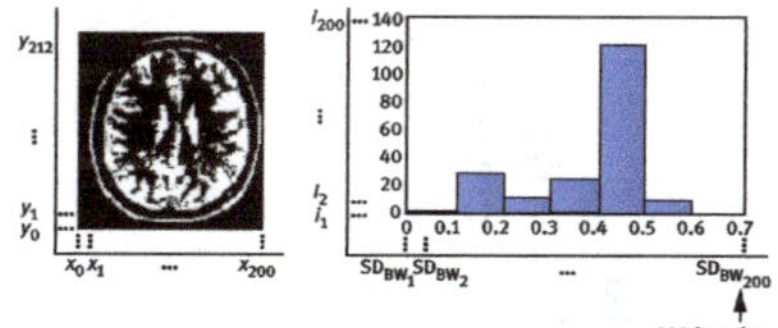

4.3.2 Statistical application with algorithms: RGB image on coordinate systems

In each pixel, the color is the function $f(x_i, y_j)$. In particular, for an RGB image in each pixel (x_i, y_j), three values $R(x_i, y_j)$, $G(x_i, y_j)$ and $B(x_i, y_j)$ are defined, each one ranging from 0 to 255; the resulting color in (x_i, y_j) is given by the combination as follows [15]:

$$\frac{R(x_i, y_j) + G(x_i, y_j) + B(x_i, y_j)}{3} \tag{4.1}$$

How the original MRI of an RRMS patient can be calculated through algorithms with statistical methods in coordinates of RGB is provided below:

Step (1) First of all, pixel values that correspond to *x*- and *y*-coordinate axes in the image used are obtained.

Step (2) Statistical methods are used along with the algorithm through the pixel values obtained (Figure 4.5).

Now, let us apply the algorithm for an image data for calculating RGB image of an RRMS patient (see Figure 4.17(b)) using statistical methods such as mean, median, range, variance and standard deviation in Examples 4.18– 4.22.

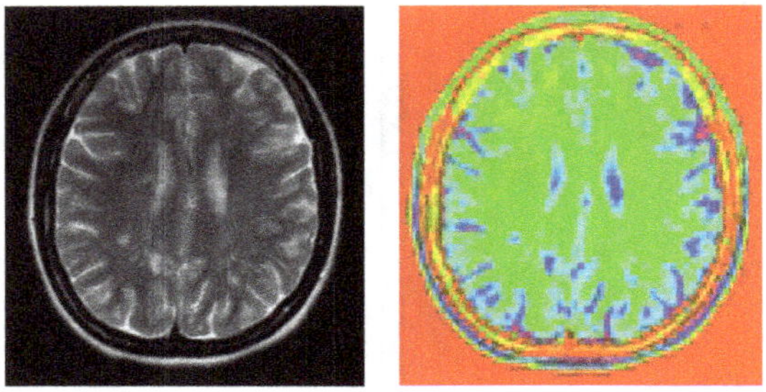

(a) (b)

Figure 4.17: RGB application of original MRI belonging to RRMS. (a) Original RRMS MRI and (b) RGB application of MRI belonging to an RRMS patient.

Example 4.18: The application for *RGB* image in Figure 4.17 for the *Mean* function algorithm has been provided step by step. The steps of the mean function algorithm as elicited in Figures 4.18 and 4.19 show how RGB is applied to the MRI image of a patient who is afflicted with RRMS.
The histogram graphs are provided owing to the application of Steps 1–8 for the *Mean()* function algorithm (see Figure 4.18) for the RGB image in Figure 4.19.

```
(1) foreach i = 1 in K//image in x coordinate value K is 196
(2) {
(3)     foreach j = 1 in L//image in y coordinate value L is 216
(4)     {
(5)         //Mean
(6)     M_RGB_i = Σ_{i=1}^{K} Σ_{j=1}^{L} (R(x_i, y_j) + G(x_i, y_j) + B(x_i, y_j)) / L
(7)     }
(8) }
```

$$M_{RGB_i} = \sum_{i=1}^{K}\sum_{j=1}^{L}\frac{R(x_i,y_j)+G(x_i,y_j)+B(x_i,y_j)}{L}$$

Figure 4.18: *RGB* image *Mean()* function algorithm.

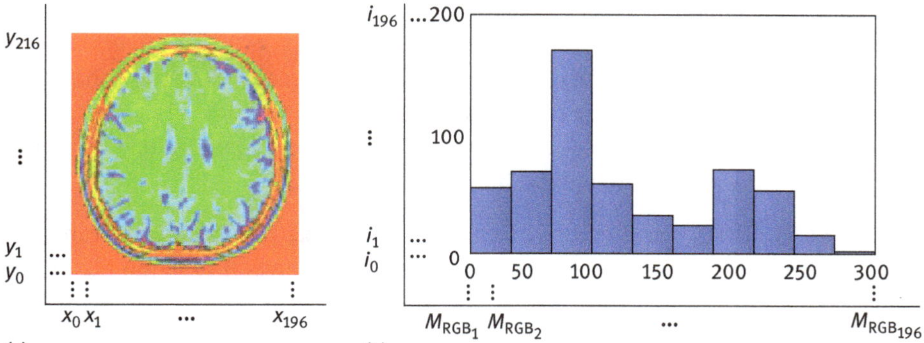

Figure 4.19: RGB image *Mean()* function algorithm. (a) RGB MRI image for pixel value and (b) RGB MRI image of histogram for *Mean()*.

Figure 4.19(a) shows in RGB MRI image, x-axis $x = (x_1, x_2, ..., x_{196})$ row and y-axis $y = (y_1, y_2, ..., y_{216})$ column as the corresponding pixel values.

In RGB MRI image of histogram (Figure 4.19(b)), x-axis represents the results as obtained by calculating the mean value in Figure 4.19(a) (for each pixel values in x- and y-axes) and y-axis represents the iteration numbers in Figure 4.19 (a) (iteration $i = 1, 2, ..., 196$) (mean values for pixel values in x- and y-axes).

In order to calculate the mean value in RGB image (196×216 dimension), the following iteration steps are applied ((i): 1–196) along with the steps of *Mean()* function algorithm (Figure 4.18) (which means while passing from Figure 4.19(a) RGB MRI image for pixel value to Figure 4.19(b) RGB MRI image of histogram for *Mean()*).

For RGB MRI image,

Iteration (i) = 1

Step (1) When ($i = 1$), with the loop to be formed for $(x_i, y_j) = (x_1, y_1, ..., y_{216})$, the pixel values that correspond to x- and y-coordinate axes are obtained.

Step (2) When ($i=1$), for $j=1, ..., 216$, with the pixel values obtained in the first iteration, the mean values are calculated.

...

...

...

Iteration *(i)* = 196

Step (1) When ($i=196$), for $(x_i, y_j)=(x_{196}, y_1, ..., y_{216})$, the pixel values that correspond to x- and y-coordinate axes are obtained.

Step (2) When ($i=196$), for the pixel values obtained in the first iteration, mean value is calculated.

The mean (M) values above ($M_{\text{RGB}_i}=M_{\text{RGB}_1}, M_{\text{RGB}_2}, ..., M_{\text{RGB}_{196}}$): ($i=1, 2, ..., 196$) for the pixel values in all columns in the y-axis ($j=1, 2, ..., 216$) corresponding to x-coordinate axis ($i=1$), the calculation of mean values is done until ($i=196$).

In Table 4.7, the mean values in Figure 4.19 are calculated along with the application of *Mean()* function algorithm for ($i=1,..., 196$).

Example 4.19: The application of *RGB* image in Figure 4.20 for the *Median* function algorithm has been provided step by step. The steps of median function algorithm elicited in Figures 4.20 and 4.21 show how RGB is applied to the MRI image of a patient who is afflicted with RRMS.

The histogram graphs are provided owing to the application of Steps (1–8) for the *Median()* function algorithm (see Figure 4.20) for the RGB image in Figure 4.21.

Figure 4.21 (b) in RGB MRI image of histogram., x-axis represents the results as obtained by calculating the median value in Figure 4.21(a) (for each pixel values in x- and y-axes). y-Axis represents the iteration numbers in Figure 4.21(a) (iteration = 1,2, ..., 196) (median values for pixel value in x- and y-axes).

In order to calculate the median value in RGB image (196 × 216 dimension), the following iteration steps are applied ((i): 1–196) along with the steps of *Median()* function algorithm (Figure 4.21) (where medians pass from Figure 4.21(a) RGB MRI image for pixel value to Figure 4.18(b) RGB MRI image of histogram for *Median()*).

For RGB MRI image,

Iteration (*i*) = 1

Step (1) When ($i=1$), with the loop to be formed for $(x_i, y_j)=(x_1, y_1, ..., y_{216})$, the pixel values that correspond to x- and y-coordinate axes are obtained.

Step (2) When ($i=1$), for $j=1, ..., 216$, with the pixel values obtained in the first iteration, the median values are calculated.

...

...

...

Table 4.7: The detailed illustration of red/green/blue MRI image *Mean()* calculation.

Iteration	Mean() Calculation for i^{th} iteration	Mean() Value for i^{th} iteration	RGB MRI Image for pixel value-RGB MRI Image of Histogram for Mean()
$i = 1$	$M_{RGB_1} = \frac{[R(x_1, y_1)+G(x_1,y_1)]+B(x_1,y_1)+\cdots +[R(x_1,y_{216})+G(x_1,y_{216})+B(x_1,y_{216})]}{212}$	$M_{RGB_1}\ [R(x_1, y_j),\ G(x_1, y_j),\ B(x_1, y_j)] = [255]$	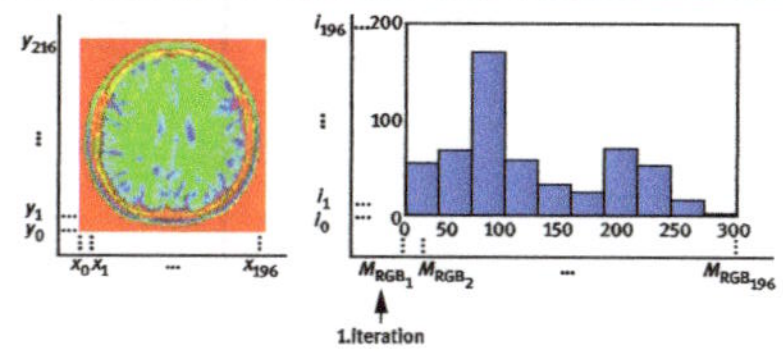
$i = 2$	$M_{RGB_2} = \frac{[R(x_2,y_1)+G(x_2,y_1)]+B(x_2,y_1)+\cdots +[R(x_2,y_{216})+G(x_2,y_{216})+B(x_2,y_{216})]}{212}$	$M_{RGB_2}\ [R(x_2, y_j),\ G(x_2, y_j),\ B(x_2, y_j)] = [71]$	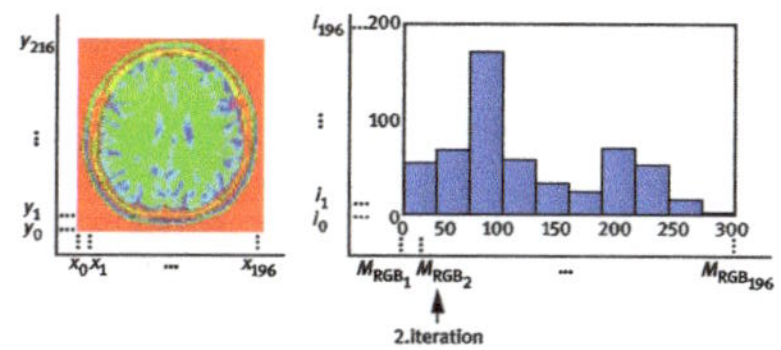
⋯ ⋯ ⋯	⋯ ⋯ ⋯	⋯ ⋯ ⋯	⋯ ⋯ ⋯
$i = 196$	$M_{RGB_{196}} = \frac{\frac{R(x_{196},y_1)+G(x_{196},y_1)]+B(x_{196},y_1)+-}{212} +[R(x_{196},y_{216})+G(x_{196},y_{216})+B(x_{196},y_{216})]}{212}$	$M_{RGB_{196}}\ [R(x_{200}, y_j),\ G(x_{200}, y_j),\ B(x_{200}, y_j)] = [0]$	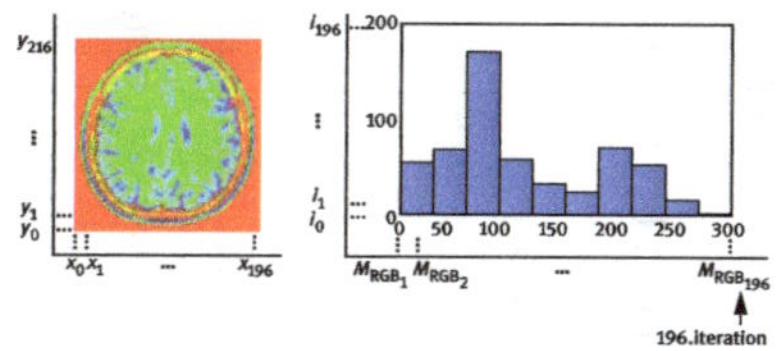

```
foreach i = 1 in K//image in x coordinate value K is 196
{
    foreach j = 1 in L//image in y coordinate value L is 216
    {
        //Median
        Mdn_RGB_i = ( Σ_{i=1}^{K} Σ_{j=1}^{L} y_j(L/2) + y_j(L/2+1) ) / 2
    }
}
```

$$Mdn_{RGB_i} = \frac{\sum_{i=1}^{K}\sum_{j=1}^{L} y_{j\left(\frac{L}{2}\right)} + y_{j\left(\frac{L}{2}+1\right)}}{2}$$

Figure 4.20: RGB image *Median()* function algorithm.

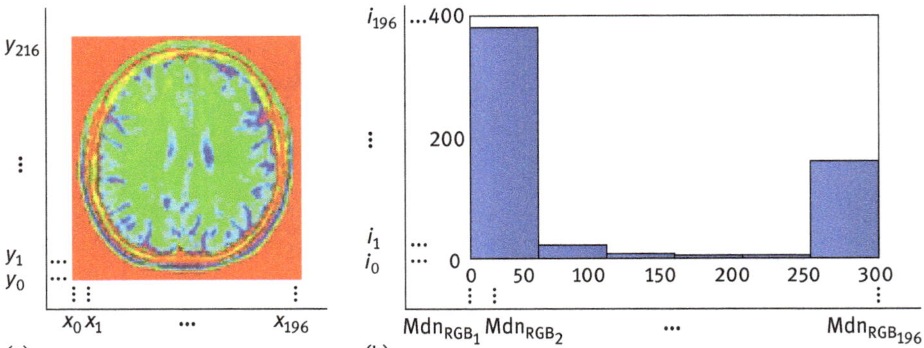

(a) (b)

Figure 4.21: The depiction of the *Median()* calculation regarding the RGB MRI image of RRMS patient through histogram. (a) RGB MRI image for pixel value and (b) RGB MRI image of histogram. for *Median()*.

Iteration (*i*) = 196

Step (1) When $(i = 196)$, for $(x_i, y_j) = (x_{196}, y_1, \ldots, y_{216})$, the pixel values that correspond to *x*- and *y*-coordinate axes are obtained.

Step (2) When $(i = 196)$, for with the pixel values obtained in the first iteration, median value is calculated.

The median (Mdn) values above $(\text{Mdn}_{\text{RGB}_i} = \text{Mdn}_{\text{RGB}_1}, \text{Mdn}_{\text{RGB}_2}, \ldots, \text{Mdn}_{\text{RGB}_{196}})$: for the pixel values in all columns in the *y*-axis $(j = 1, 2, \ldots, 216)$ corresponding to *x*-coordinate axis $(i = 1)$, the calculation of mean values is done until $(i = 196)$.

In Table 4.8, the mean values in Figure 4.20 are calculated along with the application of *Median()* function algorithm for $(i = 1, \ldots, 196)$.

Example 4.20 The application of *RGB* image in Figure 4.22 for the range function algorithm has been provided step by step. The range function algorithm is shown in Figure 4.22. Figure 4.23 shows how RGB is applied to the MRI image of a patient who is afflicted with RRMS.

Table 4.8: The detailed illustration of red/green/blue MRI image *Median()* calculation.

Iteration	Median() Calculation for i th iteration	Median() Value for i th iteration	RGB MRI Image for pixel value-RGB MRI Image of Histogram for Median()
$i=1$	$Mdn_{RGB_1} = [R(x_1,y_1)+G(x_1,y_1)] + B(x_1,y_1)+\cdots$ $+[R(x_1,y_{216})+G(x_1,y_{216}) + B(x_1,y_{216})]_{\frac{216}{2}}$ $+[R(x_1,y_1)+G(x_1,y_1)] + B(x_1,y_1)+\cdots$ $+[R(x_1,y_{216})+ G(x_1,y_{216}) + B(x_1,y_{216})]_{\frac{216}{2}+1}$	$Mdn_{RGB_1}[R(x_1,y_j),\ G(x_1,y_j),B(x_1,y_j)]$ $=[255]$	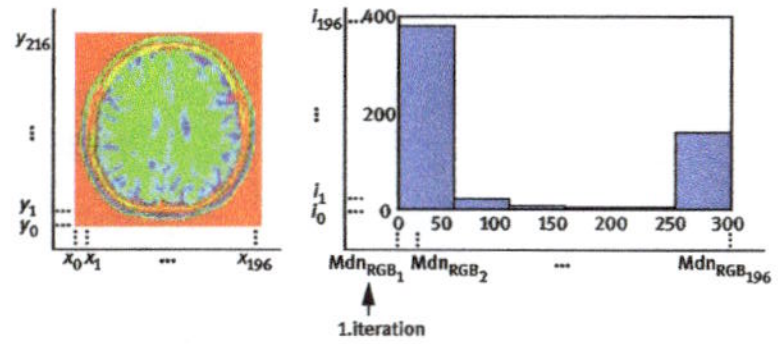
$i=2$	$Mdn_{RGB_2} = [R(x_2,y_1) + G(x_2,y_1)]+ B(x_2,y_1) +\cdots$ $+[R(x_2,y_{216})+ G(x_2,y_{216}) + B(x_2,y_{216})]_{\frac{216}{2}}$ $+[R(x_2,y_1)+ G(x_2,y_1)]+B(x_2,y_1)+\cdots$ $+[R(x_2,y_{216})+ G(x_2,y_{216})+B(x_2,y_{216})]_{\frac{216}{2}}$	$Mdn_{RGB_2}[R(x_2,y_j),\ G(x_2,y_j),B(x_2,y_j)]$ $=[255]$	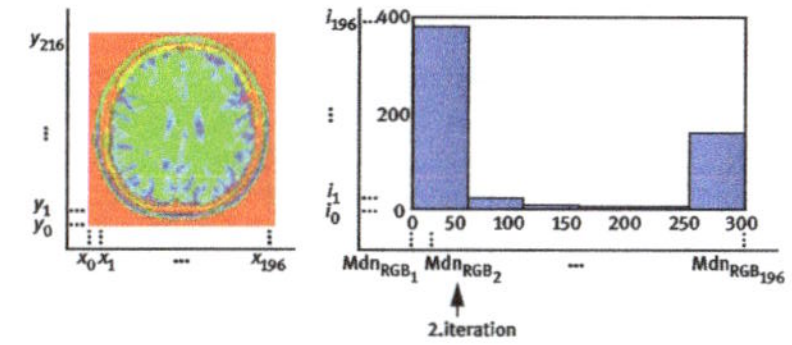
… … …	… … …	… … …	… … …
$i=196$	$Mdn_{RGB_{196}} = [R(x_{196},y_1) + G(x_{196},y_1)] + B(x_{196},y_1) +\cdots$ $+[R(x_{196},y_{216})+ G(x_{196},y_{216}) + B(x_{196},y_{216})]_{\frac{216}{2}}$ $+[R(x_{196},y_1)+ G(x_{196},y_1)] + B(x_{196},y_1) +\cdots$ $+ [R(x_{196},y_{216})+ G(x_{196},y_{216})!+B(x_{196},y_{216})]_{\frac{216}{2}+1}$	$Mdn_{RGB_{196}}[R(x_{196},y_j),G(x_{196},y_j),B(x_{196},y_j)]$ $=[255]$	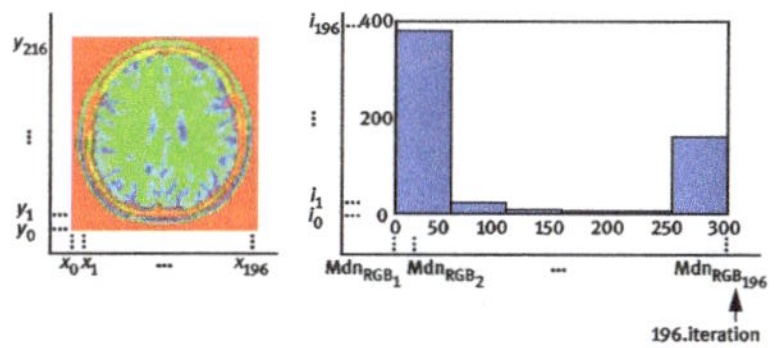

```
(1) foreach i = 1 to K//image in x coordinate value K is 196
(2) {
(3)     foreach j = 1 in L//image in y coordinate value L is 216
(4)     {
(5)         //Range
(6)         R_RGB_i = max(x_i, y_j) − min(x_i, y_j)
(7)     }
(8) }
```

Figure 4.22: RGB image *Range()* function algorithm.

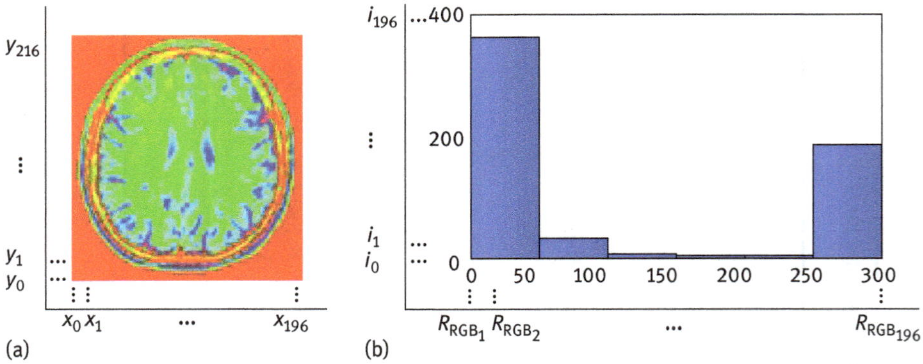

Figure 4.23: The depiction of the *Range()* calculation regarding RGB MRI image of an RRMS patient through histogram. (a) RGB MRI image for pixel value and (b) RGB MRI image of histogram for *Range()*.

The histogram graphs are provided owing to the application of Steps (1–8) for the *Range()* function algorithm (see Figure 4.22) for the RGB image in Figure 4.23.

Figure 4.23(a) shows in RGB MRI image, x-axis $x = (x_1, x_2, \ldots, x_{196})$ row and y-axis $y = (y_1, y_2, \ldots, y_{216})$ column as the corresponding pixel values.

In RGB MRI image of histogram (Figure 4.23(b)), x-axis represents the results as obtained by calculating the range value in Figure 4.24(a) (for each pixel values in x- and y-axes) and y-axis represents the iteration numbers in Figure 4.23(a) ($i = 1,2,\ldots, 196$) (range values for pixel values in x- and y-axes).

In order to calculate the range value in RGB image (196×216 dimension), the following iteration steps are applied ((i): 1–196) along with the steps of *Range()* function algorithm (Figure 4.22) (where ranges pass from Figure 4.22(a) RGB MRI image for pixel value to Figure 4.23(b) RGB MRI image of histogram for *Range()*).

For RGB MRI image,

Iteration (i) = 1

Step (1) When ($i = 1$), with the loop to be formed for $(x_i, y_j) = (x_1, y_1, \ldots, y_{216})$ the pixel values that correspond to x- and y-coordinate axes are obtained.

(1) ***foreach*** $i = 1$ **in** K//image in x coordinate value K is 196
(2) {
(3) ***foreach*** $j = 1$ **in**//image in y coordinate value L is 216
(4) {
(5) **//Variance**
(6) $\sigma^2_{RGB_i} = \frac{1}{K \times L} \sum_{i=1}^{K} \sum_{j=1}^{L} (x_i - \bar{x})^2 (y_j - \bar{y})^2$
(7) }
(8) }

Figure 4.24: *RGB* image *Variance* function algorithm.

Step (2) When ($i = 1$), for $j = 1, \ldots, 216$, with the pixel values obtained in the first iteration, the range values are calculated.

...

...

...

Iteration *(i)* = 196

Step (1) When ($i = 196$), for $(x_i, y_j) = (x_{196}, y_1, \ldots, y_{216})$ the pixel values that correspond to x- and y-coordinate axes are obtained.

Step (2) When ($i = 196$), for the pixel values obtained in the first iteration, range value is calculated.

The range values above ($R_{\mathrm{RGB}_i} = R_{\mathrm{RGB}_1}, R_{\mathrm{RGB}_2}, \ldots, R_{\mathrm{RGB}_{196}}$): for the pixel values in all columns in the y-axis ($j = 1, 2, \ldots, 216$) corresponding to x-coordinate axis ($i = 1$), the calculation of range values is done until ($i = 196$).

In Table 4.9, the range values in Figure 4.23 are calculated along with the application of *Range()* function algorithm for ($i = 1, \ldots, 196$).

Example 4.21 The application for *RGB* image in Figure 4.24 for the *Variance* function algorithm has been provided step by step. The steps of *Variance* function algorithm elicited in Figure 4.25 show how RGB is applied to the MRI image of a patient who is afflicted with RRMS.

The histogram graphs are provided owing to the application of Steps (1–8) for the *Variance()* function algorithm (see Figure 4.24) of the RGB image in Figure 4.25.

Figure 4.25(a) shows in RGB MRI image, x-axis $x = (x_1, x_2, \ldots, x_{196})$row and y-axis $y = (y_1, y_2, \ldots, y_{216})$ column as the corresponding pixel values.

In RGB MRI image of histogram (Figure 4.25(b)), x-axis represents the results as obtained by calculating the variance value in Figure 4.26(a) (for each pixel values in x- and y-axes) and y-axis represents the iteration numbers in Figure 4.26(a) (iteration $i = 1, 2, \ldots, 196$) (variance values for pixel value in x- and y-axes).

Table 4.9: The detailed illustration of red/green/blue MRI image *Range()* calculation.

Iteration	Range() Calculation for i^{th} iteration	Range() Value for i^{th} iteration	RGB MRI Image for pixel value-RGB MRI Image of Histogram for Range()
$i=1$	$R_{RGB_1} = [Max(R(x_1,y_1)+G(x_1,y_1)]+B(x_1,y_1))- Min((R(x_1,y_1)+G(x_1,y_1)+B(x_1,y_1)))]+\cdots +[Max(R(x_1,y_{216})+G(x_1,y_{216})]+B(x_1,y_{216})- [2Min((R(x_1,y_{216})+G(x_1,y_{216})+B(x_1,y_{216}))]]$	$R_{RGB_1}[(x_1,y_j),\ G(x_1,y_j),B(x_1,y_j)] = [255]$	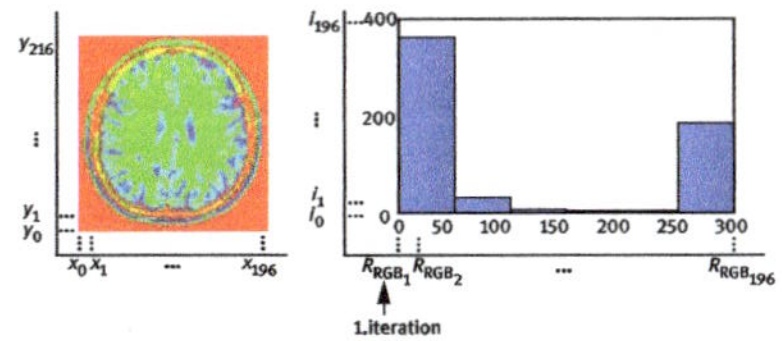
$i=2$	$R_{RGB_2} = [Max(R(x_2,y_1)+G(x_2,y_1)]+B(x_2,y_1))- Min((R(x_2,y_1)+G(x_2,y_1)+B(x_2,y_1)))]+\cdots +[[Max(R(x_{196},y_{216})+G(x_{196},y_{216})]+B(x_{196},y_{216})- [2Min((R(x_{196},y_{216})+G(x_{196},y_{216})+B(x_{196},y_{216}))]]$	$R_{RGB_2}[(x_2,y_j),\ G(x_2,y_j),B(x_2,y_j)] = [255]$	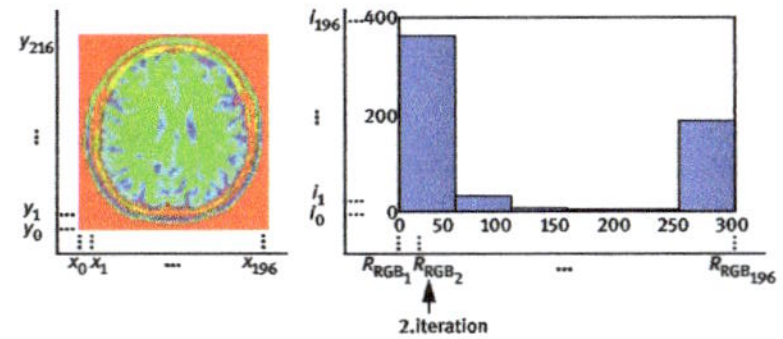
…	…	…	…
…	…	…	…
…	…	…	…
$i=196$	$R_{RGB_{196}} = [Max(R(x_1,y_1)+G(x_1,y_1)]+B(x_1,y_1))- Min((R(x_1,y_1)+G(x_1,y_1)+B(x_1,y_1)))]+\cdots +[Max(R(x_{196},y_{216})+G(x_{196},y_{216})]+B(x_{196},y_{216})- [2Min((R(x_{196},y_{216})+G(x_{196},y_{216})+B(x_{196},y_{216}))]]$	$R_{RGB_{196}}[(x_{196},y_j),\ G(x_{196},y_j),B(x_{196},y_j)] = [100]$	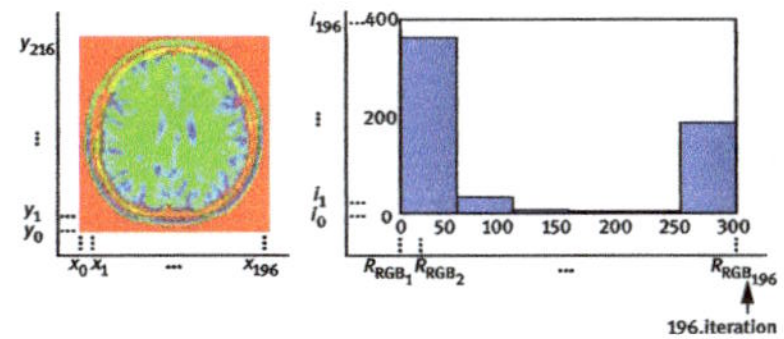

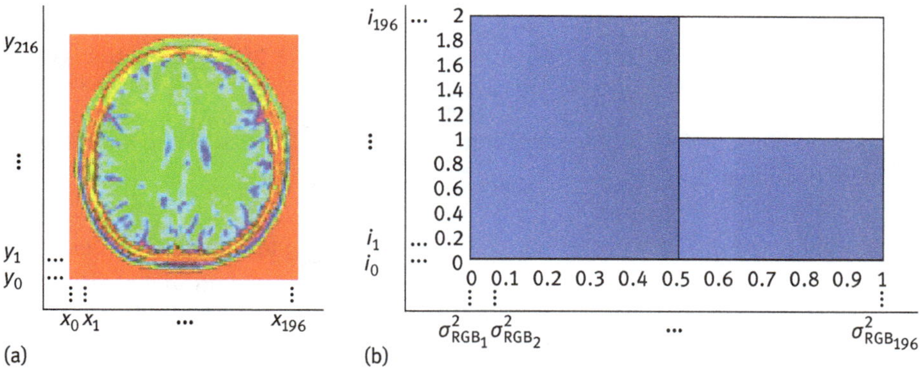

Figure 4.25: The depiction of the *Variance()* calculation regarding the RGB MRI image of an RRMS patient through histogram. (a) RGB MRI image for pixel value and (b) RGB MRI image of histogram for *Variance()*.

```
foreach i = 1 in K//image in x coordinate value K is 196
{
    foreach j = 1 in L//image in y coordinate value L is 216
    {
        //StandardDeviation
        σ² = 1/(K×L) Σ_{i=1}^{K} Σ_{j=!}^{L} (x_i − x̄)² (y_j − ȳ)²
        SD_{RGB_i} = √σ²
    }
}
```

Figure 4.26: *RGB* image *StandardDeviation* function algorithm.

In order to calculate the variance value in RGB image (196 × 216 dimension), the following iteration steps are applied ((*i*): 1–196) along with the steps of *Variance()* function algorithm (Figure 4.25) (where variances pass from Figure 4.25 (a) RGB MRI image for pixel value to Figure 4.25(b) RGB MRI image of histogram for *Variance()*).

For RGB MRI image,

Iteration (*i*) = 1

Step (1) When $(i = 1)$, with the loop to be formed for $(x_i, y_j) = (x_1, y_1, \ldots, y_{216})$, the pixel values that correspond to *x*- and *y*-coordinate axes are obtained.

Step (2) When $(i = 1)$, for $j = 1, \ldots, 216$, with the pixel values obtained in the first iteration, the variance values are calculated.

…

…

…

Iteration $(i) = 196$

Step (1) When $(i = 196)$, for $(x_i, y_j) = (x_{196}, y_1, \ldots, y_{216})$, the pixel values that correspond to x- and y-coordinate axes are obtained.

Step (2) When $(i = 196)$, for the pixel values obtained in the first iteration, variance value is calculated.

The variance values above $(\sigma^2_{\text{RGB}_i} = \sigma^2_{\text{RGB}_1}, \sigma^2_{\text{RGB}_2}, \ldots, \sigma^2_{\text{RGB}_{196}})$: for the pixel values in all the columns in the y-axis $(j = 1, 2, \ldots, 216)$ corresponding to x-coordinate axis $(i = 1)$, the calculation of variance values is done until $(i = 196)$.

In Table 4.10, the variance values in Figure 4.25 are calculated along with the application of *Variance()* function algorithm for $(i = 1, \ldots, 196)$.

Example 4.22 The application of *RGB* image in Figure 4.27 for the *standard deviation* function algorithm has been provided step by step. The steps of *standard deviation* function algorithm elicited in Figure 4.28 show how BW is applied to the MRI image of a patient who is afflicted with RRMS.

The histogram graphs are provided owing to the application of Steps (1–9) for the *StandardDeviation()* function algorithm (see Figure 4.26) for the RGB image in Figure 4.27.

In RGB MRI image of histogram (Figure 4.27(b)), x-axis represents the results as obtained by calculating the standard deviation value in Figure 4.27(a) (for each pixel values in x- and y-axes). y-Axis represents the iteration numbers in Figure 4.27(a) (iteration $i = 1, 2, \ldots, 196$) (standard deviation values for pixel values in x- and y-axes).

In order to calculate the standard deviation value in RGB image (196×216 dimension), the following iteration steps are applied ((i): 1–196) along with the algorithm steps of *StandardDeviation()* function algorithm (Figure 4.26) (where standard deviations pass from Figure 4.27(a) RGB MRI image for pixel value to Figure 4.27 (b) RGB MRI image of histogram for *StandardDeviation()*).

For RGB MRI image,

Iteration $(i) = 1$

Step (1) When $(i = 1)$, with the loop to be formed for $(x_i, y_j) = (x_1, y_1, \ldots, y_{216})$ the pixel values that correspond to x- and y-coordinate axes are obtained.

Step (2) When $(i = 1)$, for $j = 1, \ldots, 216$, with the pixel values obtained in the first iteration, the standard deviation values are calculated.

...

...

...

Iteration $(i) = 196$

Step (1) When $(i = 196)$, for $(x_i, y_j) = (x_{196}, y_1, \ldots, y_{216})$ the pixel values that correspond to x- and y-coordinate axes are obtained.

Table 4.10: The detailed illustration of red/green/blue MRI image *Variance()* calculation.

Iteration	Variance() Calculation for i^{th} iteration	Variance() Value for i^{th} iteration	RGB MRI Image for pixel value-RGB MRI Image of Histogram for Variance()
$i=1$	$\sigma^2_{RGB_1} = \left[\frac{1}{196\times 216} R \sum_{j=1}^{216} (x_1-\bar{x})^2 (y_j-\bar{y})^2 + \cdots + \frac{1}{196\times 216} G \sum_{j=1}^{216} (x_1-\bar{x})^2 (y_j-\bar{y})^2 + (\frac{1}{196\times 216} B \sum_{j=1}^{216} (x_1-\bar{x})^2 (y_j-\bar{y})^2 +) \cdots + \frac{1}{196\times 216} R \sum_{j=1}^{216} (x_1-\bar{x})^2 (y_j-\bar{y})^2 + (\frac{1}{196\times 216} G \sum_{j=1}^{216} (x_1-\bar{x})^2 (y_j-\bar{y})^2 + \cdots + \frac{1}{196\times 216} B \sum_{j=1}^{216} (x_1-\bar{x})^2 (y_j-\bar{y})^2)\right]$	$\sigma^2_{RGB_1}[R(x_1,y_j), G(x_1,y_j), B(x_1,y_j)] = [1.1]$	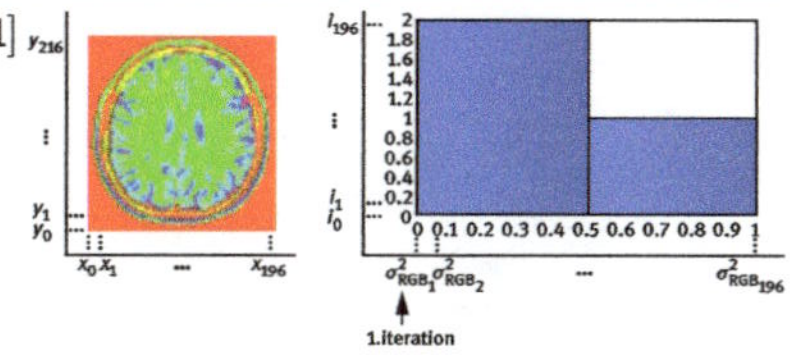
$i=2$	$\sigma^2_{RGB_2} = \left[\frac{1}{196\times 216} R \sum_{j=1}^{216} (x_2-\bar{x})^2 (y_j-\bar{y})^2 + \cdots + \frac{1}{196\times 216} G \sum_{j=1}^{216} (x_2-\bar{x})^2 (y_j-\bar{y})^2 + (\frac{1}{196\times 216} B \sum_{j=1}^{216} (x_2-\bar{x})^2 (y_j-\bar{y})^2 +) \cdots + \frac{1}{196\times 216} R \sum_{j=1}^{216} (x_2-\bar{x})^2 (y_j-\bar{y})^2 + (\frac{1}{196\times 216} G \sum_{j=1}^{216} (x_2-\bar{x})^2 (y_j-\bar{y})^2 + \cdots + \frac{1}{196\times 216} B \sum_{j=1}^{216} (x_2-\bar{x})^2 (y_j-\bar{y})^2)\right]$	$\sigma^2_{RGB_2}[R(x_1,y_j), G(x_1,y_j), B(x_1,y_j)] = [0.52]$	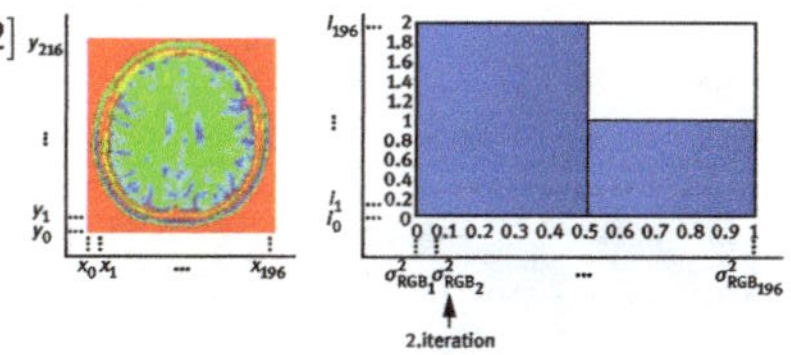

(continued)

Table 4.10 (continued)

Iteration	Variance() Calculation for i^{th} iteration	Variance() Value for i^{th} iteration	RGB MRI Image for pixel value-RGB MRI Image of Histogram for Variance()
…	…	…	…
…	…	…	…
…	…	…	…
$i=196$	$\sigma^2_{RGB_{196}} = \left[\frac{1}{196\times 216} R\sum_{j=1}^{216}(x_{196}-\bar{x})^2(y_j-\bar{y})^2 + \cdots + \frac{1}{196\times 216} G\sum_{j=1}^{216}(x_{196}-\bar{x})^2(y_j-\bar{y})^2 + (\frac{1}{196\times 216} B\sum_{j=1}^{216}(x_{196}-\bar{x})^2(y_j-\bar{y})^2 +)\cdots + \frac{1}{196\times 216} R\sum_{j=1}^{216}(x_{196}-\bar{x})^2(y_j-\bar{y})^2 + (\frac{1}{196\times 216} G\sum_{j=1}^{216}(x_{196}-\bar{x})^2(y_j-\bar{y})^2 + \cdots + \frac{1}{196\times 216} B\sum_{j=1}^{216}(x_{196}-\bar{x})^2(y_j-\bar{y})^2)\right]$	$\sigma^2_{RGB_{196}}[R(x_1,y_j),\ G(x_1,y_j),B(x_1,y_j)] = [0.13]$	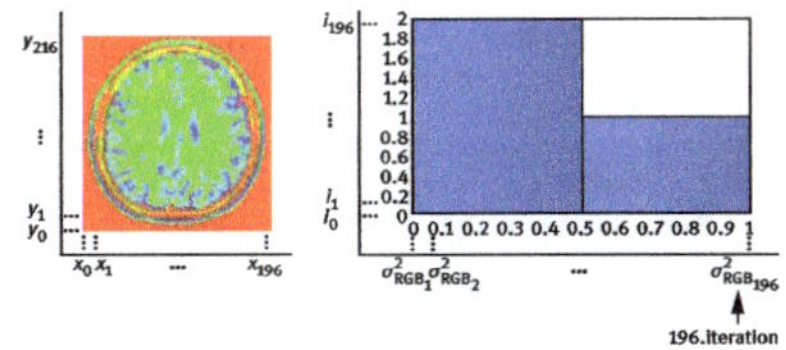

Table 4.11: The detailed illustration of red/green/blue MRI *StandardDeviation()*calculation.

Iteration	StandardDeviation() Calculation for i^{th} iteration	StandardDeviation() Value for i^{th} iteration	RGB MRI Image for pixel value-RGB MRI Image of Histogram for StandardDeviation()
$i=1$	$SD_{RGB_1} = [\sqrt{R(\frac{1}{196\times216}\sum_{j=1}^{216}(x_1-\bar{x})^2(y_j-\bar{y})^2+\cdots+\frac{1}{196\times216}\sum_{j=1}^{216}(x_1-\bar{x})^2(y_j-\bar{y})^2+G(\frac{1}{196\times216}\sum_{j=1}^{216}(x_1-\bar{x})^2(y_j-\bar{y})^2+\cdots+\frac{1}{196\times216}\sum_{j=1}^{216}(x_1-\bar{x})^2(y_j-\bar{y})^2+B(\frac{1}{196\times216}\sum_{j=1}^{216}(x_1-\bar{x})^2(y_j-\bar{y})^2+\cdots+\frac{1}{196\times216}\sum_{j=1}^{216}(x_1-\bar{x})^2(y_j-\bar{y})^2)]}$	$SD_{RGB_1}[R(x_1,y_j),\ G(x_1,y_j),B(x_1,y_j)] = [1.04]$	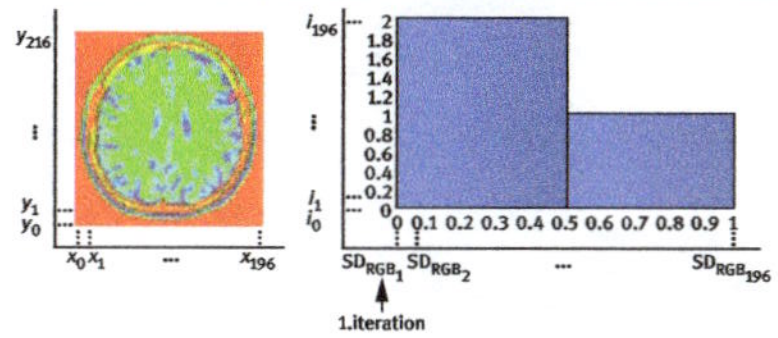
$i=2$	$SD_{RGB_2} = [\sqrt{R(\frac{1}{196\times216}\sum_{j=1}^{216}(x_2-\bar{x})^2(y_j-\bar{y})^2+\cdots+\frac{1}{196\times216}\sum_{j=1}^{216}(x_1-\bar{x})^2(y_j-\bar{y})^2+G(\frac{1}{196\times216}\sum_{j=1}^{216}(x_2-\bar{x})^2(y_j-\bar{y})^2+\cdots+\frac{1}{196\times216}\sum_{j=1}^{216}(x_2-\bar{x})^2(y_j-\bar{y})^2+B(\frac{1}{196\times216}\sum_{j=1}^{216}(x_2-\bar{x})^2(y_j-\bar{y})^2+\cdots+\frac{1}{196\times216}\sum_{j=1}^{216}(x_2-\bar{x})^2(y_j-\bar{y})^2)]}$	$SD_{RGB_2}[R(x_2,y_j),\ G(x_2,y_j),B(x_2,y_j)] = [0.72]$	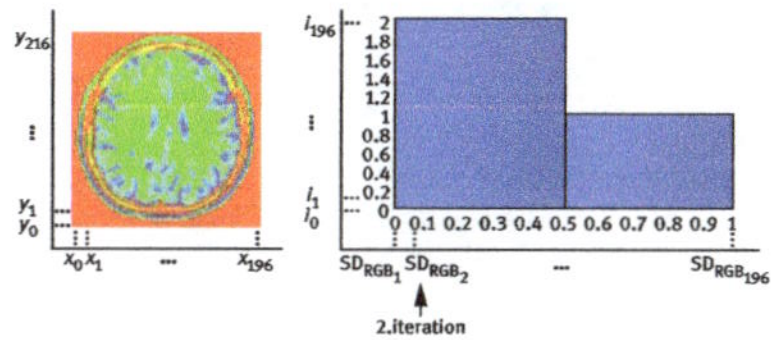
… … …	… … …	… … …	… … …
$i=196$	$SD_{RGB_{196}} = [\sqrt{R(\frac{1}{196\times216}\sum_{j=1}^{216}(x_{196}-\bar{x})^2(y_j-\bar{y})^2+\cdots+\frac{1}{196\times216}\sum_{j=1}^{216}(x_{196}-\bar{x})^2(y_j-\bar{y})^2+G(\frac{1}{196\times216}\sum_{j=1}^{216}(x_{196}-\bar{x})^2(y_j-\bar{y})^2+\cdots+\frac{1}{196\times216}\sum_{j=1}^{216}(x_{196}-\bar{x})^2(y_j-\bar{y})^2+B(\frac{1}{196\times216}\sum_{j=1}^{216}(x_{196}-\bar{x})^2(y_j-\bar{y})^2+\cdots+\frac{1}{196\times216}\sum_{j=1}^{216}(x_{196}-\bar{x})^2(y_j-\bar{y})^2)]}$	$SD_{RGB_{196}}[R(x_{196},y_j),\ G(x_{196},y_j),B(x_{196},y_j)] = [0.36]$	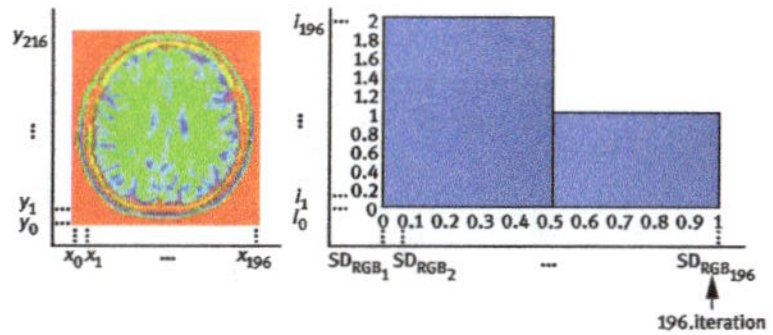

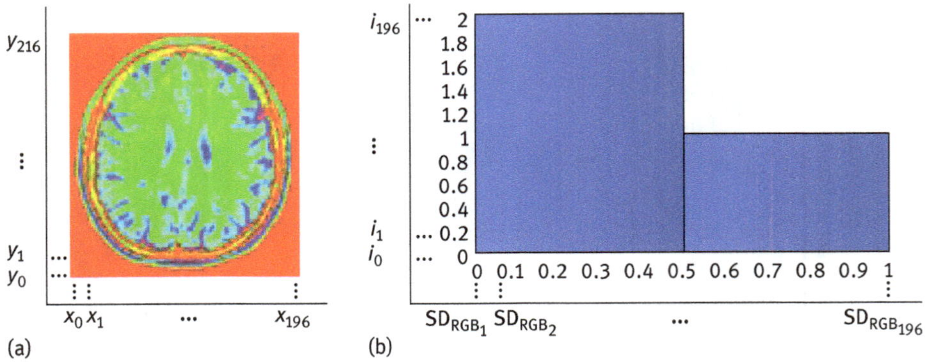

Figure 4.27: The depiction of *StandardDeviation()* calculation regarding RGB MRI image of an RRMS patient through histogram: (a) RGB MRI image for pixel value and (b) RGB MRI image of histogram for *StandardDeviation()*.

Step (2) When ($i = 196$), for the pixel values obtained in the first iteration, standard deviation value is calculated.

The standard deviation (SD) values above ($SD_{RGB_i} = SD_{RGB_1}, SD_{RGB_2}, \ldots, SD_{RGB_{196}}$): for the pixel values in all columns in the y-axis ($j = 1, 2, \ldots, 216$) corresponding to x-coordinate axis ($i = 1$), the calculation of standard deviation values is done until ($i = 196$).

In Table 4.11, the standard deviation values in Figure 4.27 are calculated along with the application of *StandardDeviation()* function algorithm for $i = 1, \ldots, 196$.

References

[1] Christofides N. Graph theory: An algorithmic approach. USA: Academic Press, Inc., 1975.

[2] Sedgewick R, Wayne K. Algorithms. USA: Pearson Education, Inc., 2011.

[3] Chaudhuri AB. The art of programming through flowcharts & algorithms. Firewall Media, 2005.

[4] Knuth DE. Computer-drawn flowcharts. Communications of the ACM, 1963, 6 (9), 555–563.

[5] Khot AS, Mishra RK . Learning functional data structures and algorithms. UK: Packt Publishing Ltd., 2017.

[6] Schriber TJ. Fundamentals of flowcharting. New York: John Wiley & Sons, Inc., 1969.

[7] Hughes, JF, Van Dam A, Foley JD, McGuire M, Feiner SK, Sklar DF, Akeley K. Computer graphics: Principles and practice. USA: Pearson Education, Inc., 2014.

[8] Goodrich MT, Tamassia R, Goldwasser MH. Data structures and algorithms in Java. USA: John Wiley & Sons, Inc., 2014.

[9] Szeliski R. Computer vision: algorithms and applications. London: Springer - Verlag, 2011.

[10] Solomon C, Breckon T. Fundamentals of Digital Image Processing: A practical approach with examples in Matlab. New York: John Wiley & Sons, Inc., 2011.

[11] Demirkaya O, Asyali MH, Sahoo PK. Image processing with MATLAB: applications in medicine and biology. USA: CRC Press. 2008.

[12] Burger W, Burge MJ. Digital image processing: an algorithmic introduction using Java. London: Springer-Verlag, 2016.
[13] Newman WM, Sproull RF. Principles of interactive computer graphics. New York, NY, USA: McGraw-Hill, Inc., 1979.
[14] Rogers DF, Adams JA. Mathematical elements for computer graphics. New York, NY, USA: McGraw-Hill, Inc. 1989.
[15] Nakamura S. Numerical analysis and graphic visualization with MATLAB. New Jersey Prentice-Hall, Inc., 2002.
[16] Wójcik W, Smolarz A. Information Technology in Medical Diagnostics. London: CRC Press. 2017.
[17] Spagnolini U. Statistical Signal Processing in Engineering. USA: John Wiley & Sons, Inc., 2018.

5 Linear model and multilinear model

This chapter covers the following items:
- Algorithm for linear modeling
- Algorithm for multilinear modeling
- Examples and applications

Sir Francis Galton (1822–1911), an English scientist and also one of the cousins of Charles Darwin, contributed significantly to the fields of not only genetics but also psychology. He came up with the concept of regression [1, 2]. He is also regarded as a pioneer in applying statistics to biology. One of the datasets he handled included the heights of fathers and their first sons. Galton focused on the prediction of heights of sons being reliant on the heights of fathers. Having analyzed the scatterplots regarding these heights, Galton discovered that the trend was linear and in increasing direction. After having fit a line to these data, he realized that the regression line estimated that taller fathers were inclined to have shorter sons, whereas shorter fathers showed the tendency to have taller sons [2, 3]. This held true for fathers whose heights were taller than the average, which showed a regression toward the mean.

In this chapter, we attempt to examine the relationship between one or more variables. Based on this, we have created a model to be utilized for predictive purposes. For instance, consider the following question: "Is there any statistical evidence to draw a conclusion that the Expanded Disability Status Scale (EDSS) score with the relapsing remitting multiple sclerosis (MS) has the greatest incidence of MS?" To answer this is of vital importance if we intent to make appropriate medical decisions, choices as well as decisions regarding lifestyles. Regression analysis will be used while we are analyzing the relationship between variables. What we aim at is creating a model and studying inferential procedures in cases when several independent variables and one dependent are existent. Y denotes the random variable to be predicted, which is also termed as the dependent variable (or response variable) and x_i denotes independent (or predictor) variables used to predict or to model Y. For instance, (x, y) represent the weight and height of a man. Let us assume that we are interested in finding the relationship between weight and height from sample measurements of $\boldsymbol{n}$ individuals. The process related to the finding of a mathematical equation that optimally fits the noisy data is termed the regression analysis. Sir Francis Galton introduced the word regression in his seminal book *Natural Inheritance* in 1894 to describe certain genetic relationships [3]. Regression is one of the most popular statistical techniques if one wishes to study the dependence of one variable in relation to another variable. There exist different forms of regression, which are simple linear, nonlinear, multiple and

https://doi.org/10.1515/9783110496369-005

others. The principal use of a regression model is prediction. When one uses a model for predicting Y for a particular set of values of $(x_1, x_2, \cdots, x_k)$, he/she may want to make out how large prediction error might be. After collecting the sample data, regression analysis broadly comprises the following steps [4]. The function $f(x_1, x_2, \cdots, x_k; w_0, w_1, w_2, \cdots, w_k)$ $(k \geq 1)$ constitutes the independent or predictor variables $x_1, x_2, \cdots, x_k$ (supposed to be nonrandom) and unknown parameters or weights $w_0, w_1, w_2, \cdots, w_k$ and e denoting the random or error variable. Subsequently, it will be possible to get on with the introduction of the simplest form of a regression model which is termed as simple linear regression [4].

Consider a random sample with $\boldsymbol{n}$ observations of the form $(x_1, y_1), (x_2, y_2), \cdots,$ (x_k, y_k). Here x is the independent variable and y is the dependent variable, both of which are scalars. The scatter diagram is a basic descriptive technique to determine the form of relationship between x and y, which can be drawn by plotting the sample observations in Cartesian coordinates. The pattern of points provides a clue of a linear or a nonlinear relationship between the relevant variables. Figure 5.1(a) presents the relationship between x and y, which is quite linear. On the other hand, in Figure 5.1(b), the relationship resembles a parabola, and in Figure 5.1(c) there exists no evident relationship between the variables [5, 6]. When the scatter diagram shows a linear relationship, the problem may be to find the linear model that fits the given data in the best way. For this purpose, we will provide a general definition of a linear statistical model, which is named as the multilinear regression model.

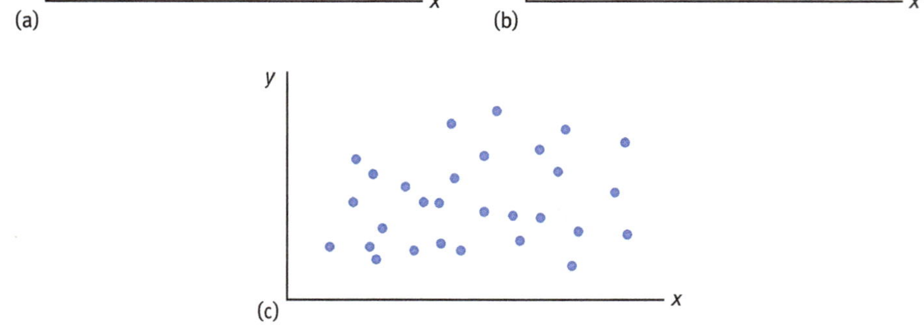

Figure 5.1: Scatter diagram: (a) linear relationship; (b) quadratic relationship; and (c) no relationship. From Ramachandran and Chris [2].

A multilinear regression model in relation to random response Y to a set of predictor variables $x_1, x_2, \cdots, x_k$ provides an equation of the form $Y = w_0 + w_1x_1 + w_2x_2 + \cdots + w_kx_k$, where $w_0, w_1, w_2, \cdots, w_k$ are unknown parameters, $x_1, x_2, \cdots, x_k$ are independent nonrandom variables and e is a random variable that signifies an error term. We suppose that $E(e) = 0$, or in an equivalent manner, $E(Y) = w_0 + w_1x_1 + w_2x_2 + \cdots + w_kx_k$ [2, 5–7]. We can handle a single dependent variable Y and a single independent nonrandom variable x to have an understanding of the basic concepts pertaining to regression analysis. If we assume that there exist no measurement errors in x_i, we can say that the possible measurement errors in y and the uncertainties in the model assumed are depicted through the random error e. There is an inability inherent to afford an exact model for a natural singularity as is expressed via the random term e. This will have a specified probability distribution (like normal) with mean zero. Hence, you can consider Y as one with a deterministic component, $E(Y)$ and a random component, e. If you take $k=1$ in the multilinear regression model, a simple linear regression model is obtained as follows [7–9]:

$$E(Y) = w_0 + w_1x + e \tag{5.1}$$

If Y is termed a simple linear regression model, w_0 is the Y-intercept of the line and w shows the line slope. Here, e is the error component. Such a basic linear model accepts the presence of a linear relationship between the variables x and y disturbed by a random error e [2, 7–9]. The data points known are the pairs $(x_1, y_1), (x_2, y_2), \cdots, (x_k, y_k)$ and the problem of simple linear regression is to fit a straight line optimal in an extent to the dataset as provided and is also shown in Figure 5.2.

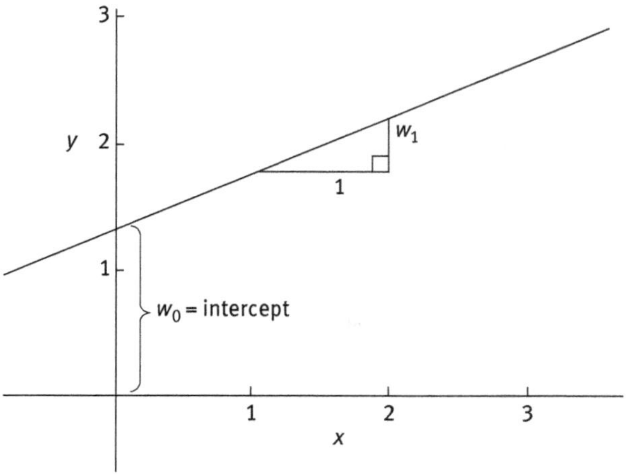

Figure 5.2: Equation regarding a straight line. From Ramachandran and Chris [2].

At this point, the problem is to find estimators for w_0 and w_1. When you obtain the "good" estimators $\hat{w}_0$ and $\hat{w}_1$ you are able to fit a line for the given data by the prediction equation $\hat{Y} = \hat{w}_0 + \hat{w}_1 x$. The question arises whether this predicted line yields the "best" description of data. The most widely used technique currently is called the method of least squares, which is described so that we can obtain the weights of the parameters or the estimators [10–12].

Many methods have been proposed for the estimation of parameters in a model. We call the method under discussion in this section *ordinary least squares (ols)*, where parameter estimates are selected to minimize the quantity which is termed as the *residual sum of squares.*

It is known that *parameters* are unknown quantities characterizing a model. *Estimates of parameters* are functions of data that can be computed; therefore, they are statistics. The fitted value for case *i* is given by $E(x_i)$. For this, $\hat{y}_i$ is used, which can be represented by the shorthand notation [10–12].

$$\hat{y}_i = w_0 + w_1 x_i \tag{5.2}$$

This should be contrasted with the equation for the statistical errors:

$$e_i = \hat{y}_i - (w_0 + w_1 x_i), \qquad i = 1, 2, \cdots, n \tag{5.3}$$

All of the least squares calculations for simple regression depend solely on averages, sums of cross-products and sums of squares [13–16]. The criterion helps to obtain estimators reliant on residuals. These are geometrically vertical distances between the fitted line and the actual values, and the illustration is shown in Figure 5.3. It is known that the residuals echo the inherent asymmetry in roles of response and predictor for the problems regarding regression.

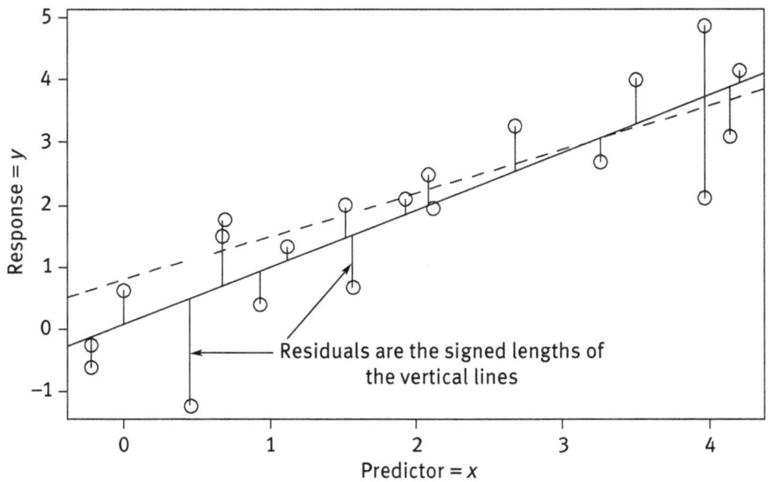

Figure 5.3: A schematic plot for *ols* fitting. From Ramachandran and Chris [2].

Figure 5.3 presents a schematic plot for *ols* fitting. A small circle indicates each data point, and the solid line is the candidate *ols* line yielded by a specific choice of slope as well as intercept. The residuals are the solid vertical lines between solid line and points. Points that are below the line have negative residuals, whereas points that are above the line have positive residuals. The *ols* estimators are values w_0 and w_1 minimizing the function as follows [13–15]:

$$\mathrm{RSS}(w_0, w_1) = \sum_{i=1}^{n} [y_i - (w_0 + w_1 x_i)]^2 \tag{5.4}$$

When we make the evaluation at $(\hat{w}_0, \hat{w}_1)$, the quantity is called $\mathrm{RSS}(\hat{w}_0, \hat{w}_1)$ (the *residual sum of squares* or the abbreviation *RSS*), which can also be used as [13–15]

$$\hat{w}_1 = \frac{\mathrm{SAB}}{\mathrm{SAA}} = r_{xy}\frac{\mathrm{SD}_y}{\mathrm{SD}_x} = r_{xy}\left(\frac{\mathrm{SBB}}{\mathrm{SAA}}\right)^2$$

$$\hat{w}_0 = \bar{y} - \hat{w}_1\bar{x} \tag{5.5}$$

Several forms of $\hat{w}_1$ are all equivalent. The variance σ^2 is basically the average squared size of the e_i^2. For this reason, it can be expected that its estimator $\hat{\sigma}^2$ is attained by averaging the squared residuals. Assuming that the errors are uncorrelated random variables that have no zero means and common variance σ^2, an unbiased estimate of σ^2 is gained through dividing $\mathrm{RSS} = \sum \hat{e}_i^{\,2}$ by its *degrees of freedom* (df). Here, residual df, in the mean function, represents the number of cases minus the number of parameters. Residual df $= n - 2$ is valid for simple regression. For this reason, the estimate of σ^2 is provided by the following equation [13–15]:

$$\hat{\sigma}^2 = \frac{\mathrm{RSS}}{n-2} \tag{5.6}$$

Another point to mention is the *F-test for regression*. In this case, if the sum of squares for regression SSreg is large, the simple regression mean function $E(x) = w_0 + w_1 x$ is supposed to be a significant improvement over the mean function $E(x) = w_0$. This is similar to stating that the additional parameter in the simple regression mean function w_1 differs from zero; or $E(x)$ is not constant as it changes. More formally, we can say that in order to be capable of judging how large is "large," we do a comparison of the regression mean square, SSreg divided by its df, with the residual mean square $\hat{\sigma}^2$. This ratio is called *F*, and the formulation is [17, 18]

$$F = \frac{(\mathrm{SBB} - \mathrm{RSS})/1}{\sigma^2} \tag{5.7}$$

The intercept is employed as in Figure 5.4 so as to show the general form of confidence intervals pertaining to the normally distributed estimates. When you perform a

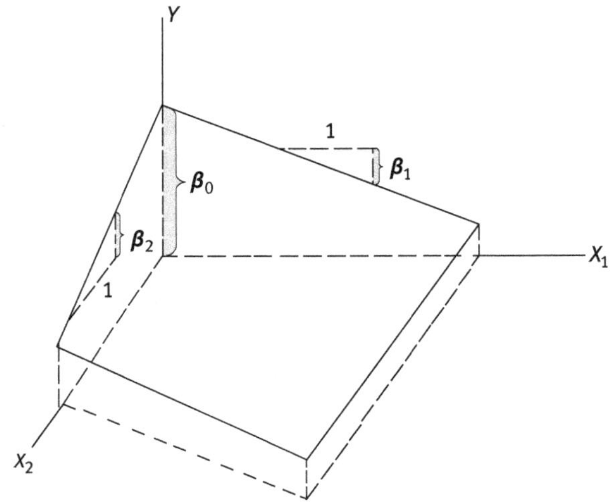

Figure 5.4: A linear regression surface $k = 2$ predictors. From Ramachandran and Chris [2].

hypothesis test in statistics, a p-value helps to determine the significance of your results. Hypothesis tests are used to test the validity of a claim put forth regarding a population. This claim that's on trial is named as the null hypothesis. p-Value is the conditional probability of observing a value of the computed statistic with appropriate assumptions. The p-value is a number between 0 and 1 which is interpreted as follows [15–17]:

- A small p-value (typically ≤ 0.05) indicates strong evidence against the null hypothesis, hence one rejects the null hypothesis.
- A large p-value (> 0.05) indicates weak evidence against the null hypothesis, hence one fails to reject the null hypothesis.
- p-Values very close to the cutoff (0.05) are considered to be marginal (that means it may go either way). It is important to report the p-value so that your readers will be able to make inferences of their own.

The value of F is large or larger, extreme or more extreme, than the observed value. We can use the estimated mean function for obtaining the values of response for given values concerning the predictor. This problem has two vital variants regarding estimation and prediction of fitted values. We aim at prediction, which is a more important component, so let us move on with prediction [17, 18]:

With response y and terms $x_1, \cdots, x_k$ regarding the general *multilinear regression* model, the following form is at stake [17–20]:

$$E(Y) = w_0 + w_1x_1 + \cdots + w_1x_k \tag{5.8}$$

$E(Y)$ shows that all the terms are being conditioned on the right-hand side of the equation. Likewise, when one is conditioning on specific values for the predictors $x_1, \cdots, x_k$, he/she will jointly call x as follows [17–20]:

$$E(Y) = w_0 + w_1 x_1 + \cdots + w_k x_k \tag{5.9}$$

where w represents unknown parameters that require to be estimated. Equation (5.4) is a linear function of the parameters; hence, it is called linear regression. If $k=1$, then Y has only one element; if $k=2$, then the mean function of eq. (5.4) corresponds to a plane in three dimensions, which are presented in Figure 5.4. If $k>t$, the fitted mean function is a hyperplane. In generalization of a k-dimensional plane in a $(k+1)$-dimensional space, it is not possible that we can draw a general k-dimensional plane in a three-dimensional world [17–20].

We have observed data for n cases or units, which means that at present there is a value of Y and all the terms for each of the n cases. We also have symbols for the response and terms that use matrices and vectors [17–21]:

$$Y = \begin{pmatrix} y_1 \\ y_2 \\ \vdots \\ y_n \end{pmatrix}, X = \begin{pmatrix} 1 & x_{11} & \cdots & x_{1k} \\ 1 & x_{21} & \cdots & x_{2k} \\ \vdots & \vdots & \cdots & \vdots \\ 1 & x_{n1} & \cdots & x_{nk} \end{pmatrix} \tag{5.10}$$

Equation (5.11) is defined, hence Y is an $n \times 1$ vector and X is an $n \times (k+1)$ matrix. w is also defined to be a $(k+1) \times 1$ vector of regression coefficients and e as the $n \times 1$ vector of statistical errors [17–20]:

$$w = \begin{pmatrix} w_1 \\ w_2 \\ \vdots \\ w_n \end{pmatrix} \text{ and } e = \begin{pmatrix} e_1 \\ e_2 \\ \vdots \\ e_n \end{pmatrix} \tag{5.11}$$

The matrix X yields all the observed values of the terms. The ith row of X is elaborated by the symbol x_i' that is a $(k+1) \times 1$ vector for mean functions including an intercept. Although x_i is row of X, the principle that all vectors are column vectors is used. For this reason, it is required to write x_i' to signify a row. The equation for the mean function evaluated at x_i is given as follows [13–21]:

$$E(x_i) = x_i' w = w_0 + w_1 x_{i1} + \cdots + w_k x_{ik} X_k \tag{5.12}$$

The multilinear regression model is written as

$$Y = Xw + e \tag{5.13}$$

for the matrix notation.

The steps of linear model based on the linear model flowchart (Figure 5.5) are given below:

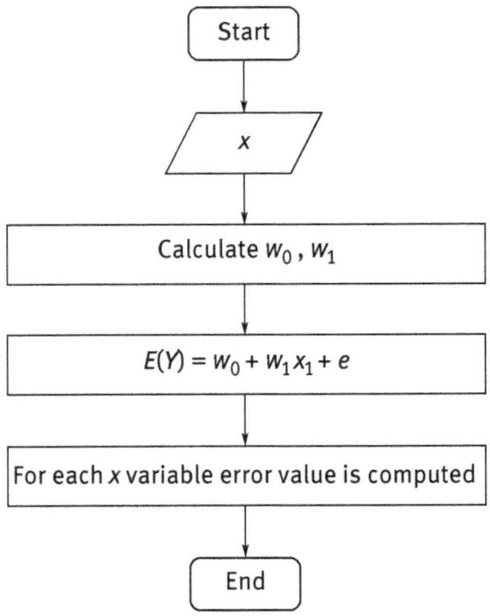

Figure 5.5: Linear model flowchart.

Algorithm: Linear Model
Input: x independent variable
Result: y dependent variable, w_0 and w_1 coefficients are identified.
Method:

(1) //Linear model
(2) //y is the dependent variable and its e value is determined.
(3) //x is the independent variable and it is assumed to be error free.
(4) ***while*** x variable ***do***{
(5) $E(Y) = w_0 + w_1 x_1 + e$
(6) ***if*** $(x = 0)$
(7) $w_0 = y$
(8) ***else***
(9) //calculate w_0, w_1, minimum error e
(10) }

Figure 5.6: General linear modeling.

For the algorithm, let us have a closer look at the steps (Figure 5.6).

The steps for the *linear model algorithm* (see Figure 5.6) in the relevant order are given below:

Steps (1–3) Simple regression model is handled with x independent variable that has direct correlation with y dependent variable.

Step (4) x independent variable is accepted as a variable that is controllable and measurable through a tolerable error.

Steps (5–7) In the linear model, y dependent variable is dependent on the x independent variable (see eq. (5.1)). If $x=0, y=w_0$. If $x \neq 0$, we move on to Steps (8–10) stage for the estimation of w_0 and w_1 parameters (see eq. (5.1)).

Steps (8–10) Least squares method is used for the estimation of w_0 and w_1 parameters. Error value (e_i) is calculated based on eq. (5.3). Afterwards, $E(Y)$ equation is obtained using eqs. (5.4) and (5.5).

If the solution of our problem is dependent on more than one independent variable, how can we do the linear model analysis?

The answer to the question can be found by applying the multilinear model flowchart in Figure 5.6 and the multilinear model algorithm in Figure 5.7.

The flowchart of *multilinear model* is shown in Figure 5.7.

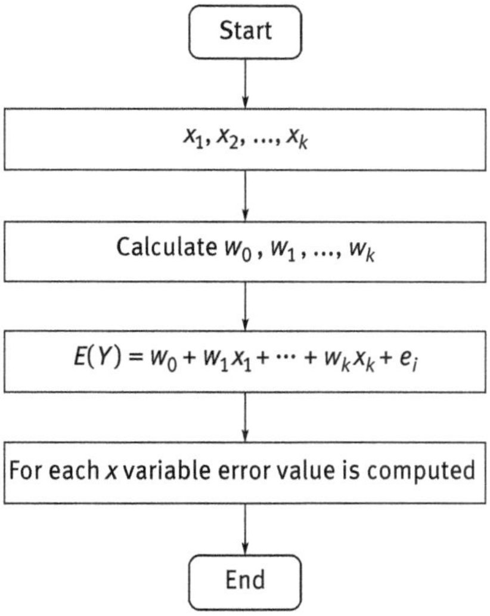

Figure 5.7: Multilinear model flowchart.

Based on Figure 5.7, the steps regarding multilinear model are given below.

The steps of multilinear model algorithm (see Figure 5.9) are as follows:

Steps (1–3) Multiregression model that has direct correlation with y dependent variable and that with more than one independent variable $X = x_1 + x_2 + \cdots + x_k$ is handled.

Step (4) x independent variable with k number can be denoted as multilinear model based on eq. (5.8). $w_0, w_1, \cdots, w_k$ parameters are accepted as the multiregression coefficients and the model to be formed will denote a hyperplane in k-dimensional space. w_k parameter shows the expected change in y dependent variable that corresponds to a unit change in x_k independent variable.

Steps (5–10) For each of the unknown multiregression coefficients $(w_0, w_1, \cdots, w_k)$ the solution of the equation will yield the estimation of the least squares method. Error value (e_i) is calculated based on eq. (5.3). Afterwards, $E(Y)$ equation is obtained. (Note: As it is more appropriate to denote multiple regression models in matrix format, eq. (5.10) can be used.)

The flowchart of the linear model and the structure of its algorithm have been provided in Figures 5.7 and 5.8, respectively. Now, let us reinforce our knowledge on linear model through the applications we shall develop.

Algorithm: Multi Linear Model

Input: x independent variable

Result: y dependent variable, $w_0, w_1, \ldots, w_k$ regression coefficients are identified.

Method:

```
(1)  //Multi linear model
(2)    //y is the dependent variable and its e_i value is determined.
(3)    //x is the independent variable and it is assumed to be error free
(4)  while x variable do{
(5)      E(Y) = w_0 + w_1x_1 +··· + w_kx_k + e_i
(6)      if (x_k = 0)
(7)          w_0 = y
(8)      else
(9)          //calculate y, minimum error e_i
(10) }
```

Figure 5.8: General multilinear modeling.

5.1 Linear model analysis for various data

The interrelationship between the attributes in our linear model dataset can be modeled in a linear manner. Linear model analysis has been applied in economy (USA, New Zealand, Sweden [U.N.I.S.]) dataset, Multiple Sclerosis (MS) dataset and Wechsler Adult Intelligence Scale Revised (WAIS-R) dataset in Sections 5.1.1, 5.1.2 and 5.1.3, respectively.

5.1.1 Application of economy dataset based on linear model

Let us see an application over an example based on linear model that analyzes the interrelationship between the attributes in our economy dataset (including economies of U.N.I.S. countries). Data for these countries' economies from 1960 to 2015 are defined based on the attributes presented in Table 2.8 economy dataset (http://data.worldeconomy.org), which is to be used in the following sections.

The attributes of data are years, time to import, time to export and so on. Values for the period 1960–2015 are known. Let's assume that we would like to develop a linear model with the data available for these four different countries. Before developing the linear model algorithmically for the economy dataset, it could be of use if we can go over Example 5.1 for the reinforcing of linear model.

Example 5.1 Let us consider the following case study: A relationship between the loan demand (in dollars) from the customer of New Zealand Bank in the second quarter of 2003 and the duration (hour) required for the preparation of loans as demanded by customers has been modeled to fit a straight line.

When eq. (5.1) is used:

$$y = 9.5 + 2.1x + e$$

The loan demand from the customer in the second quarter of 2003 is $20 < x < 80$. An anticipated time period (proposed by the New Zealand economy) for the preparation of loan for the customer whose demand for loan is 45 is estimated to be 108 h. Let's attempt to set if the loan can be offered to the customer within the anticipated time interval or not by means of linear model.

Solution 5.1
Based on eq. (5.1),

$$\begin{aligned} y &= w_0 + w_1 x \\ &= 9.5 + 2.1x \\ &= 9.5 + 2.1(45) \\ &= 104 \end{aligned}$$

Based on the linear model formed to fit a straight line, the residual error value of y is $\hat{y} = 108$ while the actual value was found to be $y = 104$. As you can see, a difference of $108 - 104 = 4$ h has been obtained between the expected value and the actual value. It can be said that the New Zealand Bank can offer the loan to the customer 4 h earlier than the time anticipated.

In order to set a linear model for the economy (U.N.I.S.) dataset, you can have a look at the steps presented in Figure 5.9.

Example 5.2 The application steps of linear modeling are as follows:

Steps (1–3) Simple regression model is handled with $x = 9.5$ independent variable that has direct correlation with y dependent variable.

Step (4) x independent variable is accepted as the variable that is controllable and measurable through a tolerable error. Based on Example 5.4, $y = 9.5 + 2.1x$ model is set up.

Algorithm: Linear Modelling for the Economy (U.N.I.S.) Data Set
Input: x independent variable for Economy (U.N.I.S.) Data Set
Result: y dependent variable w_0 and w_1 coefficients are identified.
Method:

```
(1)  //Linear model
(2)     //y is the dependent variable and its e value is determined.
(3)     // x is the independent variable and it is assumed to be error free.
(4)     while x variable do{
(5)         E(Y) = w0 + w1x1 + e
(6)     if (x = 0)
(7)         w0 = y
(8)     else
(9)         //calculate w0, w1, minimum error e
(10) }
```

Figure 5.9: Linear modeling for the analysis of the economy (U.N.I.S.) dataset.

Steps (5–7) In the linear model, y dependent variable is dependent on the x independent variable (see eq. 5.1). If $x = 0, y = w_0$. If $x \neq 0$, we move on to Steps (8–10) stage for the estimation of w_0 and w_1 parameters (see eq. (5.1)).

Steps (8–10) Least squares method is used for the estimation of w_0 and w_1 parameters (using eqs. (5.4) and (5.5)). Error value (e) is calculated based on eq. (5.3). Afterward, $E(Y)$ equation is obtained.

Figure 5.10 shows the results of the annual gross domestic product (GDP) per capita (x-coordinate) for the years between 1960 and 2015 (y-coordinate) for the Italy economy dataset, and the linear model is established so that the data are modeled to fit a straight line model.

The relationship between GDP per capita and years (1960–2015) in Figure 5.10(b) is based on the data in the linear diagonal (see Figure 5.10(a)). The statistical results for the standard error (SE) calculated and estimated values are provided below along with their interpretations.

1. The SE of the coefficients is 93.026.
2. w_1 is regression coefficient obtained as 1,667.3. The linear relation is $E(Y) = 1,667.3x$.
3. t-Statistic for each coefficient for the testing of null hypothesis that the corresponding coefficient is zero against the alternative that it is different from zero, given the other predictors in the model: t-stat = estimate/SE. The t-statistic for the intercept is 1,667.3/93.026 = 17.923. The column that is indicated as the t-stat value is the ratio of the estimate to its SE. The column labeled "t-stat" is the ratio of the estimate to its large-sample SE, which is possible to be used for a test of the null hypothesis that a particular parameter is equal to zero against either a general or one-sided alternative. The column indicated as the "p-value(>|t-stat|)"

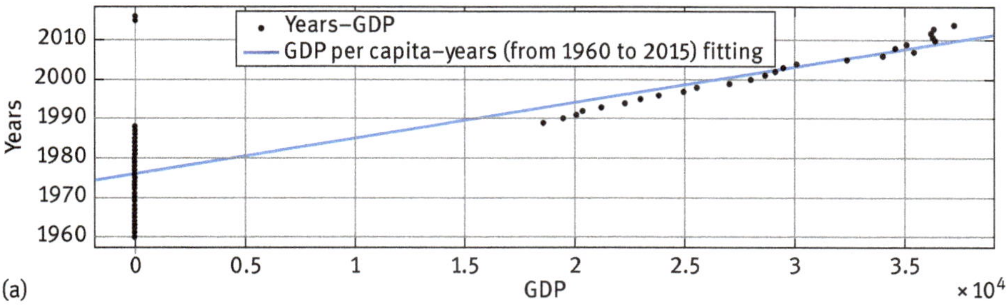

(a)

Intercept	Estimate	SE	*t*-Stat	*p*-Value
	1,667.3	93.026	17.923	2.5368e-25

(b)

Figure 5.10: Italy's GDP per capita and years based on data of linear model. (a) Data of linear model between GDP per capita (x) and years (y) for Italy. (b) Estimated coefficients between GDP per capita and years (1960–2015) for Italy based on linear regression results.

is the level of significance regarding this test, using a normal reference distribution instead of a *t*-stat.

4. The *p*-value for GDP per capita ($2.5368\text{e-}25 < 0.05$) is smaller than the common level of 0.05. This shows that it is statistically significant.

5.1.2 Linear model for the analysis of MS

At this point, let us carry out an application over an example based on linear model that analyzes the interrelationship between the attributes in our MS dataset.

As presented in Table 2.12, MS dataset has data from the following groups: 76 samples belonging to relapsing remitting MS (RRMS), 76 samples to secondary progressive MS (SPMS), 76 samples to primary progressive MS (PPMS) and 76 samples to healthy subjects of control group. The attributes of the control group are data regarding *brain stem (MRI [magnetic resonance imaging] 1), corpus callosum periventricular (MRI 2), upper cervical (MRI 3) lesion diameter size (mm) in the MRI and EDSS* score. Data are made up of a total of 112 attributes. By using these attributes of 304 individuals, we can know that whether the data belong to the MS subgroup or healthy group. How can we make the classification as to which MS patient belongs to which subgroup of MS including healthy individuals and those diagnosed with MS (based on the lesion diameters (*MRI 1, MRI 2, MRI 3*) and the number of lesion size for (*MRI 1, MRI 2, MRI 3*) as obtained from MRI and EDSS scores)? *D* matrix has a dimension of (304×112), which means it includes the MS dataset of 304 individuals along with their 112 attributes (see Table 2.12 for the MS dataset). For the classification of *D* matrix through linear model algorithm, the first step training procedure is to be employed. For the training procedure, 66.66% of the *D* matrix can be split for the training dataset (203 × 112) and 33.33% as the test dataset (101 × 112). For MS dataset, we would like to develop a linear model for MS patients and healthy individuals whose data are available.

Before developing the linear model algorithmically for the MS dataset, it will be for the benefit of the reader to study Example 5.2 for the reinforcing of linear model.

Example 5.3 Based on the research carried out in neurology clinic for an SPMS patient, the linear model set for the lesion diameter (x) in the MRI and EDSS score (y) is given below:

When eq. (5.1) is applied

$$y = 1 + 1.5x$$

Based on this model, for w_0 a 0.5 mm increase in the lesion diameter will lead to 1 unit increase in the EDSS score. This could be assumed based on the given parameters. Thus, it can also be assumed that the EDSS score value for an individual who is for the first time diagnosed with SPMS will be 1. At this point, let's address a hypothetical question to the reader. Can we predict, based on this model, what the EDSS score will be for an individual whose lesion diameter is 6 in line with the linear regression model?

Solution 5.3

$$x = 6$$
$$y = 0.5 + 1(6) = 6.5$$

Based on the linear model set, it can be estimated that an SPMS patient with lesion diameter as 6 will have an EDSS of 6.5 (see Table 2.10 for details).

In order to set a linear model for the MS dataset, you can have a look at the steps presented in Figure 5.10.

Example 5.4 The application of the linear model algorithm (Figure 5.11) is given below:

Steps (1–3) Simple regression model is handled with $x = 6$ independent variable that has direct correlation with y dependent variable.

Step (4) x independent variable is accepted as one that is controllable and measurable through a tolerable error.

Steps (5–7) In the linear model, y dependent variable is dependent on the x independent variable (see eq. (5.1)). If $x = 0, y = w_0$. If $x \neq 0$, then we move on to Steps (8–10) stage for the estimation of w_0 and w_1 parameters (see eq. (5.1)).

Steps (8–10) Least squares method is used for the estimation of w_0 and w_1 parameters (using eqs. (5.4) and (5.5)) and error value (e) is calculated based on eq. (5.3). Afterwards, $E(Y)$ equation is obtained.

Algorithm: Linear Modelling for the MS Data Set
Input: x independent variable for MS Data Set
Result: y dependent variable w_0 and w_1 coefficients are identified.
Method:

//Linear model
//y is the dependent variable and its e value is determined.
//x is the independent variable and it is assumed to be error free.
while x variable **do**{
  $E(Y) = w_0 + w_1 x_1 + e$
if $(x = 0)$
  $w_0 = y$
else
  //calculate w_0, w_1, minimum error e
}

Figure 5.11: Linear modeling for the analysis of MS dataset.

Figure 5.12(a) presents the results obtained in line with linear model set for EDSS (x-coordinate) and *SPMS, PPMS, RRMS, healthy* classes (y-coordinate). From the EDSS and classes (MS subgroups data: *SPMS* [labeled in y-axis by 1], *PPMS* [labeled in y-axis by 2], *RRMS* [labeled in y-axis by 3] and *healthy* [labeled in y axis by 4]), find the linear regression relation $E(Y) = w_1 x$ between the EDSS and classes of a least squares regression.

Figure 5.12(a) shows the linear model between EDSS attribute (x) and classes (MS subgroups and healthy) (y).

The relationship between EDSS and classes (MS subgroups, healthy) in Figure 5.12(b) is based on the data in the linear diagonal (see Figure 5.12(a)). The statistical results for the SE calculated and estimated values are provided with their interpretations.

1. SE of the coefficients is 0.33102.
2. w_1 is the regression coefficient obtained as 5.8127. The linear relation is $E(Y) = 5.8127x$.
3. t-Statistic for each coefficient to test the null hypothesis that the corresponding coefficient is zero against the alternative that it is different from zero, given the other predictors in the model. Note that t-stat = estimate/SE. The t-statistic for the intercept is $5.8127/0.33102 = 17.56$. The column indicating the t-stat value is the ratio of the estimate to its SE. The column labeled "t-stat" is the ratio of the estimate to its large-sample SE and can be used for the null hypothesis test that a particular parameter is equal to zero against either a general or one-sided alternative. The column indicating the "p-value($>|t$-stat$|$)" is the level of significance with regard to this test, using a normal reference distribution instead of a t-stat.

   The coefficient value obtained from t-statistic is used with predictors different from zero (zero against) in order to test the null hypothesis.

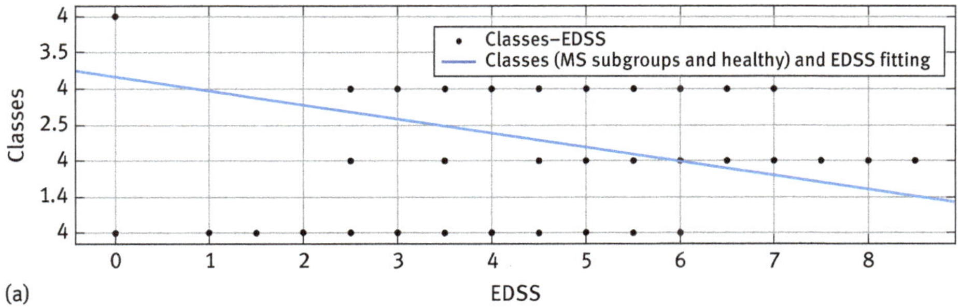

(a)

Intercept	Estimate	SE	*t*-Stat	*p*-Value
	5.8127	0.33102	17.56	4.6492e-48

(b)

Figure 5.12: Linear model for the EDSS–MS subgroups and healthy group: (a) Linear model for the EDSS attribute (*x*) and classes (MS subgroups and healthy) (*y*) and (b) estimated coefficients of EDSS attribute and classes (MS subgroups and healthy group) based on linear regression results.

Note: *t*-stat = estimate/SE is calculated. For the intercept, *t*-stat value has been calculated as 5.8127/0.33102 = 17.56. *t*-Stat row represents the SE and is used for the test of the null hypothesis (general or one dimensional). "*p*-Value" uses a normal reference distribution vis-à-vis sStat, defining the significance level of the null hypothesis.

4. The *p*-value for EDSS ($4.6492e\text{-}48 < 0.05$) is smaller than the common level of 0.05. This shows that it is statistically significant.

5.1.3 Linear model for the analysis of mental functions

Let us see an application over an example based on linear model that analyzes the interrelationship between the attributes in our WAIS-R dataset.

WAIS-R dataset in Table 2.19 includes data of 200 samples belonging to healthy group and 200 samples in control group. The attributes of data that belong to control group are *school education, gender, ..., DM* information. Using these attributes of 400 individuals, it is known which individuals are healthy or patient. For WAIS-R dataset we would like to develop a linear model for control group subjects and healthy individuals whose data are available.

Before developing the linear model algorithmically for the WAIS-R dataset, it will be for the benefit of the reader to study Example 5.5 for the reinforcing of linear model.

Example 5.5 Based on the research carried out in psychology clinic on male patients, the linear regression model set stemming from the correlation between the variables of *age* (*x*) and *assembly ability* (*y*) is described later:

When eq. (5.1) is used

$$y = 3.42 + 0.326x$$

Based on this model, a 1 unit increase in the age variable will lead to a 0.326 unit increase in the assembly ability value. Thus, it can be said that a newborn baby boy will have ($x = 0$) and assembly ability ($x = 0$) as 3.42. At this point, let's pose a related question: can we predict, based on this model, what the assembly ability value will be for an individual who is 50 years in line with the linear model?

Solution 5.5
For $x = 50$,

$$y = 3.42 + 0.326 \times (50) = 19.52$$

Based on the linear model set, it can be estimated that a man aged 50 will have the assembly ability value as 19.52.

In order to set a linear model for the WAIS-R dataset, please have a look at the steps presented in Figure 5.13.

Algorithm: Linear Model for the WAIS-R Data Set
Input: x independent variable for WAIS-R Data Set
Result: y dependent variable w_0 and w_1 coefficients are identified
Method:

(1) //Linear model
(2) //y is the dependent variable and its e value is determined.
(3) //x is the independent variable and it is assumed to be error free.
(4) **while** x variable **do**{
(5) $E(Y) = w_0 + w_1x_1 + e$
(6) **if** ($x = 0$)
(7) $w_0 = y$
(8) **else**
(9) //calculate w_0, w_1, minimum error e
(10) }

Figure 5.13: Linear modeling for the analysis of WAIS-R dataset.

Example 5.6 The application steps of linear modeling are as follows:

Steps (1–3) Simple regression model includes $x = 50$ independent variables that have direct correlation with the dependent variable.

Step (4) x independent variable is accepted as the variable that is controllable and measurable through tolerable error. Based on Example 5.5, $y = 3.42 + 0.326x$ model is set up.

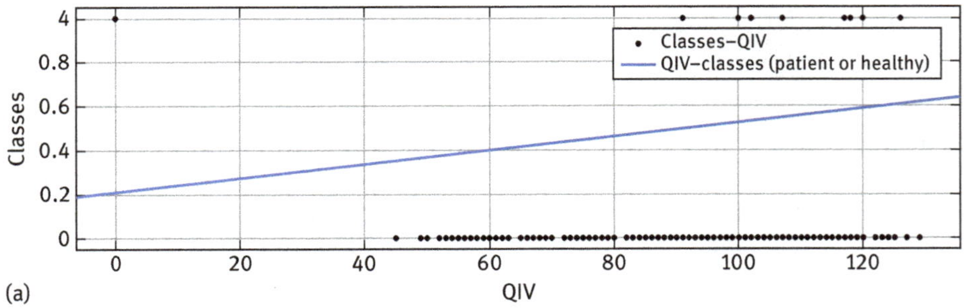

(a)

Intercept	Estimate	SE	*t*-Stat	*p*-Value
	1.2657	0.094401	1.3407	1.2261e-31

(b)

Figure 5.14: Linear model set for QIV–classes (patient and healthy). (a) Linear model between *QIV* (*x*) and classes (*patient* or *healthy*) (*y*) fitting. (b) Estimated coefficients between QIV and classes (patient and healthy) group based on linear regression results.

Steps (5–7) In the linear model, y dependent variable is dependent on the x independent variable (see eq. (4.1)). If $x = 0, y = w_0$. If $x \neq 0$, we move on to Steps (8–10) stage for the estimation of w_0 and w_1 parameters (see eq. (5.1)).

Steps (8–10) Least squares method is used for the estimation of w_0 and w_1 parameters (using eqs. (5.4) and (5.5)). Error value e is calculated based on eq. (5.3). Afterwards, $E(Y)$ equation is obtained.

In Figure 5.15, according to the linear model set for the WAIS-R dataset (x-coordinate: verbal IQ [QIV]; y-coordinate: patient or healthy class), the results obtained are presented in Figure 5.14.

Figure 5.14(a) shows the linear model between QIV and classes (patient–healthy). The relationship between QIV and classes (patient–healthy) in Figure 5.14(b) is based on the data in the linear diagonal (see Figure 5.14(a)). The statistical results for the SE calculated and estimated values are provided below along with their interpretations.

1. The SE of the coefficients is 0.094401.
2. w_1 is the regression coefficient obtained as 1.2657. The linear relation is $E(Y) = 1.2657x$.
3. t-Statistic for each coefficient to test the null hypothesis that the corresponding coefficient is zero against the alternative that it is different from zero, given the other predictors in the model. Note that t-stat = estimate/SE. The t-statistic for the intercept is 1.2657/0.094401 = 1.3407. The *pillar marked t-stat value is the ratio of the estimate to its SE. The column labeled "t-stat" is the ratio of the estimate to its large-sample SE and can be used for a test of the null hypothesis that a particular parameter is equal to zero against either a general or one-sided

alternative. The column marked "p-value(>|t-stat|)" is the significance level for this test, using a normal reference distribution rather than a t-stat.

4. The p-value for EDSS (1.2261e-31 < 0.05) is smaller than the common level of 0.05. This shows that it is statistically significant.

5.2 Multilinear model algorithms for the analysis of various data

Multilinear model is capable of modeling the relationship between the attributes in our dataset with multivariate data. Multilinear model is applied in Sections 5.2.1, 5.2.2 and 5.2.3 – economy (U.N.I.S.) dataset, MS dataset and WAIS-R dataset, respectively.

5.2.1 Multilinear model for the analysis of economy (U.N.I.S.) dataset

Let us see an application over an example based on linear model that analyzes the interrelationship between the attributes in our economy dataset (including economies of U.N.I.S.). Data for these countries' economies from 1960 to 2015 are defined based on the attributes presented in Table 2.8. Data belonging to U.N.I.S. economies from 1960 to 2015 are defined based on the attributes given in Table 2.8 economy dataset (http://data.worldEconomy.org), which is used in the following sections. The attributes of data are *years, time to import (days), time to export (days)* and others. Values for the period 1960–2015, the values over the years, are known. Let's assume that we would like to develop a linear model with the data available for four different countries.

Before developing the multivariate linear model algorithmically for the economy dataset, it will be useful if we can go over the steps provided in Figure 5.15.

For this, let us have a look at the steps provided in Figure 5.15 in order to establish the data to fit a straight line for a multivariate linear model:

Based on Figure 5.15, the steps of the multilinear model algorithm applied to economy (U.N.I.S.) dataset are as follows:

Steps (1–3) Multilinear regression model will be formed with direct correlation to the dependent variable (GDP per capita) in the dataset with Italy economy (between 1960 and 2013) including independent variables such as years and time to import (days).

Step (4) We denote the x_1, x_2, x_3 independent variables as multilinear regression model. The interrelation of the regression coefficients w_0, w_1, w_2, w_3 on three-dimensional plane will be calculated in Steps (5–10).

Steps (5–10) For each of the unknown regression coefficients(w_0, w_1, w_2, w_3), least squares model is applied and the following results have been obtained, respectively, $2.4265, 0.0005, 0.0022, 0$. In Italy for the three variables in the economy dataset (GDP per capita, years [1960–2015] and time to import [days]) multilinear model plot

Algorithm: Multi Linear Model for the Economy (U.N.I.S.) Data Set
Input: x independent variable for Economy (U.N.I.S.) Data Set
Result: y dependent variable, $w_0, w_1, \cdots, w_k$ regression coefficients are identified.
Method:

//Multi linear model
y is the dependent variable and its e_i value is determined.
x is the independent variable and it is assumed to be error free.
while x variable ***do***{
  $E(Y) = w_0 + w_1x_1 + \cdots + w_kx_k + e_i$
if $(x_k = 0)$
  $w_0 = y$
else
  calculate y, minimum error e_i
}

Figure 5.15: Multilinear modeling for the analysis of economy (U.N.I.S.) dataset.

(see Figure 5.16), the multilinear regression coefficients are presented as $w_0 = 2.4265$, $w_1 = 0.0005$, $w_2 = 0.0022$, $w_3 = 0$. The error value (e_i) of the y equation calculated is computed based on eq. (5.3).

For the Italy economy dataset, the multiple regression analysis established for the data to be modeled to fit a straight line between GDP per capita, years and time to import (days) figure is provided in Figure 5.16.

Based on Figure 5.17, a relation has been established for the data to be modeled to fit a straight line between GDP per capita, years and time to import (days) as stated in the equations in the multilinear regression model.

When eq. (5.8) is used:

$$y = 2.4265 - 0.0005x_1 + 0.0022x_2 - (0)x_3$$

$$y = 2.4265 - 0.0005x_1 + 0.0022x_2$$

All these show that each year evens on the average 0.0005 and for Italy economy it is 0.0022 to the time to import (days). w_0 coefficient (>0), GDP per capita coefficient (= −0.0005) and years coefficient affects (=0.0022) are significant attributes.

5.2.2 Multilinear model for the analysis of MS

Let us do an application over an example based on linear model that analyzes the interrelationship between the attributes in our MS dataset.

As presented in Table 2.12, MS dataset has data from the following groups: 76 samples belonging to RRMS, 76 samples to SPMS, 76 samples to PPMS and 76 samples

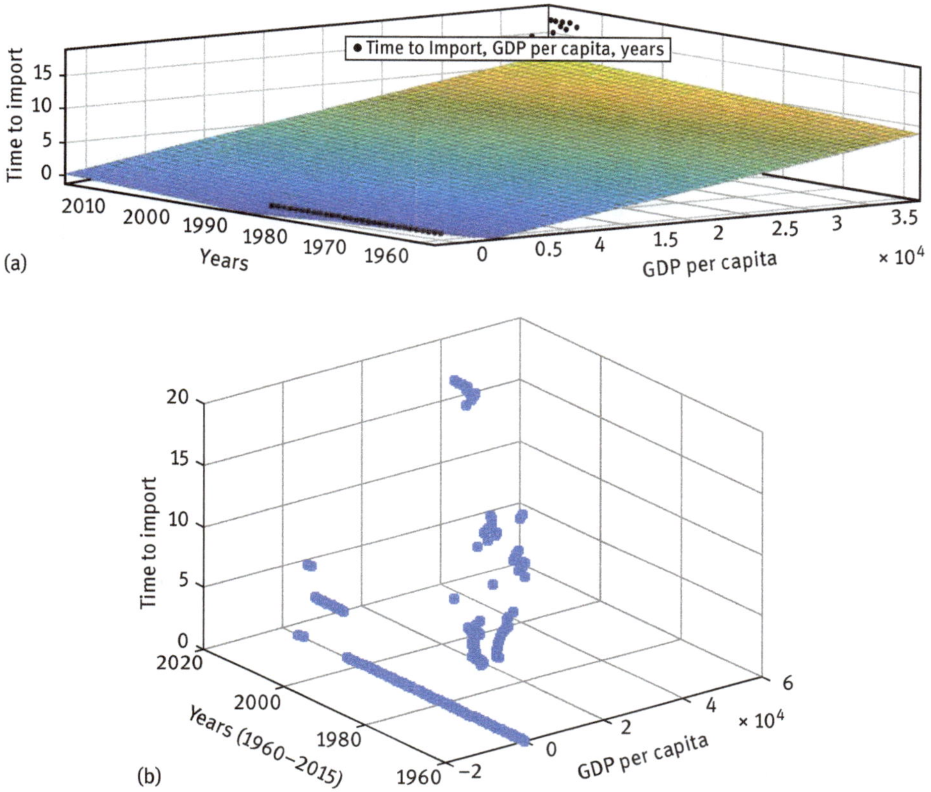

Figure 5.16: (a) Multilinear model mesh plot for GDP per capita, years and time to import attributes for Italy. (b) Multilinear model plot for GDP per capita, years and time to import attributes for Italy.

to healthy subjects of control group. The attributes of the control group are data regarding *brain stem (MRI 1), corpus callosum periventricular (MRI 2), upper cervical (MRI 3) lesion diameter size (mm) in the MRI and EDSS* score. Data are made up of a total of 112 attributes. By using these attributes of 304 individuals, we can know that whether the data belong to the MS subgroup or healthy group. How can we make the classification as to which MS patient belongs to which subgroup of MS including healthy individuals and those diagnosed with MS (based on the lesion diameters (*MRI 1, MRI 2, MRI 3*) and the number of lesion size for (*MRI 1, MRI 2, MRI 3*) as obtained from MRI and EDSS scores)? D matrix has a dimension of (304×112), which means D matrix includes the MS dataset of 304 individuals along with their 112 attributes (see Table 2.12 for the MS dataset). For the classification of D matrix through naive Bayesian algorithm, the first step training procedure is to be employed. For the training procedure, 66.66% of the D matrix can be split for the training dataset (203×112) and 33.33% as the test dataset (101×112). For this, let us have a look at the steps provided in Figure 5.17 to establish the data to fit a straight line regarding a multivariate linear model:

Algorithm: Multi Linear Model for the MS Data Set
Input: x independent variable for MS Data Set
Result: y dependent variable, $w_0, w_1, \ldots, w_k$ regression coefficients are identified.
Method:

```
(1)  //Multi linear model
(2)    //y is the dependent variable and its e_i value is determined.
(3)    //x is the independent variable and it is assumed to be error free.
(4)    while x variable do{
(5)      E(Y) = w_0 + w_1 x_1 + ··· + w_k x_k + e_i
(6)    if (x_k = 0)
(7)      w_0 = y
(8)    else
(9)      //calculate y, minimum error e_i
(10) }
```

Figure 5.17: Multilinear modeling for the analysis of MS dataset.

For the MS dataset, the multiple regression analysis established for the data to be modeled to fit a straight line between the MRI data MRI 1, MRI 2 and the individuals' EDSS three-dimensional space figure is provided in Figure 5.18.

Based on Figure 5.18, the linear model steps applied to MS dataset (see Table 2.12) linear model are given below.

Steps (1–3) Multilinear regression model will be formed with MRI that has direct correlation with dependent variable (EDSS score) with data belonging to 304 individuals, including these independent variables: lesion count (in the first and second regions of the brain).

Step (4) We denote the x_1, x_2, x_3 independent variables as the multilinear regression model. The interrelation of the regression coefficients w_0, w_1, w_2, w_3 on three-dimensional plane will be calculated in Steps (5–10).

Steps (5–10) For each of the unknown regression coefficients(w_0, w_1, w_2, w_3), least squares model is applied and the following results have been obtained, respectively: 0.0483, 0.1539, 0.1646, 0.0495. For the three variables in the MS dataset (EDSS, MRI 1, MRI 2 attributes) multilinear model plot (see Figure 5.18), the interrelations of the multilinear regression coefficients are presented as $w_0 = 0.0483, w_1 = 0.1539, w_2 = 0.1646, w_3 = 0.0495$. The error value e_i of the y equation calculated is computed based on eq. (5.3).

In the multilinear regression model established for the data modeled to fit a straight line as in Figure 5.18, the following equations show the relation between the EDSS score of the individuals, the lesion count of the first region and the lesion count of the second region.

If we use eq. (5.8),

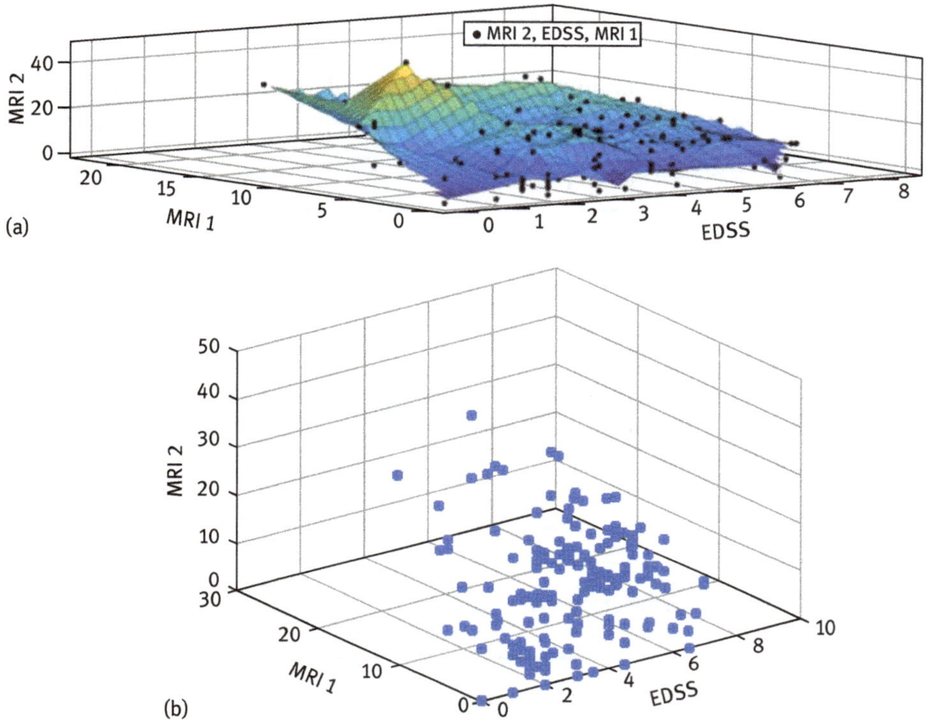

Figure 5.18: (a) Multilinear model mesh plot for EDSS, MRI 1, MRI 2 attributes. (b) Multilinear model plot for EDSS, MRI 1, MRI 2 attributes.

$$y = 0.0483 + 0.1539x_1 + 0.1646x_2 + 0.0495x_3$$

All these show that each EDSS score adds on the average 0.1539 and each lesion count of the second region corresponds to 0.1646 as the lesion count of the first region. Based on w_0 coefficient (>0), MRI 1 coefficient (=0.1646) and MRI 2 coefficient (=0.0495), the coefficients obtained for the EDSS score affects (=0.1539) reveal that they are significant attributes.

5.2.3 Multilinear model for the analysis of mental functions

Let us do an application over an example based on linear model that analyzes the interrelationship between the attributes in our WAIS-R dataset.

WAIS-R dataset as provided in Table 2.19 includes data of 200 samples belonging to healthy group and 200 samples in control group. The attributes of data that belong to control group are *school education, gender, . . ., DM* information. Using these attributes of 400 individuals, it is known which individuals are healthy or patient. For WAIS-R dataset we would like to develop a linear model for control group subjects and healthy individuals whose data are available.

Before developing the multivariate linear model algorithmically for the WAIS-R dataset, it will be for the benefit of the reader to study Figure 5.19 for the reinforcing of linear model.

Algorithm: Multi Linear Model for the WAIS-R Data Set

Input: x independent variable for the WAIS-R Data Set

Result: y dependent variable, $w_0, w_1, \cdots, w_k$ regression coefficients are identified.

Method:

```
(1)  //Multi linear model
(2)    //y is the dependent variable and its e_i value is determined.
(3)    // x is the independent variable and it is assumed to be error free.
(4)    while x variable do{
(5)       E(Y) = w_0 + w_1x_1 + ··· + w_kx_k + e_i
(6)    if (x_k = 0)
(7)       w_0 = y
(8)    else
(9)       //calculatey, minimum error e_i
(10) }
```

Figure 5.19: Multilinear modeling for the analysis of WAIS-R dataset.

According to Figure 5.19, the steps of the multilinear model applied to WAIS-R dataset can be explained as follows:

Steps (1–3) Multilinear regression model will be formed with direct correlation to the dependent variable (gender) in the WAIS-R dataset (with a total of 400 individuals) including independent variables such as QIV and information.

Step (4) We denote x_1, x_2, x_3 independent variables as multilinear regression model. The interrelation of the regression coefficients w_0, w_1, w_2, w_3 on three-dimensional plane will be calculated in Steps (5–10).

Steps (5–10) For each of the unknown regression coefficients(w_0, w_1, w_2, w_3), least squares model is applied and the following results have been obtained, respectively, 38.7848, 1.5510, 5.8622, 1.6936. For the three variables in the WAIS-R dataset (QIV, gender and information) multilinear model plot (see Figure 5.20), the interrelation of the multilinear regression coefficients is presented as $w_0 = 38.7848, w_1 = 1.5510, w_2 = 5.8622, w_3 = 1.6936$. The error value (e_i) of the y equation calculated is computed based on eq. (5.3).

For the WAIS-R dataset, the multiple regression analysis established to fit a straight line between gender, information and QIV score three-dimensional space figure is provided in Figure 5.20.

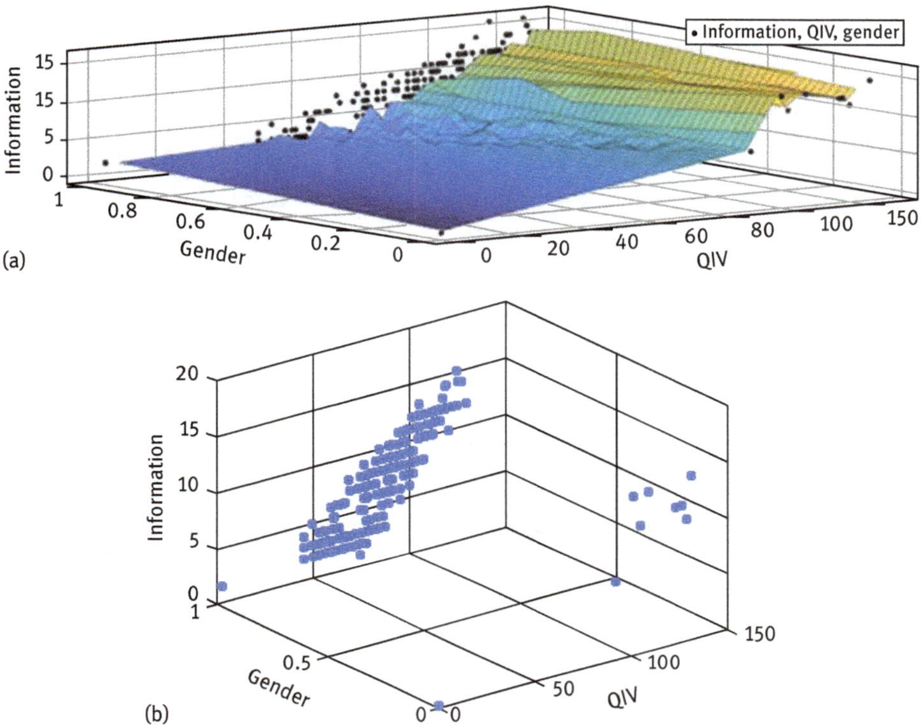

Figure 5.20: (a) Multilinear model mesh plot for QIV, gender and information attributes. (b) Multilinear model plot for QIV, gender and information attributes.

A relation has been established between gender, information and QIV as stated in the following equations based on Figure 5.20.

When eq. (5.8) is used,

$$y = 38.7848 - 1.5510x_1 - 5.8622x_2 - 1.6936x_3$$

All these reveal that each QIV score evens on the average 1.5510 and gender 5.8622 to the information. According to w_0 coefficient (>0), *QIV* coefficient (= −1.5510), *information coefficient* (= −1.6936), the coefficient obtained for *gender* affects (= −5.8622) so this shows that they are significant attributes.

References

[1] Dobson AJ, Barnett AG. An introduction to generalized linear models. USA: CRC Press, 2008.

[2] Ramachandran KM, Chris PT. Mathematical statistics with applications in R.USA: Academic Press, Elsevier, 2014.

[3] C, Zidek J, Lindsey J. An introduction to generalized linear models. New York, USA: CRC Press, 2010.

[4] Rencher AC, Schaalje GB. Linear models in statistics. USA: John Wiley & Sons, 2008.
[5] Neter J, Kutner MH, Nachtsheim CJ, Wasserman W. Applied linear statistical models. New York, NY, USA: The McGraw-Hill Compan Inc., 2005.
[6] John F. Regression diagnostics: An introduction . USA: Sage Publications, Vol. 79, 1991.
[7] John F. Applied regression analysis and generalized linear models. USA: Sage Publications, 2015.
[8] Allison PD, Event History and Survival Analysis: Regression for Longitudinal Event Data. USA: Sage Publications, Inc., 2014.
[9] Weisberg S. Applied linear regression. USA: John Wiley & Sons, Inc., 2005.
[10] Berry WD, Feldman S. Multiple regression in practice. USA: Sage Publications, Inc.,1985.
[11] Everitt BS, Dunn G. Applied multivariate data analysis. USA: John Wiley & Sons, Ltd., 2001.
[12] Agresti A. Foundations of linear and generalized linear models. Hoboken, NJ: John Wiley & Sons, 2015.
[13] Raudenbush SW, Bryk A S. Hierarchical linear models: Applications and data analysis methods (Vol. 1). USA: Sage Publications, Inc., 2002.
[14] Darlington RB, Hayes AF. Regression analysis and linear models: Concepts, applications, and implementation, New York: The Guilford Press, 2016.
[15] Stroup WW. Generalized linear mixed models: modern concepts, methods and applications. USA: CRC press, 2012.
[16] Searle SR., Gruber MH. Linear models. Hoboken, New Jersey, John Wiley & Sons, Inc., 2016.
[17] Faraway JJ. Extending the linear model with R: generalized linear, mixed effects and nonparametric regression models (Vol. 124). USA: CRC press, 2016.
[18] Hansen LP, Sargent TJ. Recursive models of dynamic linear economies. USA: Princeton University Press, 2013.
[19] Bingham NH, Fry JM. Regression: Linear models in statistics. London: Springer Science & Business Media, 2010.
[20] Karaca Y, Onur O, Karabudak R. Linear modeling of multiple sclerosis and its subgroubs. *Turkish Journal of Neurology*, 2015, 21(1), 7–12.
[21] Reddy TA, Claridge DE. Using synthetic data to evaluate multiple regression and principal component analyses for statistical modeling of daily building energy consumption. *Energy and buildings*, 1994, 21(1), 35-44.

6 Decision Tree

This chapter covers the following items:
- Algorithm for decision tree
- Algorithm for classifying with decision tree algorithms
- Examples and applications

At this point, let us now dwell on what a decision tree is. A *decision tree* is a tree structure that is similar to a flowchart tree structure. In a typical decision tree structure, each *internal node* (also known as *nonleaf node*) shows a test on an attribute [1–3]. On the other hand, each *branch* shows an outcome of the test. In decision trees, each *leaf node* (also known as *terminal node*) contains a class label. The *root* node is the top node in a tree. Figure 6.1 provides a typical illustration of a decision tree. It shows the concept of *time to export (days)*. In other words, it predicts if an economy has USA, New Zealand, Italy or Sweden.

Each internal (nonleaf) node represents a test on an attribute. Each leaf node represents a class that is likely to have the time to export (days), internal nodes are represented by rectangles and leaf nodes by ovals. Some decision tree algorithms yield only *binary* trees in which each internal node branches to exactly two other nodes, while some others can produce on binary trees [4–6].

Some questions regarding decision trees are elicited and elaborated:

1. How can we use decision trees for classification purposes?
   If the associated class label is not known for X a dataset, then the dataset attribute are tested against the decision tree. From the root to the leaf node, one follows a certain track or path. This contains the class prediction for that particular dataset. It is easy to change the decision trees into classification rules. For these reasons, decision trees are commonly used for the classification purposes.
2. What is the reason for the popularity of decision tree classifiers?
   First, there is no requirement of having field knowledge for the formation of the decision tree classifiers. There is no need to have knowledge of parameter setting, either. In addition, it is possible to deal with multi-dimensional data through decision trees. Another advantage is related to the simplicity and time-saving characters of decision trees with regard to learning and classification steps of decision tree induction. In addition, decision tree classifiers yield a good level of accuracy rate. Nevertheless, it should be noted that a successful use relies on the available data one is using for research [7]. Algorithms using decision tree induction enjoy numerous areas of applications including but not limited to medicine, production exact sciences (including biology and astronomy) and financial areas.

https://doi.org/10.1515/9783110496369-006

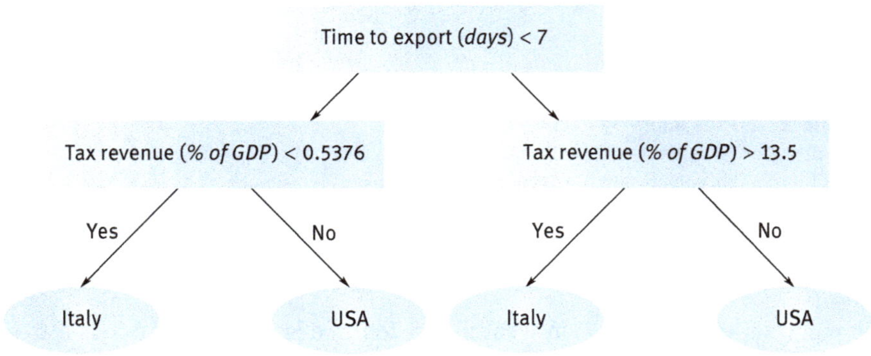

Figure 6.1: A decision tree for the concept of *time to export (days)*, indicating whether an economy is likely to have *USA, New Zealand, Italy* or *Sweden*.

Section 6.2 provides a basic algorithm for learning decision trees. *Attribute selection measures* are utilized for the selection of the attribute while building the tree. This selection is concerned with getting the best partitions for the datasets so that they can fall into distinct classes. Section 6.1 presents some of the popular measures seen in attribute selection. When decision trees are constructed, various branches are likely to reflect noise or outliers in the training data. At this stage, *tree pruning* method is used for trying to identify and remove these kinds of branches. This is particularly significant for the improvement of classification accuracy on the unseen data. Tree pruning is elaborated in Section 6.2, addressing scalability issues concerning large datasets.

6.1 Decision tree induction

Decision tree induction is referred to as learning of decision trees from the training datasets that are class labeled. In this section, it is important to mention two cornerstone algorithms that generated a flood of works carried out on decision tree induction. A decision tree algorithm called *ID3* (iterative dichotomiser (ID)) was developed by a researcher engaged in machine learning. J. Ross Quinlan started the work in the late 1970s and early 1980s and E. B. Hunt, J. Marin and P. T. Stone later expanded on the previous work, with Quinlan presenting a benchmark called *C4.5* (a successor of ID3) [8]. In 1984, a group of statisticians with L. Breiman, J. Friedman, R. Olshen and C. Stone published *Classification and Regression Trees* (*CART*) book [9]. In their book, they described the binary decision tree generation. ID3 and CART were invented independent of each other and surprisingly at around the same time periods. They track a similar approach about learning decision trees from training datasets.

Most algorithms with regard to the decision tree induction track an approach that is stop-down, which means decision trees are built in a top-down recursive divide and conquer way starting with a training set of datasets and their associated class labels.

Then one partitions the training set into smaller subsets while building the tree; Figure 6.2 provides a summarized form of a decision tree algorithm. Although the algorithm may seem very long at first glance, there is no need to worry about it since it is really straightforward. The strategy regarding the algorithm is given as follows:

- The algorithm has three parameters that are *D*, *attribute list* and *attribute selection*. *D* is referred to as data partition, the complete set of training datasets and associated class labels thereof. A list of attributes that describe the datasets is called the parameter attribute list. Finally, *attribute selection* identifies a heuristic procedure for the selection of the attribute that differentiates the "best" given datasets based on the class. This procedure uses an attribute selection measure like the *Gini* index or the information gain. The *Gini* index imposes the resulting tree to be binary. Other measures such as the information gain do not allow multiway splits to illustrate better; we can say that two or more branches are to be grown from a node.
- The starting of the tree is seen as a single node and *N* depicts the training datasets in *D* (Step 1). The partition of class-labeled training datasets at node *N* gives us the set of datasets. These sets track a path from the tree root to the node *N* while it is being processed by the tree [10].

Following is the classification of the training dataset being trained with decision tree algorithm (see Figure 6.2).

Steps (1–3) The datasets in *D* may all be of the same class. In such a case, node *N* becomes a leaf that is labeled with that class.

Steps (4–5) Note that there are conditions that are terminating. At the end of the algorithm, all the terminating conditions are explicated.

Step (6) If they are not explained, then the algorithm calls *Attribute_selection_method* so that the *splitting criterion* can be determined. It is the splitting criterion that shows which attribute one shall test at node *N* by the determination of the "best" way for separating the datasets in *D* into individual classes. There is another function of the slitting criterion that shows which branches one shall grow from node *N* with regard to the results of the test chosen. More specifically, the splitting criterion presents the *splitting attribute*, which may also demonstrate either a *split point* or a *splitting subset*. At this point, it is important to obtain the resulting partitions at each branch with highest "purity" possible. If all the datasets in a partition belong to the same class, it is possible to call a partition a *pure* one.

Steps (7–9) The splitting criterion labels the node *N*. This serves as a test at the node. From node *N* a branch is grown for each result of the splitting criterion.

Steps (10–15) The datasets in *D* are divided accordingly. Figure 6.3 shows three of the possible scenarios. Let us say, *A* is the splitting attribute. On the basis of the training data, *A* has distinct values $\{a_1, a_2,:::, a_v\}$. The three possible cases are discussed.

Algorithm: Generate decision tree from the training dataset of datasets partition D.

Input: Data partition. D a set of training datasets and associated class labels thereof attributes the set of candidate attributes;

Attribute selection (), a procedure to decide on the splitting criterion that "best" splits the datasets into individual classes. This criterion involves an attribute of splitting and probably one of the two choices, which are either split-point or splitting subset.

Result: A decision tree.

Method:

create a node N;
if dataset in D are all of the same class Y **then**
    return N as a leaf node labeled with the class Y;
if *attribute list* is empty **then**
    return N as a leaf node labeled with the majority class in D;//majority voting
    apply **attribute selection** (*D attribute list*) to find the "best" *splitting criterion*;
    label node N with splitting criterion;
if splitting attribute is discrete value **and** multiway splits enabled **then**//not restricted to binary trees
    attribute list = attribute list − splitting attribute;//remove the splitting attribute
while outcome j of splitting criterion **do**//split the dataset and grow subtrees for each partition
let D_j be the set of datasets in D satisfying outcome j;//a partition
    if D_j is empty **then**
        attach a leaf labeled with the majority class in D to node N;
    else attach the node returned by **decision tree** (D_j attribute list) to node N;
return N;//learned tree

Figure 6.2: Basic algorithm to induce a decision tree from training datasets.

In Figure 6.3(a)–(c), examples for the splitting criterion labels regarding decision tree are explained in detail.

Let us take case in which *A is discrete valued.* In this case, test outcomes at node N match directly with the known values of A. For each value that is known, a branch is created: a_j of A and labeled with that value (Figure 6.3(a)). Partition D_j is the subset of class-labeled datasets in D that has the value a_j of A. All datasets in a particular partition have the same value for A, and so A does not have to be taken into account in any of the future dataset partitioning. For this reason, it is taken out from *attribute list.*

1. The case in which *A is continuous valued.* In this second possible case, the test at node N has two possible outcomes that correspond to the conditions $A \le$ *split point* and $A >$ *split point*. In practice, the split point a is generally taken as the midpoint of the two known adjacent values of A. Thus, it may not essentially be a pre-existing value of A from the training data. Two branches are grown from N and labeled based on the outcomes previously gained as can be seen from Figure 6.3(b). The datasets are partitioned in a way in which D_1 contains the subset of class-labeled datasets in *D:* $A \le$ *split point*, but D_2 holds the rest.

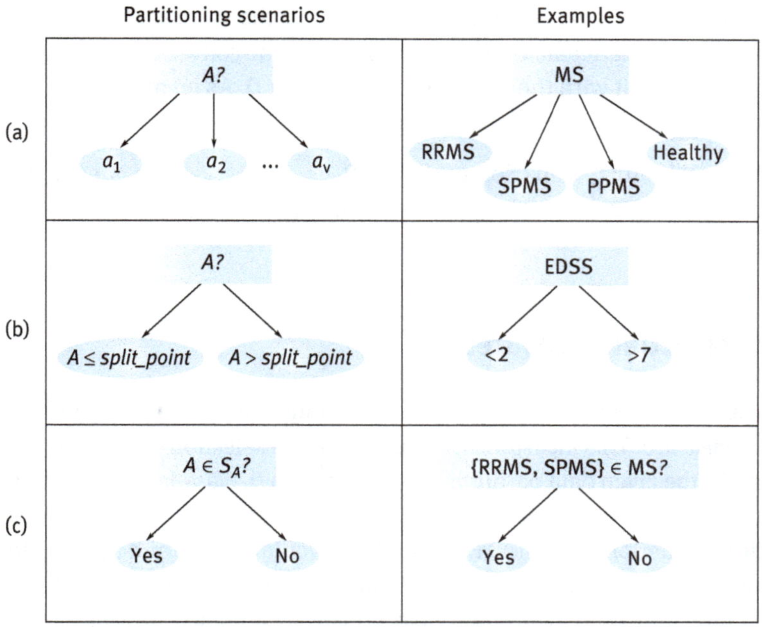

Figure 6.3: The figure displays three possibilities for the partitioning of datasets based on the splitting criterion. Each with examples, let us assume *A* be the splitting attribute. (a) If *A* is discrete valued, then one branch is grown for each *A* known value. (b) In the condition that *A* is continuous valued, two branches are grown parallel to $A \leq split\_point$ and $A >$ split_point. (c) In the condition that *A* is discrete valued and a binary tree must be generated, then the test will have the form as $A \in S_A$. Here as you can see S_A is the splitting subset for *A*.

2. The case in which *A is discrete valued* and a *binary tree* must be produced: The test at node *N* is in the form "$A \in S_A$." Here S_A is the splitting subset for *A*, a subset of the known values of *A*. If the dataset has value a_j of *A* and if $a_j \in S_A$, the test at node *N* is regarded as satisfied. Two branches are grown from *N* (which you can see in Figure 6.3(c)). As a general convention in the field, the left branch out of *N* is marked *yes* in order that D_1 corresponds to the subset of class-labeled datasets in *D* fulfilling the test. Right branch out of *N* is marked *no* in order that D_2 matches up with the subset of class-labeled datasets from *D* that do not fulfil or satisfy the test.

   The same process is iteratively used by the algorithm so as to form a decision tree for the datasets at each of the resulting partition. D_j of *D* (as you may see Step 14). The repetitive partitioning comes to an end when one of the following terminating conditions provided is true:

   a. Steps (1–3) All the datasets in *D* partition represented at node *N* are members of the same class.

b. Steps (4–11) There are no remaining attributes for the datasets to be partitioned further and one patient *majority voting* involving the conversion of N into a leaf as well as labeling it with the most common class in D. As an alternative the node class distribution of the datasets can be stored.

c. Steps (12–14) In this scenario, partition D_j is empty, which means no datasets exist for a given branch. In such a case, a leaf is created with the majority class in D.

6.2 Attribute selection measures

After having covered decision tree and decision tree induction, we can move on to *attribute selection measure*. This measure is a heuristic for the selection of the splitting criterion which splits the given data partition "best" separates D, class-labeled training datasets into individual classes. If it is necessary to divide D into smaller partitions in line with the splitting criterion outcomes. Each partition would be pure and that would be an ideal situation since all the datasets in the partition belongs to the same class. *Splitting rule* is another term used for attribute selection measures since these measures resolve how the datasets at a given node are going to be partitioned. The attribute selection measure presents a ranking for each attribute that describe the given training datasets. The attribute that has the best score for the measure is selected as the *splitting attribute* for the given datasets. If the splitting attribute has continuous value or if there is a restriction to binary trees, then either a *split point* or a *splitting subset* has to be identified as a component of the splitting criterion. The tree node that is formed for partition D is marked with the splitting criterion. Branches are grown for the each outcome of the criterion. Additionally, the datasets are partitioned in view of that. In this section, you can find the descriptions of three popular attribute selection measures: *information gain, gain ratio* and *Gini index* [11].

6.3 Iterative dichotomiser 3 (ID3) algorithm

D, the data partition, is the training set of class-labeled datasets. Assume that the class label attribute has m distinct values that define m distinct classes C_i (for $i = 1, \ldots, m$). D is the set of datasets of class C_i in D. $|D|$ denotes the number of datasets in D and $|C_{i,D}|$ in $C_{i,D}$.

As the attribute selection measure, ID3 uses *information gain*. This measure is reliant on the leading work carried out by Claude Shannon [12] on information theory, which examined the value or "information content" of messages. Node N represents or holds the datasets of partition D. The attribute with the highest information gain is selected as the splitting attribute for node N. This attribute reduces the information required to classify the datasets in the resultant partitions and reflects the least randomness or “impurity” in the partitions. This approach reduces the expected number of tests that are required to classify a given dataset to the minimum and ensures that a simple tree is found (although not always the simplest one).

The expected information that is needed for the classification of a dataset in D is provided by the following formula:

$$\text{Info}(D) = -\sum_{i=1}^{m} p_i \log_2(p_i) \tag{6.1}$$

where p_i is the nonzero probability concerning an arbitrary dataset in D that belongs to class C_i. It is estimated by $\frac{|C_{i,D}|}{|D|}$. $\text{Info}(D)$ is the average amount of information required to identify the class label of a dataset in D. The available information at this point is based merely on the proportions of datasets pertaining to each class. $\text{Info}(D)$is also recognized as the *entropy* of D.

Ideally, the partitioning process is desired to produce an exact classification of the datasets. In other words, each partition is aspired to be pure. Yet, it is probable that the partitions will not be pure as a partition might contain a collection of datasets from different classes instead of being from a single class.

The information still needed following the partitioning to reach an exact classification is measured as an amount by the following formula:

$$\text{Info}_A(D) = \sum_{j=1}^{v} \frac{|D_j|}{|D|} \times (D_j) \tag{6.2}$$

The term $\frac{|D_j|}{|D|}$ refers to the weight of the j th partition. $\text{Info}_A(D)$ is the expected information that is needed for the classification of a dataset from D, which is based on the separation by A. If the expected information required is smaller, the purity of the partitions shall be greater.

Information gain is the difference between the original information requirement (based on the proportion of classes) and the new requirement (as can be obtained following the partitioning process on A). In short,$\text{Gain}(A)$ shows how much one could gain by branching on A; the attribute A that has the highest information gain. $\text{Gain}(A)$ is selected as the splitting attribute at node N. This is comparable to saying that partition on the attribute A that would do the “best classification” is needed so that the amount of information still needed to finish categorizing the datasets is smallest minimum $\text{Info}_A(D)$.

The following formula clarifies the situation:

$$\text{Gain}(A) = \text{Info}(D) - \text{Info}_A(D) \tag{6.3}$$

If you want to calculate an attribute’s information gain with continuous value, you have an attribute A not a discrete-valued one. (For instance, suppose that we have the raw values for this attribute rather than having the discretized version of *Age* from Table 2.19.) For this case, you will have to determine the "best" *split-point* for A, where the split-point is a threshold on A.

You first sort the values of A in an increasing order. The midpoint between each pair of adjacent values is regarded as a possible split-point for given v values of A. Then v–1 possible splits are taken into consideration and evaluated [13]. For instance, the midpoint between the values a_i and a_{i+1} of A is found through the following formula:

$$\frac{a_i + a_{i+1}}{2} \tag{6.4}$$

The values of A may be arranged beforehand; next determination of the best split for A needs only one pass through the values. For each possible split-point for A, you evaluate $\mathrm{Info}_A(D)$. Here, the number of partitions is two: $v = 2$ (or $j = 1, 2$) as can be

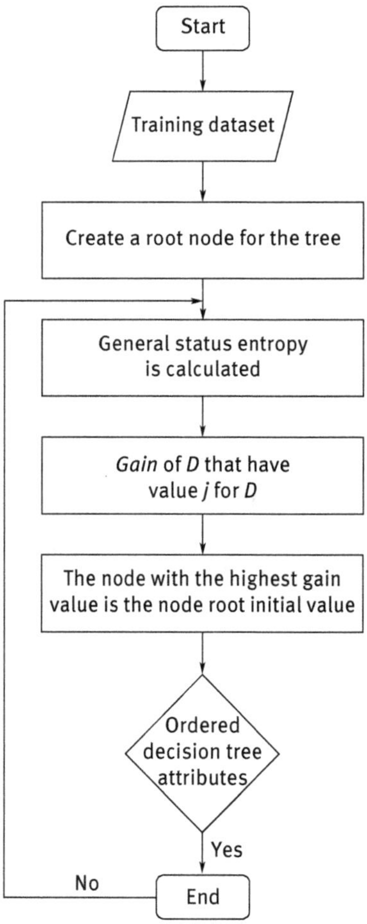

Figure 6.4: The flowchart for ID3 algorithm.

seen in eq. (6.2). The point that has the minimum expected information obligation for A is chosen as the *split point* for A. D_1 is the set of datasets in D that satisfies $A \le$ *split_point*. D_2 is the set of datasets in D, which satisfies A>*split_point*.

The flowchart of ID3 algorithm is as the one provided in Figure 6.4.

Based on ID3 algorithm flowchart (Figure 6.4), you can find the steps of ID3 algorithm. The steps applied in ID3 algorithm are as follows:

Algorithm: ID3 Algorithm

Input: Training set including m number of samples $D = [x_1, x_2, \ldots, x_m]$ $(x_i \in R^d)$
Test data including t number of samples $T = [x_1, x_2, \ldots, x_t]$ $(x_k \in R^d)$

Result: Classification information corresponding to the test and training data $C = [c_1, c_2, \ldots, c_m]$ number of trees N

Method:

(1) Create a node for the tree
(2) **while** each sample i of belonging to the training set **do** {
(3) //The general status entropy is calculated

$$Info(D) = -\sum_{i=1}^{m} p_i \log_2(p_i)$$

(4) root = j//refers to the weight of the j *th* partition
(5) **while** possible value j of D **do**

$$Info_A(D) = \sum_{j=1}^{v} \frac{|D_j|}{|D|} \times (D_j)$$

(6) //gain of D that have value j for D

$$Gain(A) = Info(D) - Info_A(D)$$

(7) **if** is empty, the leaf is the most common value in D
(8) **else** add the subtree ID3()
(9) **return** the learned tree(N);
(10) //The node with the highest gain value is the node root initial value.
(11) c_i = majority ($Gain$)}
(12) //Predict each test sample's class

Figure 6.5: General ID3 decision tree algorithm.

The following Example 6.1 shows the application of the ID3 algorithm to get a clearer picture through relevant steps.

Example 6.1 In Table 6.1 the values for the individuals belonging to WAIS-R dataset are elicited as *gender, age, verbal IQ (QIV)* and *class*. The data are provided in Chapter 2 for WAIS-R dataset (Tables 2.19 and 6.1). These three variables (*gender, age* and *QIV*) have been handled. Class (*patient* and *healthy individuals*). At this moment, let us see how we can form the ID3 tree form using these data.

We can apply the steps for ID3 algorithm in the relevant step-by-step order so as to build the tree form as shown in Figure 6.5.

Table 6.1: Selected sample from the WAIS-R dataset.

Individual (ID)	Gender	Age	QIV	Class
1	Female	48	170	Patient
2	Female	49	151	Healthy
3	Female	52	158	Patient
4	Female	56	165	Patient
5	Male	59	160	Healthy
6	Female	61	159	Patient
7	Male	62	162	Healthy
8	Male	63	174	Patient
9	Female	68	168	Patient
10	Female	69	177	Patient
11	Male	72	170	Patient
12	Male	74	165	Healthy
13	Male	85	175	Patient
14	Male	85	190	Patient
15	Male	98	190	Patient

Steps (1–4) The first step is to calculate the root node to form the tree structure. The total number of record is 15 as can be seen in Table 6.1 (first column), which can go as follows: (Individual (ID): 1, 2, 3, ..., 15). We can use the entropy formula in order to calculate the root node based on eq. (6.1).

As can be seen clearly in Table 6.1, class column 11 individuals belong to *patient* classification and 4 to *healthy*. The general status entropy can be calculated based on eq. (6.1) as follows:

$$\frac{11}{15}\log\left(\frac{15}{11}\right)+\frac{4}{15}\log\left(\frac{15}{4}\right)=0.2517$$

Table 6.1 lists the calculation of the entropy value for the *gender* variable (female–male; our first variable).

Step (5) In order to calculate the entropy value of the *gender* variable.

$$\text{Info}(D)=\frac{6}{7}\log\left(\frac{7}{6}\right)+\frac{1}{7}\log\left(\frac{7}{1}\right)=0.1780$$

$$\text{Info}_{\text{Gender}=\text{Male}}(D) = \frac{5}{8}\log\left(\frac{8}{5}\right) + \frac{3}{8}\log\left(\frac{8}{3}\right) = 0.2872$$

The weighted total value of the *gender* variable can be calculated as in Figure 6.5.

$$\text{Info}_A(D) = \sum_{j=1}^{v} \frac{|D_j|}{|D|} \times (D_j) \text{ (based on (6.2))}$$

$$\left(\frac{7}{15}\right) \times 0.1780 + \left(\frac{8}{15}\right) \times 02872 = 0.2361$$

Step (6) The gain calculation of the *gender* variable is done according to eq. (6.3). The calculation is as follows:

$$\text{Gain(Gender)} = 0.2517 - 0.2361 = 0.0156$$

Now let us do the *gain* calculation for another variable of ours, *age* variable following the same steps mentioned. In order to form the branches of the tree formed based on ID3 algorithm, the *age* variable should be split into groups: let us assume that the *age* variable is grouped as follows and labeled accordingly.

Table 6.2: The grouping of the age variable.

Age interval	Group no.
48–55	1
59–64	2
65–75	3
76–85	4
86–100	5

Steps (1–4) Let us now calculate the entropy values of the group numbers as labeled with the *age* interval values given in Table 6.2; eq. (6.2) is used for this calculation.

$$\text{Info}_{\text{Age}=1.\text{Group}} = \frac{1}{2}\log\left(\frac{3}{1}\right) + \frac{2}{3}\log\left(\frac{3}{2}\right) = 0.2764$$

$$\text{Info}_{\text{Age}=2.\text{Group}} = \frac{2}{5}\log\left(\frac{5}{2}\right) + \frac{3}{5}\log\left(\frac{5}{3}\right) = 0.2923$$

$$\text{Info}_{\text{Age}=3.\text{Group}} = \frac{1}{4}\log\left(\frac{4}{1}\right) + \frac{2}{4}\log\left(\frac{4}{2}\right) + \frac{1}{4}\log\left(\frac{4}{1}\right) = 0.4515$$

$$\text{Info}_{\text{Age}=4.\text{Group}} = \frac{1}{2}\log\left(\frac{2}{1}\right) + \frac{1}{2}\log\left(\frac{2}{1}\right) = 0.3010$$

$$\text{Info}_{\text{Age}=5.\text{Group}} = \frac{1}{1}\log\left(\frac{1}{1}\right) = 1$$

Step (5) The calculation of the weighted total value for the *age* variable is as follows:

$$\left(\frac{3}{15}\right) \times 0.2764 + \left(\frac{5}{15}\right) \times 0.2923 + \left(\frac{4}{15}\right) \times 0.4515 + \left(\frac{2}{15}\right) \times 0.3010 + \left(\frac{1}{15}\right) \times 1 = 0.3799$$

Step (6) The gain calculation for the age variable Gain(Age) is calculated according to eq. (6.3):

$$\text{Gain}(\text{Age}) = 0.2517 - 0.3799 = -0.1282.$$

At this point the *gain* value of our last variable, namely, the QIV variable, is to be calculated.

We shall apply the same steps likewise for the QIV variable as well. In order to form the branches of the tree to be formed, it is required to split the QIV variable into groups. If we assume that the QIV variable is grouped as follows:

Table 6.3: Grouping the QIV variable.

QIV	Group no.
151–159	1
160–164	2
165–174	3
175–184	4
>185	5

Steps (1–4) Let us calculate the entropy value of the QIV variable based on Table 6.3.

$$\text{Info}_{\text{QIV}=1.\text{Group}} = \frac{1}{3}\log\left(\frac{3}{1}\right) + \frac{2}{3}\log\left(\frac{3}{2}\right) = 0.2764$$

$$\text{Info}_{\text{QIV}=2.\text{Group}} = \left(\frac{2}{2}\right)\log\left(\frac{2}{2}\right) = 1$$

$$\text{Info}_{\text{QIV}=3.\text{Group}} = \left(\frac{1}{6}\right)\log\left(\frac{6}{1}\right) + \left(\frac{5}{6}\right)\log\left(\frac{6}{5}\right) + \left(\frac{1}{4}\right)\log\left(\frac{4}{1}\right) = 0.1957$$

$$\text{Info}_{\text{QIV}=4.\text{Group}} = \left(\frac{1}{2}\right)\log\left(\frac{2}{1}\right) + \left(\frac{1}{2}\right)\log\left(\frac{2}{1}\right) = 0.3010$$

$$\text{Info}_{\text{QIV}=5.\text{Group}} = \left(\frac{2}{2}\right)\log\left(\frac{2}{2}\right) = 1$$

Step (5) Let us use eq. (6.2) in order to calculate the weighted total value of the QIV variable.

$$\left(\frac{3}{15}\right)\times 0.2764 + \left(\frac{2}{15}\right)\times 1 + \left(\frac{6}{15}\right)\times 0.1957 + \left(\frac{2}{15}\right)\times 0.3010 + \left(\frac{2}{15}\right)\times 1 = 0.4404$$

Step (6) Let us apply the *gain* calculation for the QIV variable based on eq. (6.3):

$$Gain(QIV) = 0.2517 - 0.4404 = -0.1887$$

Step (7) We have been able to calculate the *gain* values of all the variables in Table 6.1. These variables are *age, gender* and *QIV*. The calculation has yielded the following results for the variables:

$$\text{Gain}(\text{Gender}) = 0.0156; \text{Gain}(\text{Age}) = -0.1282; \text{Gain}(\text{QIV}) = -0.1887$$

When ranked from the smallest to the greatest, the greatest *gain* value has been obtained from *gain* (*gender*). Based on the greatest values, gender will be selected as the root node and it is presented as shown in Figure 6.6 Gender variable constitutes the root of the tree in line with the procedural steps of the ID3 algorithm as shown in Figure 6.6. The branches of the tree are shown as *female* and *male*.

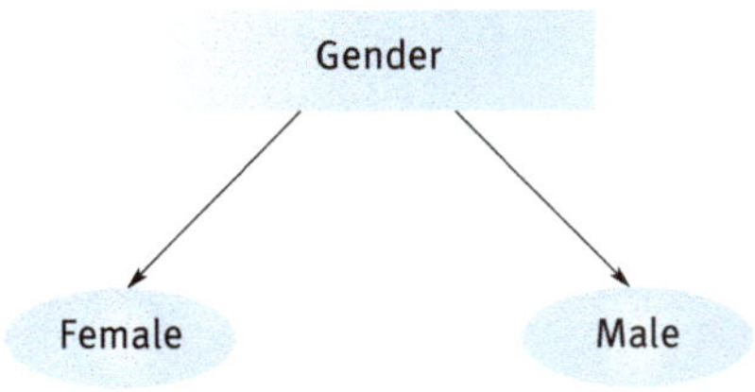

Figure 6.6: The root node initial value based on Example 6.1 ID3 algorithm.

6.3.1 ID3 algorithm for the analysis of various data

ID3 algorithm has been applied on economy (U.N.I.S.) dataset, multiple sclerosis (MS) dataset and WAIS-R dataset in Sections 6.3.1.1, 6.3.1.2 and 6.3.1.3, respectively.

6.3.1.1 ID3 Algorithm for the analysis of economy (U.N.I.S.)

As the first set of data, in the following sections, we will use some data related to economies for U.N.I.S. countries. The attributes of the these countries' economies are data regarding *years, unemployment youth male (% of male labor force ages 15–24) (national estimate), gross value added at factor cost ($), tax revenue,..., gross domestic product (GDP) growth (annual %)*. Economy (U.N.I.S.) dataset consists of a total of 18 attributes. Data belonging to USA, New Zealand, Italy and Sweden economies from 1960 to 2015 are defined based on the attributes given in Table 2.7 economy (U.N.I.S.)

dataset (http://data.worldbank.org) [15], to be used in the following sections. For the classification of D matrix through ID3 algorithm, the first step training procedure is to be employed. For the training procedure, 66.66% of the D matrix can be split for the training dataset (151 × 18), and 33.33% as the test dataset (77 × 18).

Following the classification of the training dataset being trained with ID3 algorithm, we can classify the test dataset. Although the procedural steps of the algorithm may seem complicated at first glance, the only thing you have to do is to concentrate on the steps and grasp them. For this, let us have a close look at the steps provided in Figure 6.7.

Algorithm: ID3 Algorithm for Economy (U.N.I.S.) Data Set

Input: Training set including 151 number of samples $D = [x_1, x_2, \ldots, x_{151}]$ $(x_i \in R^d)$
Test data including 77 number of samples $T = [x_1, x_2, \ldots, x_{77}]$ $(x_i \in R^d)$

Result: Classification information corresponding to the test and training data C = [USA, New Zealand, Italy, Sweden] number of trees N

Method:

(1) Create a node for the tree
(2) **while** each sample i of belonging to the training set **do** {
(3) //The general status entropy is calculated

$$Info(D) = -\sum_{i=1}^{151} p_i \log_2(p_i)$$

(4) root = j
(5) **while** possible value j of D **do**

$$Info_A(D) = \sum_{j=1}^{2} \frac{|D_j|}{|D|} \times (D_j)$$

(6) //gain of D that have value j for D

$$Gain(A) = Info(D) - Info_A(D)$$

(7) **if** is empty, the leaf is the most common value in D
(8) **else** add the subtree ID3()
(9) **return** the learned tree(N);
(10) //The node with the highest gain value is the node root initial value.
(11) c_i = majority ($Gain$)}
(12) //Predict each test sample's class

Figure 6.7: ID3 algorithm for the analysis of the economy (U.N.I.S.) data set.

How can we find the class of a test data belonging to economy (U.N.I.S.) dataset provided in Table 2.7.1 according to ID3 algorithm? We can find the class label of a sample data belonging to economy (U.N.I.S.) dataset through the following steps (Figure 6.9):

Steps (1–3) Let us split the U.N.I.S. dataset as follows: 66.6% as the training data ($D = 151 \times 18$) and 33.3% as the test data ($T = 77 \times 18$).

Steps (4–10) Create random subset. The training of the training set ($D = 228 \times 18$) through the ID3 algorithm has yielded the following results: number of nodes: 15. Sample ID3 decision tree structure is elicited in Figure 6.10.

Steps (11–12) The splitting criterion labels the node.

The training of the economy (U.N.I.S.) dataset is carried out by applying ID3 algorithm (Steps 1–12 as specified in Figure 6.8.). Through this, the decision tree made up of 15 nodes (the learned tree ($N = 15$)) has been obtained. The nodes chosen via the ID3 learned tree (gross value added at factor cost($), tax revenue, unemployment youth male) and some of the rules pertaining to the nodes are presented in Figure 6.8.

Based on Figure 6.8(a), some of the rules obtained from the ID3 algorithm applied to economy (U.N.I.S.) dataset can be seen as follows.

Figure 6.8(a) and (b) shows economy (U.N.I.S.) dataset sample decision tree structure based on ID3, obtained as such; **IF-THEN** rules are as follows:

…

…

…

Rule 1
IF ((gross value *added at factor cost*($) (current US ($) < $2.17614e + 12) **and** (tax revenue (*% of GDP*) < %12.20) **and** (unemployment youth *male* (*% labor force ages 15–24*)< % 3.84))
THEN class = Italy

Rule 2
IF ((gross value added at factor cost($) (*current US ($)*) < $ 2.17614e + 12) **and** (tax revenue (*% of GDP*) < % 12.20) **and (**unemployment youth male (*% labor force ages 15–24*)> % 3.84)
THEN class = USA

Rule 3
IF ((gross value added at factor cost($) (*current US ($)* < $ 2.17614e + 12) **and** (tax revenue (*% of GDP)* ≥ % 12.20)
THEN class = Italy

Rule 4
IF ((Gross value added at factor cost($)(*Current US ($))* ≥ $ 2.17614e + 12)
THEN class = USA

…

…

…

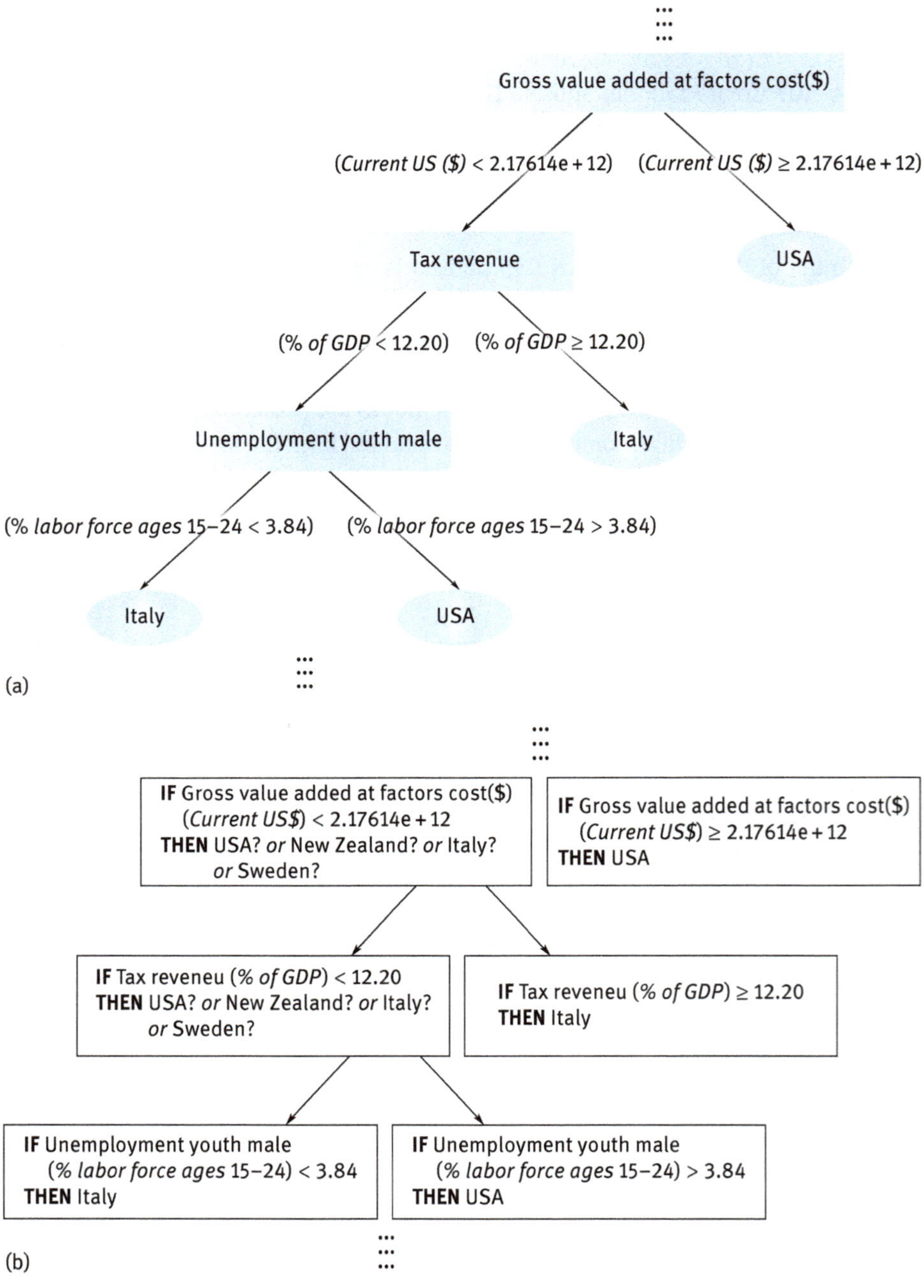

Figure 6.8: Sample decision tree structure for the economy (U.N.I.S.) dataset based on ID3. (a) ID3 algorithm flowchart for the sample economy (U.N.I.S.) dataset. (b) Sample decision tree rule space for the economy (U.N.I.S) dataset.

As a result, the data in the economy (U.N.I.S.) dataset with 33.33% portion allocated for the test procedure and classified as *USA, New Zealand, Italy* and *Sweden* have been classified with an accuracy rate of *95.2%* based on ID3 algorithm.

6.3.1.2 ID3 algorithm for the analysis of MS

As presented in Table 2.10, MS dataset has data from the following groups: 76 sample belonging to RRMS, 76 sample to SPMS, 76 sample to PPMS and 76 sample belonging to healthy subjects of control group. The attributes of the control group are data regarding *brain stem (MRI 1), corpus collasum periventricular (MRI 2), upper cervical (MRI 3) lesion diameter size (milimetric (mm)) in the MR image and EDSS* score. Data consist of a total of 112 attributes. It is known that using these attributes of 304 individuals, the data whether they belong to the MS subgroup or healthy group can be known. How can we make the classification as to which MS patient belongs to which subgroup of MS including healthy individuals and those diagnosed with MS (based on the lesion diameters (*MRI 1, MRI 2* and *MRI 3*) and number of lesion size for (*MRI 1, MRI 2* and *MRI 3*) as obtained from MRI images and EDSS scores)? *D* matrix has a dimension of (304 × 112). This means *D* matrix includes the MS dataset of 304 individuals along with their 112 attributes (see Table 2.12 for the MS dataset). For the classification of *D* matrix through Naive Bayesian algorithm, the first step training procedure is to be employed. For the training procedure, 66.66% of the *D* matrix can be split for the training dataset (203 × 112) and 33.33% as the test dataset (101 × 112).

Following the classification of the training dataset being trained with ID3 algorithm, we can classify the test dataset. Although the procedural steps of the algorithm may seem complicated at first glance, the only thing you have to do is to concentrate on the steps and grasp them. For this let us have a close look at the steps provided in Figure 6.9.

In order to be able to do classification through ID3 algorithm, the test data has been selected randomly from the MS dataset as 33.33%.

Let us now think for a while about how to classify the MS dataset provided in Table 2.10.1 in accordance with the steps of ID3 algorithm (Figure 6.9).

Steps (1–3) Let us identify 66.66% of the MS dataset as the training data ($D = 203 \times 112$) and 33.33% as the test data ($T = 101 \times 112$).

Steps (4–10) Create random subset. As a result of training the training set with ID3 algorithm. The following results have been yielded: number of nodes – 39.

Steps (11–12) The splitting criterion labels the node.

The training of the MS dataset is done by applying ID3 algorithm (Steps 1–12 as specified in Figure 6.7). Through this, the decision tree comprised of 39 nodes (the learned tree ($N = 39$)) has been obtained. The nodes chosen via the ID3 learned tree (MRI 1) and some of the rules pertaining to the nodes are presented in Figure 6.9.

Algorithm: ID3 Algorithm for MS Data Set

Input: MS Training set including 203 number of samples $D = [x_1, x_2, \ldots, x_{203}]$ $(x_i \in R^d)$ Test data including 101 number of samples $T = [x_1, x_2, \ldots, x_{101}]$ $(x_i \in R^d)$

Result: Classification information corresponding to the test and training data C = [PPMS, RRMS, SPMS, Healthy] number of trees N

Method:

(1) Create a node for the tree
(2) **while** each sample i of belonging to the training set **do** {
(3) //The general status entropy is calculated

$$Info(D) = -\sum_{i=1}^{203} p_i \log_2(p_i)$$

(4) root = j
(5) **while** possible value j of D **do**

$$Info_A(D) = \sum_{j=1}^{2} \frac{|D_j|}{|D|} \times (D_j)$$

(6) //gain of D that have value j for D

$$Gain(A) = Info(D) - Info_A(D)$$

(7) **if** is empty, the leaf is the most common value in D
(8) **else** add the subtree ID3()
(9) **return** the learned tree(N);
(10) //The node with the highest gain value is the node root initial value.
(11) c_i = majority $(Gain)$}
(12) //Predict each test sample's class

Figure 6.9: ID3 algorithm for the analysis of the MS data set.

Based on Figure 6.10(a), some of the rules obtained from the ID3 algorithm applied to MS dataset can be seen as follows.

Figure 6.10(a) and (b) shows MS dataset sample decision tree structure based on ID3, obtained as such; *if–then* rules are as follows:

…

…

…

Rule 1
If ((1st region lesion count) < (*number of lesion* is 10.5))
Then CLASS = RRMS

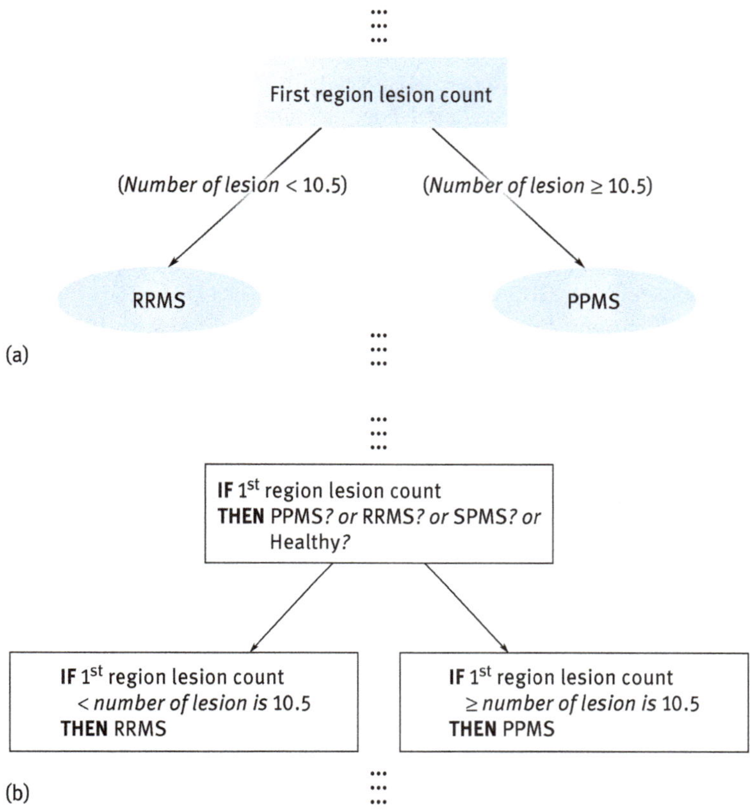

Figure 6.10: Sample decision tree structure for the MS dataset based on ID3. (a) ID3 algorithm flowchart for sample MS dataset. (b) Sample decision tree rule space for the MS dataset.

Rule 2
If ((*1st* region lesion count) ≥ (*number of lesion* is 10.5))
Then CLASS = *PPMS*

...

...

...

As a result, the data in the MS dataset with 33.33% portion allocated for the test procedure and classified as *RRMS, SPMS, PPMS* and *healthy* have been classified with an accuracy rate of *88.3%* based on ID3 algorithm.

6.3.1.3 ID3 algorithm for the analysis of mental functions

As presented in Table 2.16 the WAIS-R dataset has data of 200 belonging to patient group and 200 sample to healthy control group. The attributes of the control group are data regarding *school education, gender, three-dimensional modeling, memory, ordering ability, age,..., DM*. WAIS-R dataset is consists of a total of 21 attributes. It is known that using these attributes of 400 individuals, the data whether they belong to patient or healthy group are known. How can we make the classification as to which individual belongs to which patient or healthy individuals and those diagnosed with WAIS-R test (based on the *school education, gender, three-dimensional modeling, memory, ordering ability, age,..., DM*)? D matrix has a dimension of 400×21. This means D matrix includes the WAIS-R dataset of 400 individuals along with their 21 attributes (see Table 2.16) for the WAIS-R dataset. For the classification of D matrix through ID3 algorithm, the first step training procedure is to be employed. For the training procedure, 66.66% of the D matrix can be split for the training dataset (267×21) and 33.33% as the test dataset (133×21).

Following the classification of the training dataset being trained with ID3 algorithm, we can classify the test dataset. Although the procedural steps of the algorithm may seem complicated at first glance, the only thing you have to do is to concentrate on the steps and grasp them. We can have a glance at the steps shown in Figure 6.11:

Algorithm: ID3 Algorithm for WAIS-R Data Set

Input: Training set including 267 number of samples $D = [x_1, x_2, \ldots, x_{267}]$ $(x_i \in R^d)$ Test data including 133 number of samples $T = [x_1, x_2, \ldots, x_{133}]$ $(x_i \in R^d)$

Result: Classification information corresponding to the test and training data $C =$ [Patient, Healthy] number of trees N

Method:

(1) Create a node for the tree
(2) **while** each sample i of belonging to the training set **do** {
(3) //The general status entropy is calculated

$$Info(D) = -\sum_{i=1}^{267} p_i \log_2(p_i)$$

(4) root = j
(5) **while** possible value j of D **do**

$$Info_A(D) = \sum_{j=1}^{2} \frac{|D_j|}{|D|} \times (D_j)$$

(6) //gain of D that have value j for D

$$Gain(A) = Info(D) - Info_A(D)$$

(7) **if** is empty, the leaf is the most common value in D
(8) **else** add the subtree ID3()
(9) **return** the learned tree(N);
(10) //The node with the highest gain value is the node root initial value.
(11) y_i = majority ($Gain$)}
(12) //Predict each test sample's class

Figure 6.11: ID3 algorithm for the analysis of WAIS-R data set.

How can we classify the WAIS-R dataset in Table 2.16.1 in accordance with the steps of ID3 algorithm (Figure 6.11)?

Steps (1–3) Let us identify the 66.66% of WAIS-R dataset as the training data ($D = 267 \times 21$) and 33.33% as the test data ($T = 133 \times 21$).

Steps (4–10) Create random subset. As a result of training of the training set ($D = 267 \times 21$) through the ID3 algorithm. The following results have been obtained: Number of nodes – 27. The sample decision tree structure is provided in Figure 6.12.

Steps (11–12) The splitting criterion labels the node.

The training of the WAIS-R dataset is conducted by applying ID3 algorithm (Steps 1–12 as specified in Figure 6.12.). Through this, the decision tree made up of 27 nodes (the learned tree ($N = 27$)) has been obtained. The nodes chosen via the ID3 learned tree (*three- dimensional modeling, memory, ordering ability, age*) and some of the rules pertaining to the nodes are presented in Figure 6.12.

Based on Figure 6.12(a), some of the rules obtained from the ID3 algorithm applied to WAIS-R dataset can be seen as follows.

Figure 6.12(a) and (b) shows WAIS-R dataset sample decision tree structure based on ID3, obtained as such; *if–then* rules are as follows:

…

…

…

Rule 1
IF (three-dimensional modeling of *test score result* < 13.5 **and** ordering ability of *test score result* < 0.5)
THEN class = healthy

Rule 2
IF (three-dimensional modeling of *test score result* ≥ 13.5 **and** memory of *test score result* < 16.5 **and** *age* < 40
THEN class = patient

Rule 3
IF (three-dimensional modeling of *test score result* ≥ 13.5 **and** memory of *test score result* < 16.5 **and** ≥ 40
THEN class = healthy

Rule 4
IF (three-dimensional modeling *of test score result*) ≥ 13.5 **and** memory of *test score result* ≥ 16.5
THEN class = patient

…

…

…

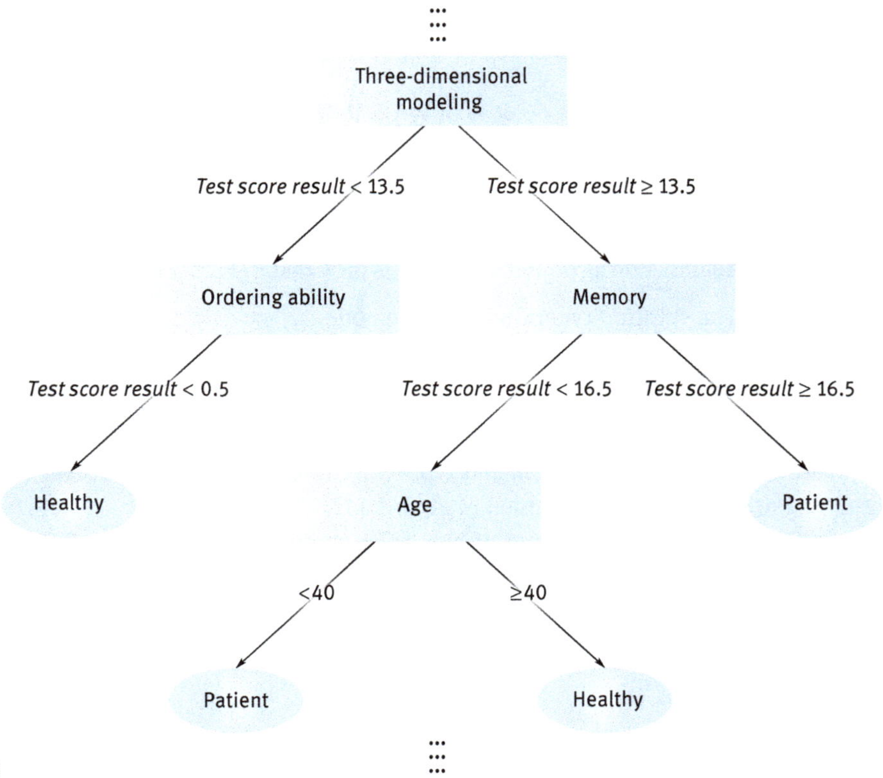

(a)

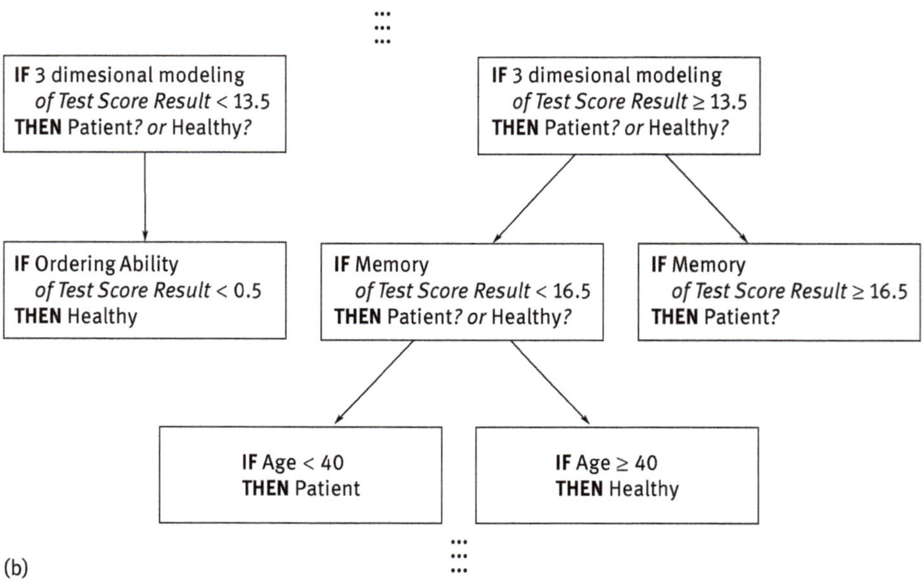

(b)

Figure 6.12: Sample decision tree structure for the WAIS-R dataset based on ID3.(a) ID3 algorithm flowchart for sample WAIS-R dataset. (b) Sample decision tree rule space for the WAIS-R dataset.

As a result, the data in the WAIS-R dataset with 33.33% portion allocated for the test procedure and classified as *patient* and *healthy individuals* have been classified with an accuracy rate of *51%* based on ID3 algorithm.

6.4 C4.5 algorithm

Following some information on ID3, we can have a look at C4.5. Information gain measure prefers to choose attributes with a large number of values. For instance, you can consider an attribute acting as a unique identifier such as *ID*. A split on *ID* would bring about a large number of partitions (as many number as there are values) each one contains just one dataset.

As the example shows each partition is pure and the information needed for the classification of classify dataset D based on this partitioning shall be $\text{Info}_{\text{individual\_ID}}(D) = 0$. For this very reason, the information gained by partitioning on this attribute is highest and this classification could prove to be useless at the end [14].

ID3 is a precursor of C4.5. An extension to information gain is known as *gain ratio*. *Gain ratio* tries to overcome the bias toward tests with numerous outcomes. Implementing a sort of normalization to information gain (using a "split information" value defined in an analogous manner with Info(D) as given as follows):

$$\text{SplitInfo}_A(D) = -\sum_{j=1}^{v} \frac{|D_j|}{|D|} \times \log_2\left(\frac{|D_j|}{|D|}\right) \tag{6.5}$$

The value stands for the potential information produced by splitting the training dataset D into v partitions. This corresponds to the v outcomes of a test on attribute A. It considers the number of datasets with that outcome for the total number of datasets in D for each outcome. Unlikely, information gain measures the information relating to classification acquired as derived from the same partitioning, but the gain ratio is defined through the calculation as follows [15]:

$$\text{GainRatio}(A) = \frac{\text{Gain}(A)}{\text{SplitInfo}_A(D)} \tag{6.6}$$

The attribute that has the maximum gain ratio is selected as the splitting attribute. Here, as the split information comes close to 0, the ratio gets unstable. To evade this situation, a constraint is included. And then the information gain of the test selected has to be large.

Figure 6.13: presents the flowchart of C4.5 algorithm.

Based on C4.5 algorithm flowchart, you can find the steps of C4.5 algorithm according to Figure 6.13.

Example 6.2 shows the application of the C4.5 algorithm to get a clearer picture through relevant steps (the same procedure was followed for ID3 algorithm Figure 6.14).

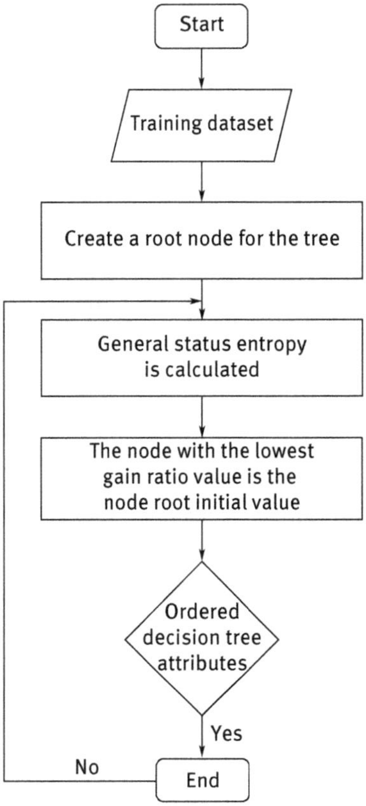

Figure 6.13: The flowchart for C4.5 algorithm.

Example 6.2 In Table 6.1, the values for the individuals belonging to WAIS-R dataset are elicited as *gender, age, QIV* and *class* label values. The data in Section 6.2 for WAIS-R dataset (indicated in Table 2.16.1 and Table 6.1) has been handled.

We can step-by-step apply the steps for C4.5 algorithm in the relevant order so as to build the tree form as in Figure 6.1.

Steps (1–4) The first step is to calculate the root node in order to form the tree structure. The total number of record is 15, as can be seen in Table 6.1 (first column), which goes as follows: (Individual (ID): 1, 2, 3,..., 15). We can use the entropy formula to calculate the root node based on eq. (6.1).

As can be seen clearly in Table 6.1 class column, 11 individuals belong to *patient* classification and four to *healthy*. The general status entropy can be calculated based on eq. (6.1):

$$\frac{11}{15}\log\left(\frac{15}{11}\right)+\frac{4}{15}\log\left(\frac{15}{4}\right)=0.2517$$

Algorithm: C4.5 Algorithm

Input: Training set including m number of samples $D = [x_1, x_2, \ldots, x_m]$ $(x_i \in R^d)$
Test data including n number of samples $T = [x_1, x_2, \ldots, x_k]$ $(x_k \in R^d)$

Result: Label information corresponding to test and training sample $C = [c_1, c_2, \ldots, c_m]$ Number of trees N

Method:

(1) Create a Root node for the tree
(2) **while** each sample m of belonging to the training set **do** {
(3) //General status entropy is calculated

$$Info(D) = -\sum_{i=1}^{m} p_i \log_2(p_i)$$

(4) root = j
(5) **for each** possible value j of D

$$SplitInfo_A(D) = -\sum_{j=1}^{v} \frac{|D_j|}{|D|} \times \log_2\left(\frac{|D_j|}{|D|}\right)$$

(6) gain of D that have value j for D

$$GainRatio(A) = \frac{Gain(A)}{SplitInfo_A(D)}$$

(7) **if** root is empty in this case leaf will be the most common value in D
(8) **else** add the subtree C4.5();
(9) **return** the learned tree(N);
(10) //The node root with the lowest *Gain* Ratio value will be the initial value
(11) c_i = minimum(*GainRatio*())}
(12) //Predict Sample Test Data Class

Figure 6.14: General C4.5 decision tree algorithm.

Step (5) Table 6.1 lists the calculation of the entropy value for the gender variable (female–male; our first variable) based on eq. (6.4).

$$\text{SplitInfo}_{\text{Gender}}\left(\frac{7}{15}; \frac{8}{15}\right) = \frac{7}{15}\log\left(\frac{15}{7}\right) + \frac{8}{15}\log\left(\frac{15}{8}\right) = 0.3001$$

Step (6) The *gain* ratio calculation for the *gender* variable can be calculated based on eq. (6.5).

The calculation will be as follows:

$$\text{GainRatio(Gender)} = \frac{0.2517}{0.3001} = 0.8387$$

At this point, let us do the *gain ratio* calculation for the *age* variable following the same steps. In order to form the branches of the tree formed based on C4.5 algorithm, the *age* variable should be split into groups: let us assume that the age variable is grouped as follows and labeled accordingly.

Step (3) Let us now calculate the entropy values of the group numbers as labeled with the age interval values in Table 6.2; eq. (6.2) is used for this calculation.

$$\text{SplitInfo}_{\text{Age}}\left(\frac{3}{15},\frac{5}{15},\frac{4}{15},\frac{2}{15},\frac{1}{15}\right)=\frac{3}{15}\log\left(\frac{15}{3}\right)+\frac{5}{15}\log\left(\frac{15}{5}\right)+\frac{4}{15}\log\left(\frac{15}{4}\right)$$
$$+\frac{2}{15}\log\left(\frac{15}{2}\right)+\frac{1}{15}\log\left(\frac{15}{1}\right)=0.6470$$

Step (6) The gain calculation for the age variable *Gain*(*Age*) is calculated according to eq. (6.5):

$$\text{GainRatio}(\text{Age})=\frac{0.2517}{0.6470}=0.3890$$

Now, let us calculate the *gain ratio* value of the QIV variable, which is our last variable.

Let us follow the same steps and similarly apply them for the QIV variable as well.

In order to form the branches of the tree to be formed, it is required to split the QIV variable into groups. Let us assume that the QIV variable is grouped and are given as follows.

Step (3) Let us calculate the entropy value of the QIV variable based on Table 6.3:

$$\text{Info}_{\text{QIV}=1.\text{Group}}=\frac{1}{3}\log\left(\frac{3}{1}\right)+\frac{2}{3}\log\left(\frac{3}{2}\right)=0.2764$$

Step (5) We can use eq. (6.5) to calculate the total weighted value of the QIV variable:

$$\text{SplitInfo}_{\text{QIV}}\left(\frac{3}{15},\frac{2}{15},\frac{6}{15},\frac{2}{15},\frac{2}{15}\right)=\frac{3}{15}\log\left(\frac{15}{3}\right)+\frac{2}{15}\log\left(\frac{15}{2}\right)+\frac{6}{15}\log\left(\frac{15}{6}\right)$$
$$+\frac{2}{15}\log\left(\frac{15}{2}\right)+\frac{2}{15}\log\left(\frac{15}{2}\right)=0.6490$$

Step (6) Let us calculate the *gain ratio* value for the QIV variable.

$$\text{GainRatio}(\text{QIV})=\frac{0.2517}{0.6490}=0.3878$$

Step (7) We have been able to calculate the *gain ratio* values of all the variables in Table 6.1. These variables are *age, gender* and *QIV*. The calculation has yielded the following results for the variables:

Gain Ratio(Gender) = 0.8387; Gain Ratio(Age) = 0.3890; Gain Ratio(QIV) = 0.3878
When ranked from the smallest to the greatest, the greatest *gain ratio* value has been obtained from *Gain Ratio(Gender)*. Based on the values from the greatest to the smallest, *gender* is to be selected as the root node and it is presented as can be seen in Figure 6.15.

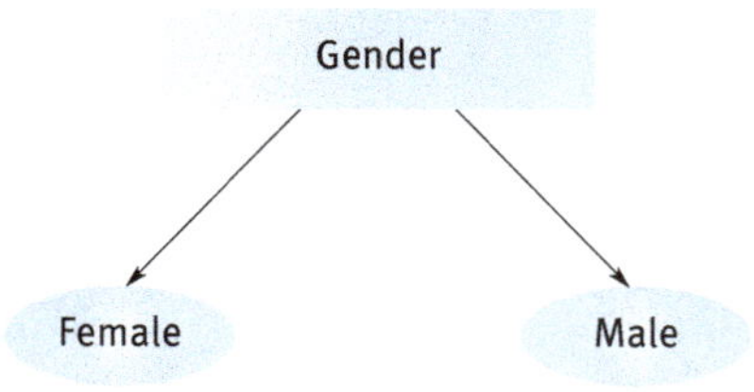

Figure 6.15: Root node initial value based on Example 6.1 C4.5 algorithm.

Gender variable constitutes the root of the tree in line with the procedural steps of C4.5 algorithm as can be seen in Figure 6.15. The branches of the tree are depicted as *female* and *male*.

6.4.1 C4.5 Algorithm for the analysis of various data

C4.5 algorithm has been applied on economy (U.N.I.S.) dataset, multiple sclerosis (MS) dataset and WAIS-R dataset in Sections 6.4.1.1, 6.4.1.2 and 6.4.1.3, respectively.

6.4.1.1 C4.5 algorithm for the analysis of economy (U.N.I.S.)

As the first set of data, we will use, in the following sections, some data related to the U.N.I.S. countries' economies. The attributes of the these countries' economies are data regarding *years, unemployment youth female (% of male labor force ages 15–24) (national estimate), account at a financial institution, male(% age 15+), gross value added at factor cost ($), tax revenue..., GDP growth (annual %)*. Data consist of a total of 18 attributes. Data belonging to USA, New Zealand, Italy and Sweden economies from 1960 to 2015 are defined based on the attributes given in Table 2.7.1 (economy dataset (http://data.worldbank.org)) [15], to be used in the following sections. For the classification of *D* matrix through C4.5 algorithm, the first step training procedure is to be employed. For the training procedure, 66.66% of the *D* matrix can be split for the training dataset (151 × 18) and 33.33% as the test dataset (77 × 18).

Following the classification of the training dataset being trained with C4.5 algorithm, we can classify the test dataset. Although the procedural steps of the algorithm may seem complicated at first glance, the only thing you have to do is to concentrate on the steps and grasp them. We can take a close glance at the steps presented in Table 6.16.

Algorithm: C4.5 Algorithm for Economy (U.N.I.S.) Data Set

Input: Training set including 151 number of samples $D = [x_1, x_2, \ldots, x_{151}]$ $(x_i \in R^d)$ Test data including 77 number of samples $T = [x_1, x_2, \ldots, x_{77}]$ $(x_i \in R^d)$

Result: Label information corresponding to test and training sample

$$C = [USA, NewZealand, Italy, Sweden]$$

Number of trees N

Method:

(1) Create a Root node for the tree
(2) **while** each sample i of belonging to the training set **do** {
(3) //General status entropy is calculated

$$Info(D) = -\sum_{i=1}^{151} p_i \log_2(p_i)$$

(4) root = j
(5) **for each** possible value j of D

$$SplitInfo_A(D) = -\sum_{j=1}^{2} \frac{|D_j|}{|D|} \times \log_2\left(\frac{|D_j|}{|D|}\right)$$

(6) gain of D that have value j for D

$$GainRatio(A) = \frac{Gain(A)}{SplitInfo_A(D)}$$

(7) **if** root is empty in this case leaf will be the most common value in D
(8) **else** add the subtree C4.5();
(9) **return** the learned tree(N);
(10) //The node root with the lowest *Gain* Ratio value will be the initial value
(11) c_i = minimum($GainRatio$())}
(12) //Predict Sample Test Data Class

Figure 6.16: C4.5 algorithm for the analysis of the economy (U.N.I.S.) data set.

How can we find the class of a test data belonging to economy (U.N.I.S.) dataset provided in Table 2.7.1 according to C4.5algorithm? We can find the class label of a sample data belonging to economy (U.N.I.S.) dataset through the following steps (Figure 6.16):

Steps (1–3) Let us split the U.N.I.S. dataset as follows: 66.6% as the training data ($D = 151 \times 18$) and 33.33% as the test data ($T = 77 \times 18$).

Steps (4–10) Create random subset: The training of the training set ($D = 228 \times 8$) through C4.5 algorithm has yielded the following results: number of nodes – 15. Sample decision tree structure for C4.5 is elicited in Figure 6.17.

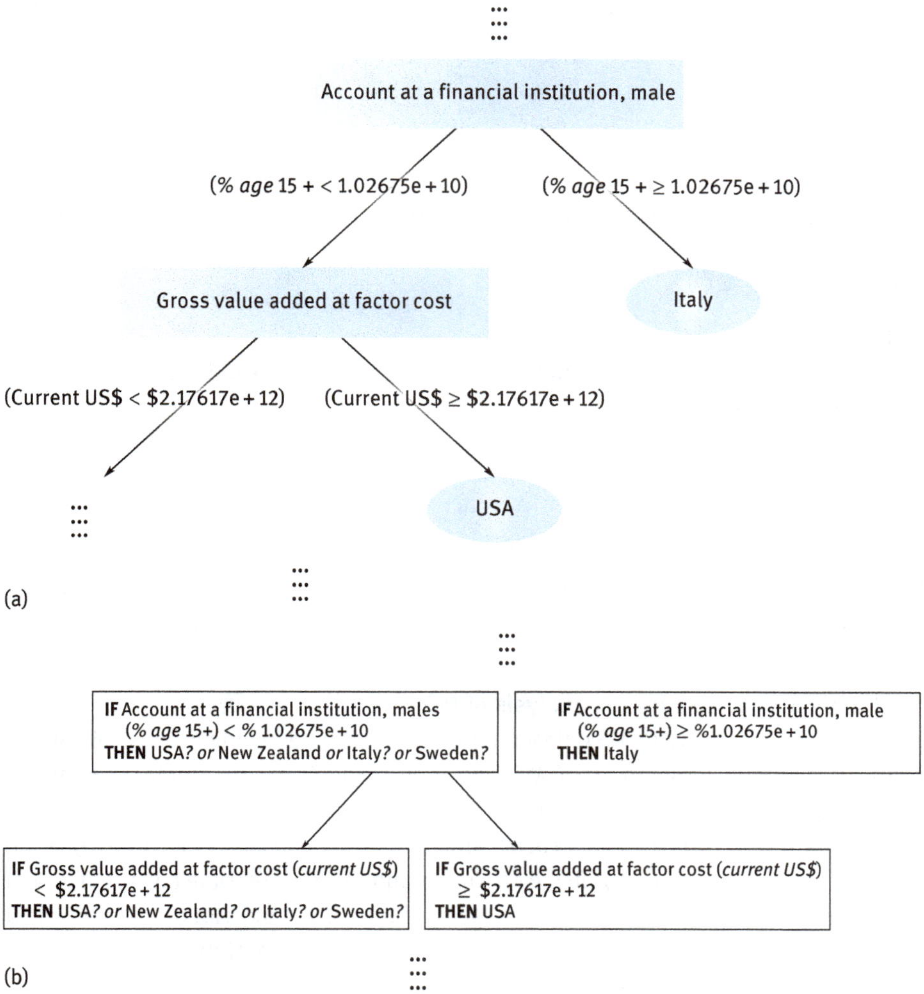

Figure 6.17: Sample decision tree structure for economy (U.N.I.S.) dataset based on C4.5. (a) C4.5 algorithm graph for the sample economy (U.N.I.S.) dataset. (b) Sample C4.5 decision tree rule space for economy (U.N.I.S.) dataset.

Steps (11–12) The splitting criterion labels the node.

The training of the economy (U.N.I.S.) is done by applying C4.5 algorithm (Steps 1–12 as specified in Figure 6.16.). Through this, the decision tree made up of 15 nodes (the learned tree ($N = 15$)) has been obtained. The nodes chosen via the C4.5 learned tree nodes (account at a financial institution (male), gross value added at factor cost) and some of the rules regarding the nodes are provided in Figure 6.17.

Based on Figure 6.17, some of the rules obtained from the C4.5 algorithm applied to economy (U.N.I.S.) dataset can be seen as follows.

Figure 6.17(a) and (b) shows economy (U.N.I.S.) dataset sample decision tree structure based on C4.5 algorithm, obtained as such; *if–then* rules are as follows:

...

...

...

Rule 1
IF (account at a financial institution for male (*% age 15+*) < %1.02675e+10) **and** (gross value added at factor cost (*current US($)*) ≥ *$*2.17617e+12)
THEN class = USA

Rule 2
IF (account at a financial institution, male (*%age 15+*) ≥ %1.02675e+10)
THEN class = Italy

...

...

...

As a result, the data in the economy (U.N.I.S.) dataset with 33.33% portion allocated for the test procedure and classified as *USA, New Zealand, Italy* and *Sweden* have been classified with an accuracy rate of *86.4%* based on C4.5 algorithm.

6.4.1.2 C4.5 algorithm for the analysis of MS

As presented in Table 2.10, MS dataset has data from the following groups: 76 sample belonging to RRMS, 76 sample to SPMS, 76 sample to PPMS and 76 sample belonging to healthy subjects of control group. The attributes of the control group are data regarding *brain stem (MRI 1), corpus collasum periventricular (MRI 2), upper cervical (MRI 3) lesion diameter size (milimetric (mm)) in the MR image* and *EDSS* score. Data consist of a total of 112 attributes. It is known that using these attributes of 304 individuals, we can know whether the data belong to the MS subgroup or healthy group. How can we make the classification as to which MS patient belongs to which subgroup of MS including healthy individuals and those diagnosed with MS (based on the lesion diameters (*MRI 1, MRI 2, MRI 3*), number of lesion size for (*MRI 1, MRI 2, MRI 3*) as obtained from MRI images and EDSS scores? *D* matrix has a dimension of (304 × 112). This means *D* matrix includes the MS dataset of 304 individuals along with their 112 attributes (see Table 2.10 for the MS dataset). For the classification of *D* matrix through Naive Bayesian algorithm, the first step training procedure is to be employed. For the training procedure, 66.66% of the *D* matrix can be split for the training dataset (203 × 112) and 33.33% as the test dataset (101 × 112).

Following the classification of the training dataset being trained with C4.5 algorithm, we can classify the test dataset. Although the procedural steps of the algorithm may seem complicated at first glance, the only thing you have to do is to concentrate on the steps and grasp them. For this, let us have a close look at the steps provided in Figure 6.18.

Algorithm: C4.5 Algorithm for MS Data Set

Input: Training set including 203 number of samples $D = [x_1, x_2, \ldots, x_{203}]$ $(x_i \in R^d)$ Test data including 101 number of samples $T = [x_1, x_2, \ldots, x_{101}]$ $(x_i \in R^d)$

Result: Label information corresponding to test and training sample C = [PPMS, RRMS, SPMS, Healthy] Number of trees N

Method:

(1) Create a Root node for the tree
(2) **while** each sample *m* of belonging to the training set **do** {
(3) //General status entropy is calculated

$$Info(D) = -\sum_{i=1}^{203} p_i \log_2(p_i)$$

(4) root = *j*
(5) **for each** possible value *j* of *D*

$$SplitInfo_A(D) = -\sum_{j=1}^{2} \frac{|D_j|}{|D|} \times \log_2\left(\frac{|D_j|}{|D|}\right)$$

(6) gain of *D* that have value *j* for *D*

$$GainRatio(A) = \frac{Gain(A)}{SplitInfo_A(D)}$$

(7) **if** root is empty in this case leaf will be the most common value in *D*
(8) **else** add the subtree C4.5();
(9) **return** the learned tree(*N*);
(10) //The node root with the lowest *Gain* Ratio value will be the initial value
(11) c_i = minimum (*GainRatio()*)}
(12) //Predict Sample Test Data Class

Figure 6.18: C4.5 algorithm for the analysis of the MS data set.

In order to be able to do classification through C4.5 algorithm, the test data has been randomly selected as 33.33% from the MS dataset.

Let us now think for a while about how to classify the MS dataset provided in Table 2.10.1 in accordance with the steps of C4.5 algorithm (Figure 6.18.)

Steps (1–3) Let us identify 66.66% of the MS dataset as the training data ($D = 203 \times 112$) and 33.33% as the test data ($T = 101 \times 112$).

Steps (4–10) Create random subset: The training of the training set ($D = 203 \times 112$) through C4.5 algorithm has yielded the following results: number of nodes – 23. Sample decision tree structure for C4.5 is elicited in Figure 6.19.

Steps (11–12) The splitting criterion labels the node.

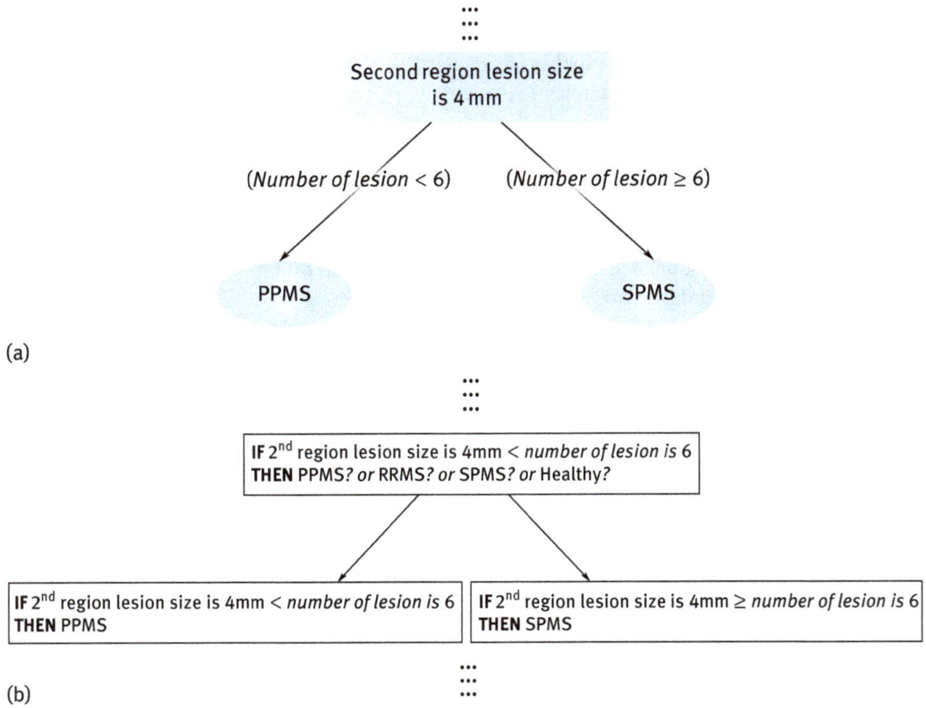

Figure 6.19: Sample decision tree structure for MS dataset based on C4.5. (a) C4.5 algorithm flowchart for the sample MS dataset. (b) Sample decision tree rule space for MS dataset.

The training of the MS dataset is done by applying C4.5 algorithm (Steps 1–12 as specified in Figure 6.18.). Through this, the decision tree made up of 23 nodes (the learned tree ($N = 23$)) has been obtained. The nodes chosen via the C4.5 learned tree (second region lesion size for 4 mm) and some of the rules pertaining to the nodes are presented in Figure 6.19.

Based on Figure 6.19(a), some of the rules obtained from C4.5 algorithm applied to MS dataset can be seen as follows.

Figure 6.19(a) and (b) shows MS dataset sample decision tree structure based on C4.5 algorithm, obtained as such; *if–then* rules are as follows:

…

…

…

Rule 1
IF (second region lesion size is 4 mm < (*number of lesion* is 6))
THEN class = PPMS

Rule 2
IF (second region lesion size is 4 mm ≥ (*number of lesion* is 6))
THEN class = SPMS

...

...

...

As a result, the data in the MS dataset with 33.33% portion allocated for the test procedure and classified as *RRMS, SPMS, PPMS* and *healthy* have been classified with an accuracy rate of *67.28%* based on C4.5 algorithm.

6.4.1.3 C4.5 algorithm for the analysis of mental functions

As presented in Table 2.16, the WAIS-R dataset has data of 200 belonging to patient group and 200 sample to healthy control group. The attributes of the control group are data regarding *school education, gender,..., DM*. Data consist of a total of 21 attributes. It is known that using these attributes of 400 individuals, we can know if the data belong to patient or healthy group. How can we make the classification as to which individual belongs to which patient or healthy groups and those diagnosed with WAIS-R test (based on the *school education, gender, ...,DM*)? *D* matrix has a dimension of 400×21. This means *D* matrix includes the WAIS-R dataset of 400 individuals along with their 21 attributes (see Table 2.16) for the WAIS-R dataset. For the classification of *D* matrix through C4.5 algorithm, the first step training procedure is to be employed. For the training procedure, 66.66% of the *D* matrix can be split for the training dataset (267×21) and 33.33% as the test dataset (133×21).

Following the classification of the training dataset being trained with C4.5 algorithm, we can classify the test dataset. Although the procedural steps of the algorithm may seem complicated at first glance, the only thing you have to do is to concentrate on the steps and grasp them. We can take a close look at the steps presented in Figure 6.20.

How can we classify the WAIS-R dataset in Table 2.16.1 in accordance with the steps of C4.5 algorithm (Figure 6.21)?

Steps (1–3) Let us identify the 66.66% of WAIS-R dataset as the training data ($D = 267 \times 21$) and 33.33% as the test data ($T = 133 \times 21$).

Steps (4–10) Create random subset: training of the training set ($D = 267 \times 21$) with C4.5algorithm. The following results have been obtained: number of nodes – 15. Sample decision tree structure is shown in Figure 6.21.

Steps (11–12) The splitting criterion labels the node.

The training of the WAIS-R dataset is done by applying C4.5 algorithm (Steps 1–12 as specified in Figure 6.21). Through this, the decision tree consisting of 15 nodes (the learned tree ($N = 15$)) has been obtained. The nodes chosen via the ID3 learned tree nodes (*gender, ordering ability*) and some of the rules regarding the nodes are shown in Figure 6.21.

Algorithm: C4.5 Algorithm for WAIS-R Data Set

Input: Training set including 267 number of samples $D = [x_1, x_2, \ldots, x_{267}]$ $(x_i \in R^d)$

Test data including 133 number of samples $T = [x_1, x_2, \ldots, x_{133}]$ $(x_i \in R^d)$

Result: Label information corresponding to test and training sample C = [Patient, Healthy]

Number of trees N

Method:

(1) Create a Root node for the tree
(2) **while** each sample i of belonging to the training set **do** {
(3) //General status entropy is calculated

$$Info(D) = -\sum_{i=1}^{267} p_i \log_2(p_i)$$

(4) root = j
(5) **for each** possible value j of D

$$SplitInfo_A(D) = -\sum_{j=1}^{2} \frac{|D_j|}{|D|} \times \log_2\left(\frac{|D_j|}{|D|}\right)$$

(6) gain of D that have value j for D

$$GainRatio(A) = \frac{Gain(A)}{SplitInfo_A(D)}$$

(7) **if** root is empty in this case leaf will be the most common value in D
(8) **else** add the subtree C4.5();
(9) **return** the learned tree(N);
(10) //The node root with the lowest *Gain* Ratio value will be the initial value
(11) c_i = minimum($GainRatio()$)}
(12) //Predict Sample Test Data Class

Figure 6.20: C4.5 algorithm for the analysis of the WAIS-R data set.

Based on Figure 6.21(a), some of the rules obtained from the C4.5 algorithm applied to the WAIS-R dataset can be seen as follows.

Figure 6.21(a) and (b) show WAIS-R dataset sample decision tree structure based on C4.5 algorithm, obtained as such; **IF-THEN** rules are as follows:

...

...

...

Rule 1

IF ((Gender = *Female*) **and** ordering ability of *test score result* < 0.5)
THEN CLASS = patient

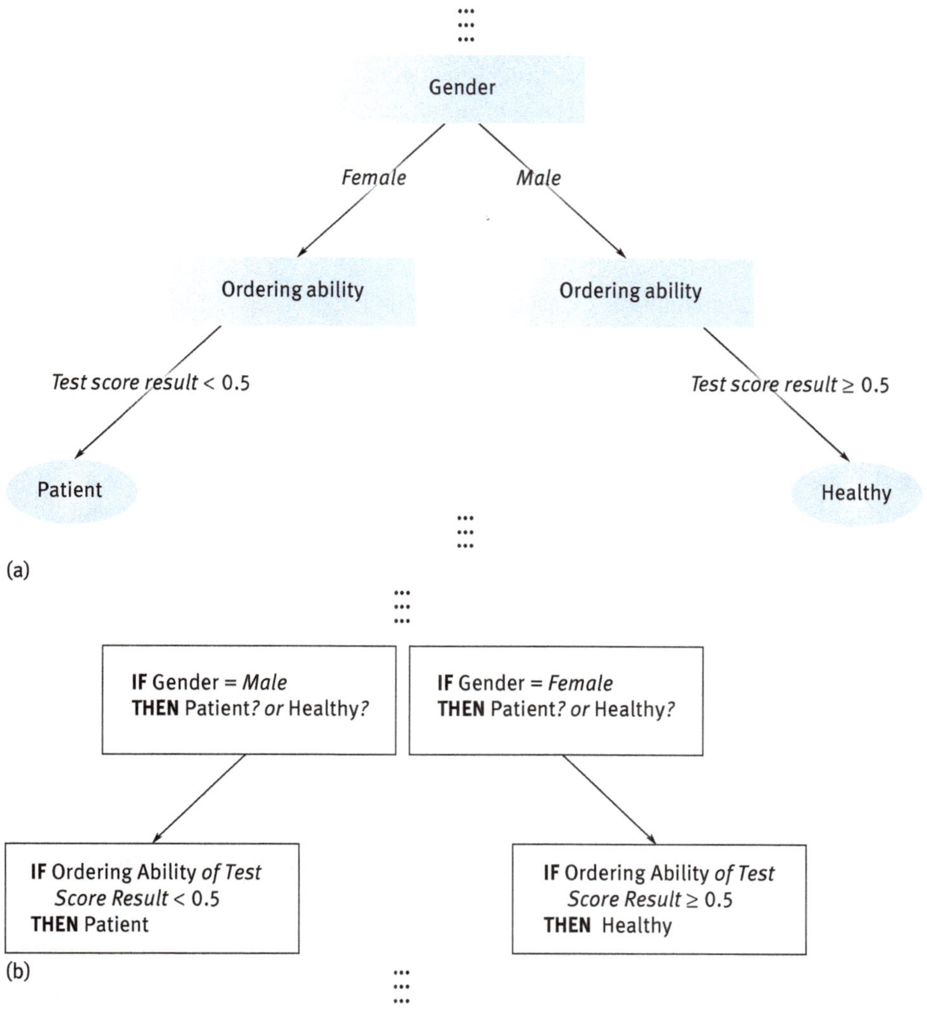

Figure 6.21: Sample decision tree structure for the WAIS-R dataset based on C4.5 algorithm. (a) C4.5 algorithm flowchart for the sample WAIS-R dataset. (b) Sample decision tree rule space for the WAIS-R dataset.

Rule 2

IF ((Gender = *Male*) **and** ordering ability of *test score result* ≥ 0.5)
THEN CLASS = healthy

...

...

...

As a result, the data in the WAIS-R dataset with 33.33% portion allocated for the test procedure and classified as *patient* and *healthy individuals* have been classified with an accuracy rate of *99%* based on C4.5 algorithm.

6.5 CART algorithm

As the third algorithm in this section, let us elaborate on CART. We can use the notation described previously. The *Gini* index measures the impurity of D, a data partition or set of training datasets:

$$\text{Gini}(D) = 1 - \sum_{i=1}^{m} p_i^2 \tag{6.7}$$

where p_i is the probability a dataset in D and belongs to class C_i. It is estimated by$|C_{i,D}|/|D|$. We calculate the sum over m classes [15–18].

The *Gini* index considers a binary split for each of the attributes. If A is a discrete-valued attribute with v distinct values, $\{a_1, a_2, ..., a_v\}$ occurs in D. If one wants to determine the best binary split on A, he/she should analyze all the possible subsets that can be generated using the known values of A. Each subset, S_A, is considered to be a binary test for A attribute of the form "$A \in S_A$." For a dataset this test will yield satisfying result if the value of A for the dataset is one of the values listed in S_A. If A has v possible values, then there will be 2^v possible subsets.

For example, if *school education* has three possible values that are {*primary school, junior high school, vocational school*}, then the possible subsets will be {*primary school, junior high school, vocational school*}, {*primary school, junior high school*}, {*primary school, vocational school*}, {*primary school, vocational school*}, {*primary school*}, {*junior high school*}, {*vocational school*} and {}. We exclude the power set {*primary school, junior high school, vocational school*}, and the empty set from consideration. They do not conceptually represent a split. This example illustrates that there are $2^v - 2$ possible ways to form two partitions of the D data depending a binary split on A. For the binary split, you calculate a weighted sum of the impurity of each resultant partition. For instance, the binary split on A splits D into D_1 and D_2. The *Gini* index of D with that partitioning is denoted as school education (*primary school, junior high school, vocational school, bachelor, master's degree, Phd*).

$$\text{Gini}_A(D) = \frac{|D_1|}{|D|}\text{Gini}(D_1) + \frac{|D_2|}{|D|}\text{Gini}(D_2) \tag{6.8}$$

Here, for each attribute, each of the possible binary splits is considered. If the attribute is one of discrete valued, then the subset yielding the minimum *Gini* index for the attribute is chosen as the splitting subset thereof [9].

If you wonder what happens with the attributes with continuous values, then each possible split-point has to be taken into consideration. A similar strategy is employed as has been described previously for information gain (with the midpoint between each pair sorted adjacent values is treated as a possible split-point). The point giving the least or minimum *Gini* index for a given (this case continuous valued) attribute is treated as the attribute's split-point. Note that D_1 is the set of datasets in D satisfying $A \leq split\_point$ and D_2 the set of datasets in D satisfying $A > split\_point$ for a possible split-point of A as has been mentioned earlier. Impurity reduction occurs by a binary split on a discrete- or continuous-valued attribute with A being:

$$\Delta\text{Gini}(A) = \text{Gini}(D) - \text{Gini}_A(D) \tag{6.9}$$

The attribute that has the minimum Gini index or the one that maximizes the reduction in impurity is chosen as the splitting attribute.

Figure 6.22 presents the flowchart of the CART algorithm.

Based on the aforementioned flowchart for CART algorithm, you can find the steps of CART algorithm in Figure 6.23. The steps applied in CART algorithm are as follows.

Example 6.3 shows the application of the CART algorithm (Figure 6.23) to get a clearer picture through relevant steps in the required order.

Example 6.3 In Table 6.3, the four variables for nine individuals as chosen from the sample WAIS-R dataset are elicited as *school education, gender, age* and *class.* In the example, let us consider how we can form the tree structure by calculating the *Gini* index value of the CART algorithm (Figure 6.23) for three variables that are (*school education, age* and *gender*) and for class these are *class* (*patient* and *healthy individuals*).

Let us now apply the steps for the CART algorithm (Figure 6.25) in order to form the tree structure for the variables in the sample dataset given in Table 6.4.

Iteration 1 The procedural steps in Table 6.4 and in Figure 6.25 regarding the CART algorithm will be used to form the tree structure for all the variables.

Steps (1–4) Regarding the variables provided (*school education, age, gender* and *class*) in Table 6.4, binary grouping is performed.

Table 6.5 is based on the class number that belongs to the variables classified as *class* (patient and healthy individuals).

Variables are as follows:

School education (primary school, vocational school, high school)

Age $(20 \leq young \leq 25)$, $(40 \leq middle \leq 50)$, $(80 \leq old \leq 90)$

Gender (female, male)

Step (5) $\text{Gini}_{\text{left}}$ and $\text{Gini}_{\text{right}}$ values (eq. (6.6)) are calculated for the variables given in Table 6.5. *School education, age* and *gender* for the CART algorithm are shown in Figure 6.25.

Gini calculation for *school education* is as follows:

$$\text{Gini}_{\text{left}}(\text{Primary School}) = 1 - \left[\left(\frac{1}{3}\right)^2 + \left(\frac{2}{3}\right)^2\right] = 0.444$$

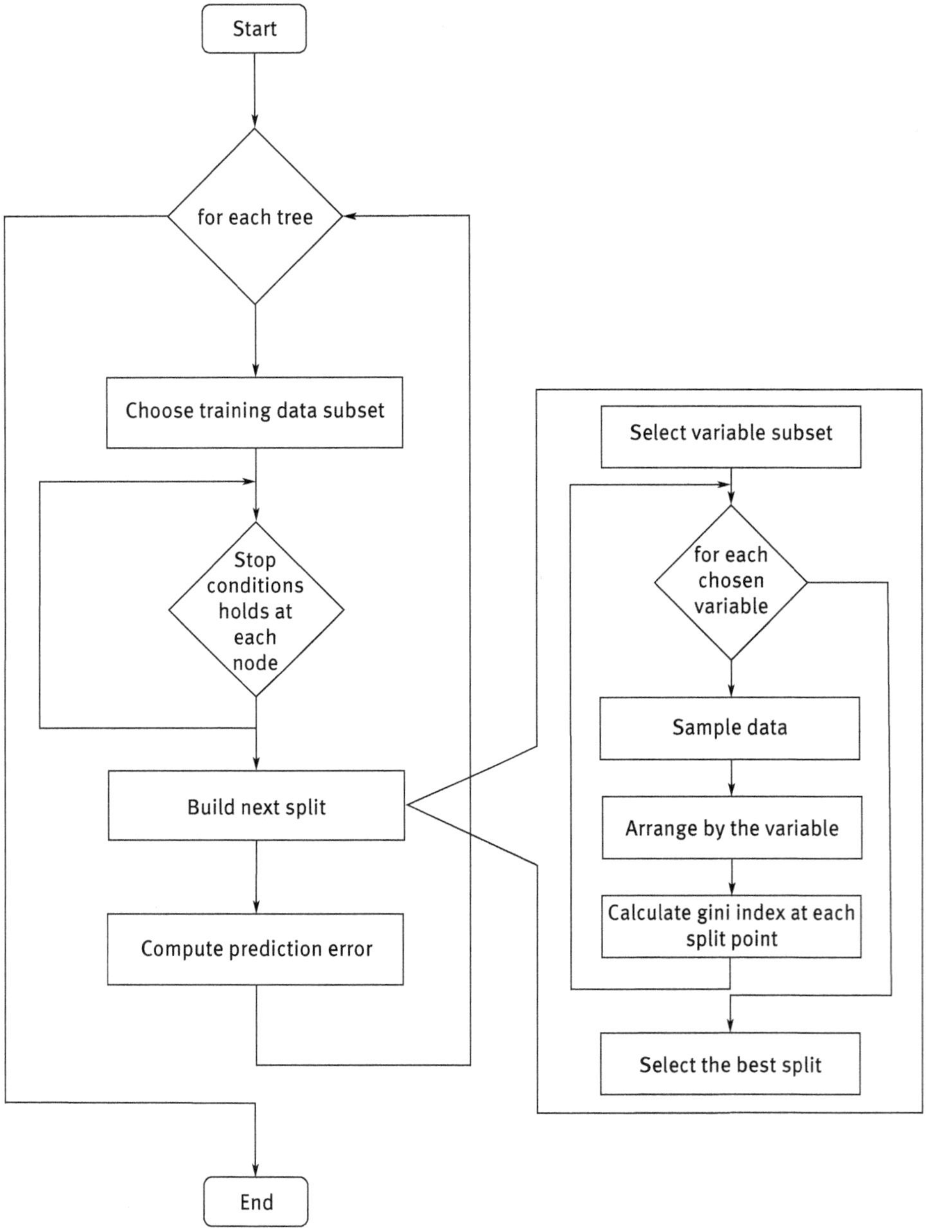

Figure 6.22: The flowchart for CART algorithm.

$$\text{Gini}_{\text{right}}(\text{Vocational School, High School}) = 1 - \left[\left(\frac{3}{4}\right)^2 + \left(\frac{1}{4}\right)^2\right] = 0.375$$

Gini calculation for *age* is as follows:

Algorithm: CART Algorithm

Input: Training set including m number of samples $D = [x_1, x_2, \ldots, x_m]$ $(x_i \in R^d)$
Test data including n number of samples $T = [x_1, x_2, \ldots, x_k]$ $(x_t \in R^d)$

Result: Classification information corresponding to the test and training data $C = [c_1, c_2, \ldots, c_m]$
Number of trees N

Method:

(1) Create a node for the tree
(2) **while** each sample i of belonging to the training set **do** {
(3) //*Gini* index value is calculated

$$Gini(D) = 1 - \sum_{i=1}^{m} p_i^2$$

(4) root = j
(5) **while** possible value j of D **do**

$$Info_A(D) = \sum_{j=1}^{2} \frac{|D_j|}{|D|} x(D_j)$$

(6) //*Gini* index for the attribute is chosen as the splitting subset

$$Gini_A(D) = \frac{|D_1|}{|D|} Gini(D_1) + \frac{|D_2|}{|D|} Gini(D_2)$$

(7) **if** is empty in this case leaf will be the most common value in D
(8) **else** add the subtree CART()
(9) **return** the learned tree(N);
(10) //maximizes the reduction in impurity is chosen as the splitting attribute.
(11)

$$\Delta Gini(A) = Gini(D) - Gini_A(D)$$

(12) //Predict each test sample's class

Figure 6.23: General CART decision tree algorithm.

$$\text{Gini}_{\text{left}}(\text{Young}) = 1 - \left[\left(\frac{0}{2}\right)^2 + \left(\frac{2}{2}\right)^2 \right] = 0$$

$$\text{Gini}_{\text{right}}(\text{Middle, Old}) = 1 - \left[\left(\frac{4}{5}\right)^2 + \left(\frac{1}{5}\right)^2 \right] = 0.320$$

Gini calculation for *gender* is as follows:

Table 6.4: Selected sample from the WAIS-R dataset.

Individual (ID)	School education	Age	Gender	Class
1	Vocational school	90	Male	Patient
2	Primary school	20	Male	Healthy
3	High school	40	Female	Healthy
4	Vocational school	45	Male	Patient
5	Vocational school	50	Male	Patient
6	High school	80	Female	Patient
7	Primary school	25	Female	Healthy

Table 6.5: The number of classes that variables belong to identified as *class* (*patient, healthy*).

Class	School education		Age		Gender	
	Primary school	Vocational school, high school	Young	Middle, old	Female	Male
Patient	1	3	0	4	1	3
Healthy	2	1	2	1	2	1

$$\text{Gini}_{\text{left}}(\text{Female}) = 1 - \left[\left(\frac{1}{3}\right)^2 + \left(\frac{2}{3}\right)^2\right] = 0.444$$

$$\text{Gini}_{\text{right}}(\text{Male}) = 1 - \left[\left(\frac{3}{4}\right)^2 + \left(\frac{1}{4}\right)^2\right] = 0.375$$

Step (6) The Gini_j value of the variables (*school education, age, gender*) is calculated (see eq. (6.7)) using the results obtained in Step 5.

$$\text{Gini}_{\text{SchoolEducation}} = \frac{3 \times (0.444) + 4 \times (0375)}{7} = 0.405$$

$$\text{Gini}_{\text{Age}} = \frac{2 \times (0) + 5 \times (0.320)}{7} = 0.229$$

$$\text{Gini}_{\text{Gender}} = \frac{3 \times (0.444) + 4 \times (0.375)}{7} = 0.405$$

Table 6.6 lists the results obtained from the steps applied for Iteration 1 to the CART algorithm as in Steps 1–6.

Table 6.6: The application of the CART algorithm for Iteration 1 to the sample WAIS-R dataset.

Class	School education		Age		Gender	
	Primary school	Vocational high school	Young	Middle, old	Female	Male
Patient	1	3	0	4	1	3
Healthy	2	1	2	1	2	1
$Gini_{left}$	0.444		0		0.444	
$Gini_{right}$		0.375		0.320		0.375
$Gini_j$	0.405		0.229		0.405	

Steps (7–11) The Gini values calculated for the variables in Table 6.6 are as follows:

$$\text{Gini}_{\text{SchoolEducation}} = 0.405$$

$$\text{Gini}_{\text{Age}} = 0.229$$

$$\text{Gini}_{\text{Gender}} = 0.405$$

The smallest value among the Gini_j values has been obtained as $\text{Gini}_{\text{Age}} = 0.229$. Branching will begin starting from the *age* variable (the smallest value), namely, the root node.

In order to start the root node, we take the registered individual (ID) values that correspond to the *age* variable classifications (right-left) in Table 6.2.

$$\text{Age}(20 \le \text{young} \le 25) \in \text{Indiviadual}(\text{ID})(2, 7)$$

$$\text{Age}(40 \le \text{middle, old} \le 90) \in \text{Indiviadual}(\text{ID})(1, 3, 4, 5, 6)$$

Given this, the procedure will start with the individuals that have the second and seventh *ID*s registered to the classification on the decision. As for the individuals registered as first, third, fourth, fifth and sixth, we can say that they will form the splitting of the other nodes in the tree. The results obtained from the first split in the tree are shown in Figure 6.24).

Iteration 2 The tree structure should be formed for all the variables in Table 6.4 in the CART algorithm. Thus, for the other variables, it is also required to apply the procedural steps of CART algorithm in Iteration 1 in Iteration 2.

The new sample WAIS-R dataset in Table 6.7 has been formed by excluding the second and seventh individuals registered in the first branching of the tree structure in Iteration 1.

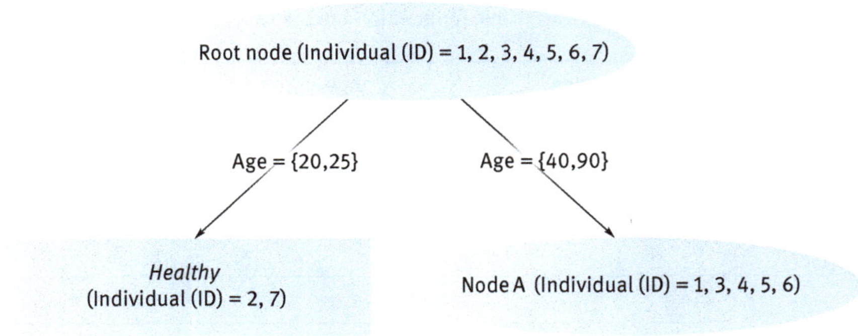

Figure 6.24: The decision tree structure of the sample WAIS-R dataset obtained from CART algorithm for Iteration 1.

Table 6.7: Sample WAIS-R dataset for the Iteration 2.

Individual (ID)	School education	Age	Gender	Class
1	Vocational school	90	Male	Patient
3	High school	40	Female	Healthy
4	Vocational school	45	Male	Patient
5	Primary school	50	Male	Patient
6	High school	80	Female	Patient

Steps (1–4) Regarding the variables provided (*school education, age, gender* and *class*) in Table 6.7 binary grouping is done.

Table 6.8 is based on the class number that belongs to the variables classified as *class* (*patient, healthy*).

Variables are as follows:

School education (primary school, vocational school, high school)
Age $(40 \leq \text{middle} \leq 50)$ $(80 \leq \text{old} \leq 90)$
Gender (female, male)

Table 6.8: The number of classes that variables belong to identified as *class* (*patient, healthy*) in Table 6.7 (for Iteration 1).

Class	School education		Age		Gender	
	Primary school	Vocational high school	Middle	Old	Female	Male
Patient	1	3	2	2	1	3
Healthy	0	1	1	0	1	0

Step (5) $\text{Gini}_{\text{left}}$ and $\text{Gini}_{\text{right}}$ values (eq. (6.6)) are calculated for the variables listed in Table 6.7: *school education, age, gender* for the CART (Figure 6.24) algorithm.

Gini calculation for *school education* is as follows:

$$\text{Gini}_{\text{left}}(\text{Primary School}) = 1 - \left[\left(\frac{1}{1}\right)^2 + \left(\frac{0}{1}\right)^2\right] = 0$$

$$\text{Gini}_{\text{right}}(\text{Vocational School, High School}) = 1 - \left[\left(\frac{3}{4}\right)^2 + \left(\frac{1}{4}\right)^2\right] = 0.375$$

Gini calculation for *age* is as follows:

$$\text{Gini}_{\text{left}}(\text{Middle}) = 1 - \left[\left(\frac{2}{3}\right)^2 + \left(\frac{1}{3}\right)^2\right] = 0.444$$

$$\text{Gini}_{\text{right}}(\text{Old}) = 1 - \left[\left(\frac{2}{2}\right)^2 + \left(\frac{0}{2}\right)^2\right] = 0$$

Gini calculation for *gender* is as follows:

$$\text{Gini}_{\text{left}}(\text{Female}) = 1 - \left[\left(\frac{1}{2}\right)^2 + \left(\frac{1}{2}\right)^2\right] = 0.5$$

$$\text{Gini}_{\text{left}}(\text{Male}) = 1 - \left[\left(\frac{3}{3}\right)^2 + \left(\frac{0}{3}\right)^2\right] = 0$$

Step (6) The Gini_j value of the variables (*school education, age, gender*) is calculated (see eq. (6.7)) using the results obtained in Step 5.

$$\text{Gini}_{\text{SchoolEducation}} = \frac{1 \times (0) + 4 \times (0.375)}{5} = 0.300$$

$$\text{Gini}_{\text{Age}} = \frac{3 \times (0.444) + 2 \times (0)}{5} = 0.267$$

$$\text{Gini}_{\text{Gender}} = \frac{3 \times (0.500) + 3 \times (0)}{5} = 0.200$$

Table 6.9: The application of the CART algorithm for Iteration 2 to the sample WAIS-R dataset.

Class	School education		Age		Gender	
	Primary school	Vocational school, high school	Middle	Old	Female	Male
Patient	1	3	2	2	1	3
Healthy	0	1	1	0	1	0
$Gini_{left}$	0		0.444		0.500	
$Gini_{right}$		0.375		0		0.000
$Gini_j$	0.300		0.267		0.200	

Table 6.9 lists the results obtained from the steps applied for Iteration 2 to the CART algorithm as in Steps 1–6.

Steps (7–11) The Gini values calculated for the variables listed in Table 6.9 are listed as follows:

$$Gini_{SchoolEducation} = 0.300$$

$$Gini_{Age} = 0.267$$

$$Gini_{Gender} = 0.200$$

The smallest value among the $Gini_j$ values has been obtained as $Gini_{Gender} = 0.200$.

Branching will begin starting from the *gender* variable (the smallest value), namely, the root node.

In order to start the root node, we take the registered *female* and *male* values that correspond to the *gender* variable classifications (right-left) in Table 6.7.

Given this the procedure will start with the individuals that have the first, fourth and fifth *IDs* registered to the classification on the decision. As for the individuals registered as first, third, fourth, fifth and sixth, we can say that they will form the splitting of the other nodes in the tree. The results obtained from the first split in the tree (see Figure 6.25) are given as follows:

Iteration 3 The tree structure should be formed for all the variables in Table 6.4 in CART algorithm. Thus, for the other variables, it is required to apply the procedural steps of CART algorithm in Iteration 1 and Iteration 2 in Iteration 3 as well.

Regarding the variables provided (*school education, age, gender* and *class*) in Table 6.10 binary grouping is done.

Table 6.10 is based on the class number that belongs to the variables classified as *class* (*patient, healthy*; see Figure 6.26).

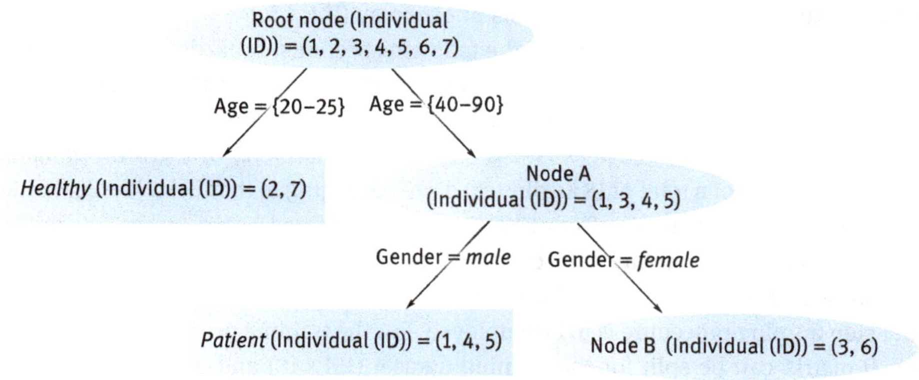

Figure 6.25: The decision tree structure of a sample WAIS-R dataset obtained from CART algorithm for Iteration 2.

Table 6.10: Iteration 3 – sample WAIS-R dataset.

Individual (ID)	School education	Age	Gender	Class
3	high school	40	Female	Healthy
6	high school	80	Female	Patient

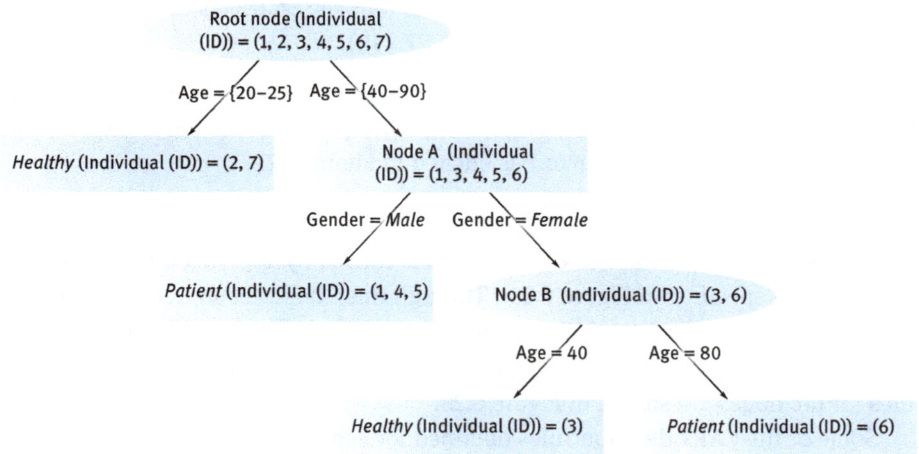

Figure 6.26: The decision tree structure of a sample WAIS-R dataset obtained from CART algorithm for Iteration 3.

6.5.1 CART algorithm for the analysis of various data

CART algorithm has been applied on economy (U.N.I.S.) dataset, multiple sclerosis (MS) dataset and WAIS-R dataset in Sections 6.5.1.1, 6.5.1.2 and 6.5.1.3, respectively.

6.5.1.1 CART algorithm for the analysis of economy (U.N.I.S.)

As the first set of data, we will use in the following sections, some data related to the U.N.I.S. countries' economies. The attributes of the these countries' economies are data regarding *years, unemployment, GDP per capita (current international $), youth male (% of male labor force ages 15–24) (national estimate), ..., GDP growth (annual %)*. (Data consist of a total of 18 attributes). Data belonging to the USA, New Zealand, Italy and Sweden economies from 1960 to 2015 are defined based on the attributes given in Table 2.7. Economy dataset (http://data.worldbank.org) [15] is used in the following sections. For the classification of D matrix through CART algorithm, the first step training procedure is to be employed. For the training procedure, 66.66% of the D matrix can be split for the training dataset (151 × 18) and 33.33% as the test dataset (77 × 18).

Following the classification of the training dataset being trained with CART algorithm, we can classify the test dataset. Although the procedural steps of the algorithm may seem complicated at first glance, the only thing you have to do is to concentrate on the steps and grasp them. For this, let us have a close look at the steps provided in Figure 6.27.

How can we find the class of a test data belonging to economy (U.N.I.S.) dataset provided in Table 2.7.1 according to CART algorithm? We can find the class label of a sample data belonging to economy (U.N.I.S.) dataset through the following steps (Figure 6.26):

Steps (1–3) Let us split the U.N.I.S. dataset as follows: 66.6% as the training data ($D = 151 \times 18$) and 33.3% as the test data ($T = 77 \times 18$).

Steps (4–10) Create random subset. The training of the training set ($D = 228 \times 18$) through the CART algorithm has yielded the following results: number of nodes – 20. Sample CART decision tree structure is elicited in Figure 6.27.

Steps (11–12) The splitting criterion labels the node.

The training of the economy (U.N.I.S.) dataset is carried by applying CART algorithm (Steps 1–12 as specified in Figure 6.28). Through this, the decision tree consisting of 20 nodes (the learned tree ($N = 20$)) has been obtained. The nodes chosen via the CART learned tree nodes (*GDP per capita (current international $)*) and some of the rules for the nodes are shown in Figure 6.28.

Some of the CART decision rules obtained for economy (U.N.I.S.) dataset based on Figure 6.30 can be listed as follows.

Figure 6.28(a) and (b) shows economy (U.N.I.S.) dataset sample decision tree structure based on CART, obtained as such; *if–then* rules are provided below:

...

...

...

Algorithm: CART Algorithm for Economy (U.N.I.S.) Data Set

Input: Training set including 151 number of samples $D = [x_1, x_2, \ldots, x_{151}]$ $(x_i \in R^d)$

Test data including 77 number of samples $T = [x_1, x_2, \ldots, x_{77}]$ $(x_i \in R^d)$

Result: Classification information corresponding to the test and training data

$C =$ [USA, New Zealand, Italy, Sweden] number of trees N

Method:

Create a node for the tree

while each sample i of belonging to the training set **do** {

//Gini value is calculated

$$Gini(D) = 1 - \sum_{i=1}^{151} p_i^2$$

root = j

while possible value j of **do** D

//*Gini* index for the attribute is chosen as the splitting subset

$$Info_A(D) = \sum_{j=1}^{2} \frac{|D_j|}{|D|} x(D_j)$$

if is empty. in this case leaf will be the most common value in D

    else add the subtree ID3()

return the learned tree(N);

//maximizes the reduction in impurity is chosen as the splitting attribute.

$$\Delta Gini(A) = Gini(D) - Gini_A(D)$$

//Predict sample test data

Figure 6.27: CART algorithm for the analysis of the economy (U.N.I.S.) data set.

Rule 1

IF (GDP per capita (*current international ($)*) < $144992e + 06)

THEN class = USA

Rule 2

IF (GDP per capita *(current international ($))* ≥ $144992e + 06)

THEN class = Italy

...

...

...

As a result, the data in the economy (U.N.I.S.) dataset with 33.33% portion allocated for the test procedure and classified as *USA, New Zealand, Italy* and *Sweden* have been classified with an accuracy rate of *95.2%* based on CART algorithm.

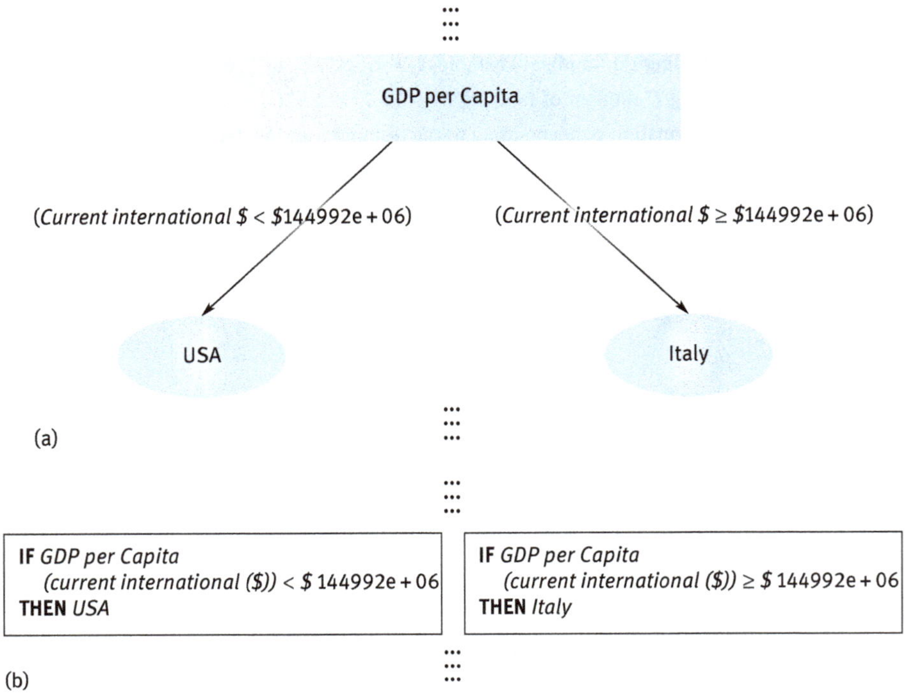

Figure 6.28: Sample decision tree structure for economy (U.N.I.S.) dataset based on CART. (a) CART algorithm graph for the sample economy (U.N.I.S.) dataset.(b) Sample decision tree rule space for economy (U.N.I.S.) dataset.

6.5.1.2 CART algorithm for the analysis of MS

As listed in Table 2.10 MS dataset has data from the following groups: 76 sample belonging to RRMS, 76 sample to SPMS, 76 sample to PPMS and 76 sample belonging to healthy subjects of control group. The attributes of the control group are data regarding *brain stem* (*MRI 1*), *corpus collasum periventricular* (*MRI 2*), *upper cervical* (*MRI 3*), *lesion diameter size (milimetric (mm)) in the MR image* and *EDSS* score. Data consist of a total of 112 attributes. It is known that using these attributes of 304 individuals, we can know whether the data belong to the MS subgroup or healthy group. How can we make the classification as to which MS patient belongs to which subgroup of MS including healthy individuals and those diagnosed with MS (based on the lesion diameters (*MRI 1, MRI 2, MRI 3*), number of lesion size for (*MRI 1, MRI 2, MRI 3*) as obtained from MRI images and EDSS scores)? *D* matrix has a dimension of (304 × 112). This means *D* matrix includes the MS dataset of 304 individuals along with their 112 attributes (see Table 2.12.1 for the MS dataset). For the classification of *D* matrix through Naive Bayesian algorithm, the first step training procedure is to be employed. For the training procedure, 66.66% of the *D* matrix can be split for the training dataset (203 × 112) and 33.33% as the test dataset (101 × 112).

Algorithm: CART Algorithm for MS Data Set

Input: MS Training set including 203 number of samples $D = [x_1, x_2, \ldots, x_{203}]$ $(x_i \in R^d)$
Test data including 101 number of samples $T = [x_1, x_2, \ldots, x_{101}]$ $(x_i \in R^d)$

Result: Classification information corresponding to the test and training data
C = [PPMS, RRMS, SPMS, Healthy]] number of trees *N*

Method:

(1) Create a node for the tree
(2) **while** each sample *i* of belonging to the training set **do** {
(3) //Gini value is calculated

$$Gini(D) = 1 - \sum_{i=1}^{203} p_i^2$$

(4) root = *j*
(5) **while** possible value *j* of **do** *D*
(6) //*Gini* index for the attribute is chosen as the splitting subset

$$Info_A(D) = \sum_{j=1}^{2} \frac{|D_j|}{|D|} x(D_j)$$

(7) **if** is empty in this case leaf will be the most common value in *D*
(8) **else** add the subtree ID3()
(9) **return** the learned tree(*N*);
(10) //maximizes the reduction in impurity is chosen as the splitting attribute.
(11)

$$\Delta Gini(A) = Gini(D) - Gini_A(D)$$

(12) //Predict sample test data

Figure 6.29: CART algorithm for the analysis of MS data set.

Following the classification of the training dataset being trained with CART algorithm, we can classify the test dataset. Although the procedural steps of the algorithm may seem complicated at first glance, the only thing you have to do is to concentrate on the steps and grasp them. We can take a close look at the steps presented in in Figure 6.29.

In order to be able to do classification through CART algorithm, the test data has been selected randomly from the MS dataset as 33.33%.

Let us now think for a while about how to classify the MS dataset provided in Table 2.10.1 in accordance with the steps of CART algorithm (Figure 6.29.)

Steps (1–3) Let us identify 66.66% of the MS dataset as the training data ($D = 203 \times 112$) and 33.33% as the test data ($T = 101 \times 112$).

Steps (4–10) As a result of training the training set with CART algorithm, the following results have been obtained: number of nodes –31.

Steps (11–12) The splitting criterion labels the node.

The training of the MS dataset is done by applying CART algorithm (Steps 1–12 as specified in Figure 6.29) . Through this, the decision tree consisting of 39 nodes (the learned tree ($N = 39$)) has been obtained. The nodes chosen via the CART learned tree nodes (EDSS, MRI 1, first region lesion size) and some of the rules pertaining to the nodes are shown in Figure 6.30.

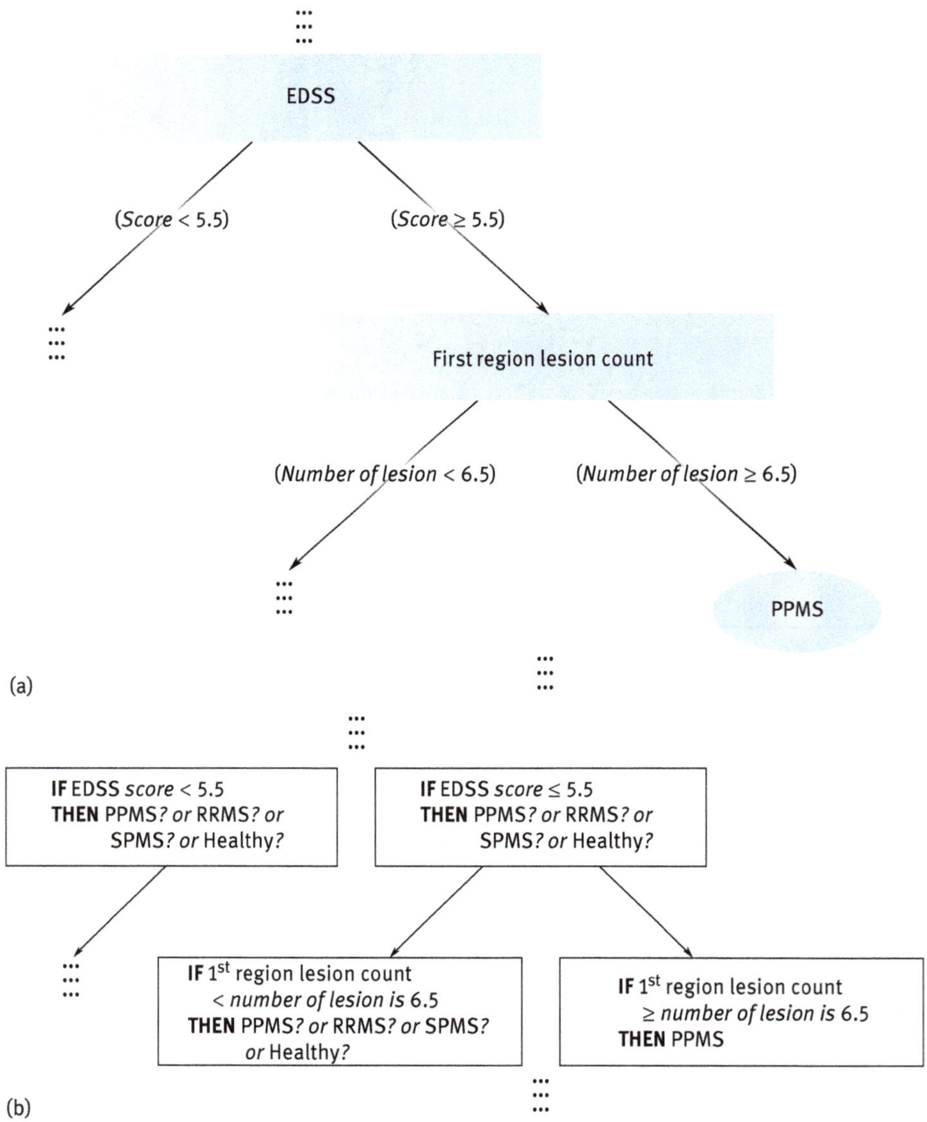

Figure 6.30: Sample decision tree structure for the MS dataset based on CART algorithm. CART algorithm graph for the sample MS dataset. (b) Sample decision tree rule space for the MS dataset.

Based on Figure 6.30, some of the decision rules obtained from the CART decision tree obtained from MS dataset can be seen as follows.

Figure 6.30(a) and (b) shows MS dataset sample decision tree structure based on CART, obtained as such; **IF-THEN** rules are as follows:

...

...

...

Rule 1
IF (*first region lesion count* (*number of lesion*) ≥ 6.5)
THEN class = PPMS

...

...

...

As a result, the data in the MS dataset with 33.33% portion allocated for the test procedure and classified as *RRMS, SPMS, PPMS* and *healthy* have been classified with an accuracy rate of *78.21%* based on CART algorithm.

6.5.1.3 CART algorithm for the analysis of mental functions

As listed in Table 2.16, the WAIS-R dataset has data of 200 people belonging to *patient* group and 200 sample to *healthy* control group. The attributes of the control group are data regarding *school education, gender, ..., DM*. Data consist of a total of 21 attributes. It is known that using these attributes of 400 individuals, we can know whether the data *patient* or *healthy* group. How can we make the classification as to which individual belongs to which patient or healthy group and those diagnosed with WAIS-R test (based on the *school education, gender, ..., DM*)? D matrix has a dimension of 400 × 21. This means D matrix includes the WAIS-R dataset of 400 individuals along with their 21 attributes (see Table 2.16) for the WAIS-R dataset. For the classification of D matrix CART algorithm, the first step training procedure is to be employed. For the training procedure, 66.66% of the D matrix can be split for the training dataset (267 × 21) and 33.33% as the test dataset (133 × 21). Following the classification of the training dataset being trained with CART algorithm, we can classify the test dataset.

Although the procedural steps of the algorithm may seem complicated at first glance, the only thing you have to do is to concentrate on the steps and grasp them. For this, let us have a close look at the steps provided in Figure 6.31.

How can we classify the WAIS-R dataset in Table 2.16.1 in accordance with the steps of CART algorithm (Figure 6.31)?

Steps (1–3) Let us identify the 66.66% of WAIS-R dataset as the training data ($D = 267 \times 21$) and 33.33% as the test data ($T = 133 \times 21$).

Algorithm: CART Algorithm for WAIS-R Data Set

Input: Training set including 267 number of samples $D = [x_1, x_2, \ldots, x_{267}]$ $(x_i \in R^d)$
Test data including 133 number of samples $T = [x_1, x_2, \ldots, x_{133}]$ $(x_i \in R^d)$

Result: Classification information corresponding to the test and training data
$C =$ [Patient, Healthy] number of trees N

Algorithm:

(1) Create a node for the tree
(2) **while** each sample i of belonging to the training set **do** {
(3) //*Gini* value is calculated

$$Gini(D) = 1 - \sum_{i=1}^{18} p_i^2$$

(4) root $= j$
(5) **while** possible value j of **do** D

$$Info_A(D) = \sum_{j=1}^{2} \frac{|D_j|}{|D|} x(D_j)$$

(6) //*Gini* index for the attribute is chosen as the splitting subset
(7) **if** is empty. in this case leaf will be the most common value in D
(8) **else** add the subtree ID3()
(9) **return** the learned tree(N);
(10) //maximizes the reduction in impurity is chosen as the splitting attribute.
(11)

$$\Delta Gini(A) = Gini(D) - Gini_A(D)$$

(12) //Predict sample test data

Figure 6.31: CART algorithm for the analysis of WAIS-R data set.

Steps (4–10) As a result of training the training set ($D = 267 \times 21$) with random forest algorithm, the following results have been obtained: number of nodes –30. Sample decision tree structure is shown in Figure 6.32.

Steps (11–12) *Gini* index for the attribute is chosen as the splitting subset.

The training of the WAIS-R dataset is done by applying CART algorithm (Steps 1–12 as specified in Figure 6.31). Through this, the decision tree consisting of 20 nodes (the learned tree ($N = 20$)) has been obtained. The nodes chosen via the CART learned tree nodes ((*logical deduction, vocabulary*) and some of the rules for the nodes are given in Figure 6.31.

Based on Figure 6.32, some of the rules obtained from the CART algorithm obtained from WAIS-R dataset can be seen as follows.

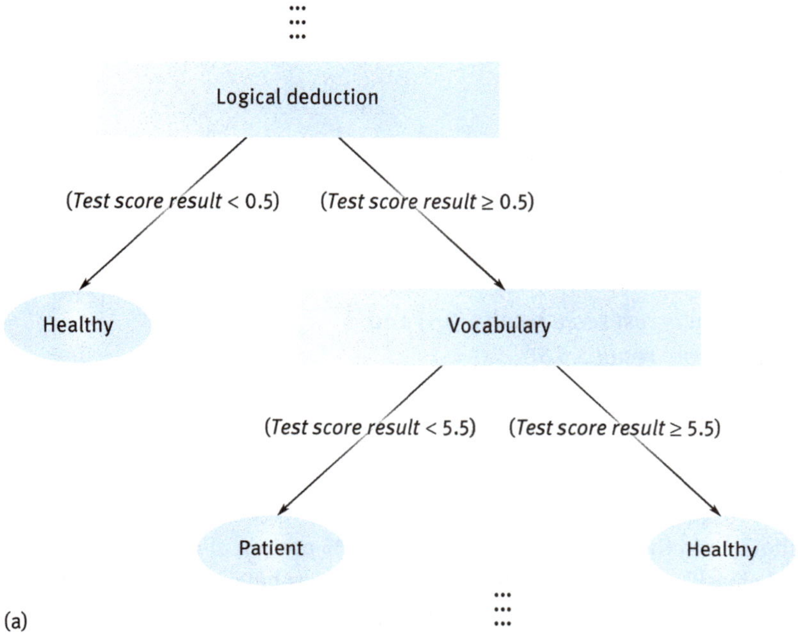

(a)

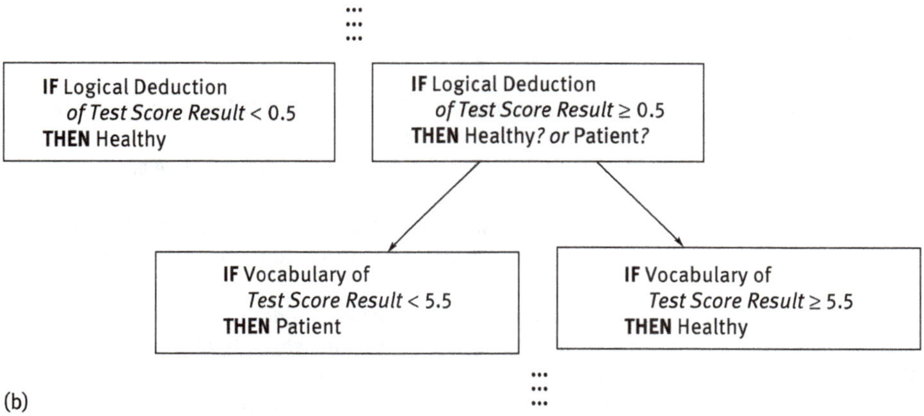

(b)

Figure 6.32: Sample decision tree structure for the WAIS-R dataset based on CART. (a) CART algorithm graph for sample WAIS-R dataset. (b) Sample decision tree rule space for the WAIS-R dataset.

Figure 6.32(a) and (b) shows WAIS-R dataset sample decision tree structure based on CART, obtained as such; *if–then* rules are as follows:

...

...

...

Rule 1
IF (logical deduction *of test score result* < 0.5)
THEN class = healthy

Rule 2
IF ((logical deduction *of test score result* ≥ 0.5) **and**
(*vocabulary of test score result* < 5.5))
THEN class = patient

Rule 3
IF ((logical deduction *of test score result* ≥ 0.5) **and**
(*vocabulary* of *test score result* ≥ 5.5))
THEN class = healthy

...

...

...

As a result, the data in the WAIS-R dataset with 33.33% portion allocated for the test procedure and classified as *patient* and *healthy individuals* have been classified with an accuracy rate of *99.24%* based on CART algorithm.

References

[1] Han J, Kamber M, Pei J, Data mining Concepts and Techniques. USA: The Morgan Kaufmann Series in Data Management Systems, Elsevier, 2012.
[2] Buhrman, H, De Wolf R. Complexity measures and decision tree complexity: A survey. *Theoretical Computer Science*, 2002, 288(1), 21–43.
[3] Du W, Zhan Z. Building decision tree classifier on private data. *In Proceedings of the IEEE international conference on Privacy, security and data mining. Australian Computer Society,* Inc., 14, 2002, 1–8.
[4] Buntine W, Niblett T. A further comparison of splitting rules for decision-tree induction. *Machine Learning, Kluwer Academic Publishers-Plenum Publishers,* 1992, 8 (1), 75–85.
[5] Chikalov I. Average Time Complexity of Decision Trees. Heidelberg: Springer. 2011.
[6] Banfield RE, Hall LO, Bowyer KW, Kegelmeyer WP. A comparison of decision tree ensemble creation techniques. *IEEE Transactions on Pattern Analysis and Machine Intelligence,* 2007, 29(1), 173–180.
[7] Barros R C, De Carvalho AC, Freitas AA. Automatic design of decision-tree induction algorithms. Germany: Springer International Publishing, 2015.
[8] Quinlan, JR. Induction of decision trees. *Machine Learning, Kluwer Academic Publishers,* 1986, 1 (1), 81–106.
[9] Breiman L, Friedman J, Charles JS , Olshen RA. Classification and regression trees. New York: Routledge, 1984.
[10] Freitas AA. Data mining and knowledge discovery with evolutionary algorithms. Berlin Heidelberg: Springer Science & Business Media, 2013.
[11] Valiente G. Algorithms on Trees and Graphs. Berlin Heidelberg: Springer, 2002.

[12] Shannon CE. A mathematical theory of communication. *ACM SIGMOBILE Mobile Computing and Communications Review*, 2001, 5,1, 3–55.

[13] Kubat M. An Introduction to Machine Learning. Switzerland: Springer International Publishing, 2015.

[14] Quinlan JR. C4. 5: Programs for machine learning. Elsevier: New York, NY, USA, 1993

[15] Quinlan JR. Bagging, boosting, and C4. 5. *AAAI/IAAI*, 1996, 1, 725–730.

[16] Quinlan JR. Improved use of continuous attributes in C4. 5. *Journal of Artificial Intelligence Research*, 1996, 4, 77–90.

[17] Rokach L, Maimon O. Data mining with decision trees: Theory and applications. Singapore: World Scientific Publishing, 2014 (2nd Ed.), 77–81.

[18] Reiter JP. Using CART to generate partially synthetic public use microdata. *Journal of Official Statistics*, 2005, 21 (3), 441–462.

7 Naive Bayesian classifier

This chapter covers the following items:

- Algorithm for *maximum posteriori* hypothesis
- Algorithm for classifying with probability modeling
- Examples and applications

This chapter provides brief information on the Bayesian classifiers that are regarded as statistical classifiers. They are capable of predicting class membership probabilities such as the probability (with a given dataset belonging to a particular class). Bayesian classification is rooted in Bayes' theorem. The *naive Bayesian classifier* has also been revealed as a result of the studies that performed the classification of algorithms. In terms of performance, simple Bayesian classifier is known to be equivalent to decision tree and some other neural network classifiers. Bayesian classifiers have proven to yield high speed as well as accuracy when they are applied to large datasets as well. Naive Bayesian classifiers presume that the value effect of an attribute pertaining to a given class is independent of the other attributes' values. Such an assumption is named as the *class conditional independence*. It makes the computations involved more simple; for this reason it has been named as "naive". First, we shall deal with the basic probability notation and Bayes' theorem. Later, we will provide the way of doing a naive Bayesian classification [1–7].

Thomas Bayesis the mastermind of *Bayes' theorem*. He was an English clergy working on subjects such as decision theory as well as probability in the eighteenth century [8]. As for the explanation, we can present the following: $\boldsymbol{X}$ is a dataset. According to Bayesian terminology, $\boldsymbol{X}$ is regarded as the "evidence" and described by measurements performed on a set of n attributes. H is a sort of hypothesis like the dataset $\boldsymbol{X}$ belongs to a particular specific class C. One may like to agree on $P(H|X)$, the probability held by the hypothesis H with the "evidence" or observed dataset $\boldsymbol{X}$. One seeks the probability that dataset $\boldsymbol{X}$ belongs to class C, given that the attribute description of $\boldsymbol{X}$ is known.

$P(H|X)$ is the *posterior probability* or it is also known as *a posteriori probability*, of H that is conditioned on $\boldsymbol{X}$. For example, based on the dataset presented in Table 2.8 *Year and Inflation (consumer prices [annual %])* attributes are the relevant ones defined and limited by the economies. Suppose that $\boldsymbol{X}$ has an inflation amounting to \$ 575.000 (belonging to the year 2003 for New Zealand). Let us also suppose that H will have profit based on the inflation of the economy in the next 2 years. Since we know the *Inflation (consumer prices [annual %])* value of the New Zealand economy belonging to 2003 $P(H|X)$, we can calculate the possibility of the profit situation. Similarly, $P(H|X)$ is the posterior possibility of $\boldsymbol{X}$ over $\boldsymbol{H}$. This means that as we know

https://doi.org/10.1515/9783110496369-007

that the New Zealand economy is in a profitable situation within that year, $\boldsymbol{X}$ is the possibility of getting an *Inflation (consumer prices [annual %])* to \$ 575.000 in 2003. $P(X)$ is the preceding possibility of $\boldsymbol{X}$ [9].

For the estimation of the *probabilities,* $P(X)$, $P(\boldsymbol{X})$ may well be predicted from the given data. It is acknowledged that Bayes' theorem is useful because it provides individuals with a way of calculating the posterior probability $P(H|X)$, from $P(H)$, $P(H|X)$and $P(\boldsymbol{X})$. Bayes' theorem is denoted as follows:

$$P(X|H) = \frac{P(X|H)P(H)}{P(X)} \tag{7.1}$$

Now let us have a look at the way Bayes' theorem is used in *naive Bayesian classification* [10]. As we have mentioned earlier, it is known as the simple Bayesian classifier. The working thereof can be seen as follows:

Let us say D is a training set of datasets and their class labels associated. As always is the case, each dataset is depicted by an n-dimensional attribute vector, $\boldsymbol{X}$, D. $x_1, x_2,::: x_n$, showing n measurements that are made on the dataset from n attributes, in the relevant order A_1, $A_2,\ldots$, A_n.

Let us think that there are m classes, C_1, $C_2,\ldots$, C_m. There is dataset denoted as $\boldsymbol{X}$. Here the classifier will make the prediction that $\boldsymbol{X}$ belongs to the class that has the maximum posterior probability conditioned on $\boldsymbol{X}$. In other words, the naive Bayesian classifier makes the prediction that dataset $\boldsymbol{X}$ belongs to the C_i class only if the following conditions are fulfilled: first of all, D being a training set of datasets and their associated class labels, each dataset is denoted by an n-dimensional attribute vector, $\boldsymbol{X} = (x_1, x_2, \ldots, x_n)$. This depicts n measurements made on the dataset derived from corresponding n attributes, A_1, $A_2,\ldots$, A_n. Second, let us assume that there exist m classes, C_1, C_2, $\ldots$, C_m. With the dataset, $\boldsymbol{X}$, the classifier will make the prediction that $\boldsymbol{X}$ belongs to the class that has the maximum posterior probability conditioned on $\boldsymbol{X}$. In this case, the naive Bayesian classifier estimates that dataset $\boldsymbol{X}$ belongs to the C_i class and if $P(C_i|X) > P(C_j|X)$ for $1 \le j \le m, j \ne i$. This is a requirement for the $\boldsymbol{X}$ to belong to C_i class. In this case, $P(C_i|X)$ is maximized and is called the maximum posterior hypothesis. Through Bayes' theorem (eq. 7.2), the relevant calculation is done as follows [10–12]:

$$P(C_i|X) = \frac{P(X|C_i)P(C_i)}{P(X)} \tag{7.2}$$

It is necessary only to maximize $P(X|C_i)P(C_i)$ since $P(X)$ is constant for all of the classes, if you do not know the class prior probabilities, it is common for one to suppose that the classes are likely to be equal with $P(C_1) = P(C_2) = \ldots = P(C_m)$; for this reason, one would maximize $P(X|C_i)$. If it is not maximized, then one may maximize $P(X|C_i)P(C_i)$. Class prior probabilities can be estimated by $P(C_i) = |C_{i,D}|/D$, in which $|C_{i,D}|$ is the number of training datasets pertaining to class C_i *in D*.

If datasets have many attributes, then the process of computation of $P(X|C_i)$ could be expensive. For the reduction of computation in the evaluation of $P(X|C_i)$, the naive assumption of *class-conditional independence* is performed, which supposes that the values of the attribute values are independent of each other on conditional bases, showing that dependence relationship does not exist among the attributes.

We obtain the following:

$$P(X|C_i) = \prod_{k=1}^{n} P(x_k|C_i) = P(x_1|C_i) \times P(x_2|C_i) \times \ldots \times P(x_n|C_i) \tag{7.3}$$

It is easy to estimate the probabilities $P(x_1|C_i) \times P(x_2|C_i) \times \ldots \times P(x_n|C_i)$ from the training datasets. For each attribute, one examines if the attribute is categorical or one with a continuous value. For example, if you would like to calculate $P(X|C_i)$, you take into account the following:

(a) If A_k is categorical, $P(x_k|C_i)$ is the number of class datasets C_i in D that has the value x_k for A_k. It is divided by $|C_{i,D}|$ which is the number of class datasets C_i in D.

(b) If A_k has a continuous value, there is some more work to be done (with calculation being relatively straightforward, though). It is assumed that a continuous-valued attribute has a Gaussian distribution with a mean μ and standard deviation, σ, you can define by using the following formulae given below [13]:

$$g(x, \mu, \sigma) = \frac{1}{\sqrt{2\pi\sigma}} e^{-\frac{(x-\mu)^2}{2\sigma^2}} \tag{7.4}$$

to get the following:

$$P(x_k|C_i) = g(x_k, \mu C_i, \sigma C_i) \tag{7.5}$$

For some, such equation might be intimidating; however, you have to calculate μC_i and σC_i, which are the mean or average; and standard deviation, respectively, of the values of A_k attribute for the training of the datasets pertaining to class C_i. Afterward, these two quantities are capped into eq. (7.6), along with x_k, for the estimation of $P(x_k|C_i)$.

For the prediction of $\boldsymbol{X}$, the class label, $P(x_k|C_i)$, is taken into consideration and assessed for each class C_i. The classifier makes the prediction that the class label of dataset $\boldsymbol{X}$ is the class C_i only if the following is satisfied (predicted class label is the class C_i for which $P(X|C_i)P(C_i)$ is the highest value) [14]:

$$P(X|C_i)P(C_i) > P(X|C_j)P(C_j) \text{ for } 1 \le j \le m, j \ne i \tag{7.6}$$

There is a question about the effectiveness of Bayesian classifiers. To answer the question to what extent the Bayesian classifiers is, some empirical studies have been conducted for making comparisons. Studies on decision tree and neural network classifiers have revealed Bayesian classifiers are comparable in some scopes. Theoretically speaking, Bayesian classifiers yield the minimum error rate when compared with all of the other classifiers. Yet, in real life or this is not always applicable in practice due to the inaccuracies in the assumptions pertaining to its use.

If we return back to the uses of Bayesian classifiers, they are believed to be beneficial since they supply a theoretical validation and rationalization for the other classifiers that do not employ Bayes' theorem explicitly. In some assumptions, many curve-fitting algorithms and neural network algorithms output the *maximum posteriori* hypothesis, as is the case with the naive Bayesian classifier [15].

The flowchart of Bayesian classifier algorithm has been provided in Figure 7.1.

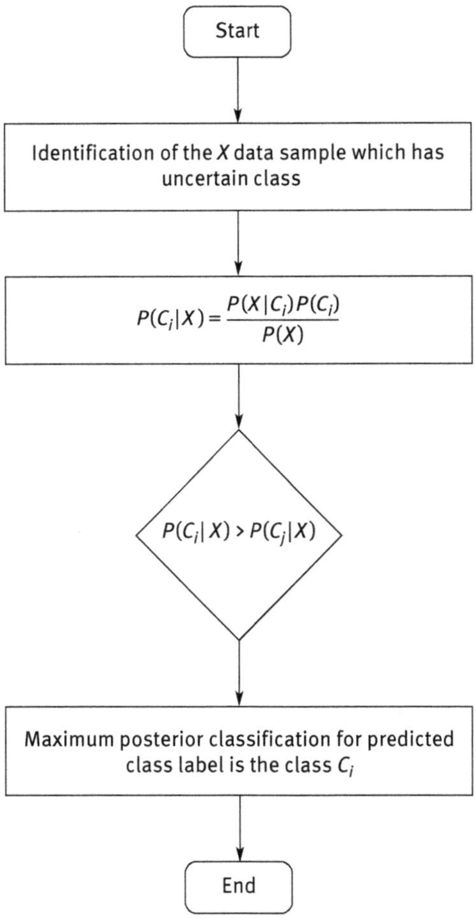

Figure 7.1: The flowchart for naive Bayesian algorithm.

Algorithm: Naive Bayesian Classifier

Input: Training set including N number of samples $D = [x_1, x_2, \ldots, x_n](x_i \in R^d)$
n attributes, correspondingly, $A_1, A_2, \ldots, A_n$.
Test data including t number of samples $T = [x_1, x_2, \ldots, x_k](x_t \in R^d)$

Result: Label data corresponding to the test and training data $C = [c_1, c_2, \ldots, c_m]$

Method:

(1) ***for*** each attribute $j \neq i$ {
(2) $//P(C_i|X)$value is calculated for all the classes
(3) ***if*** $(P(C_i|X) > P(C_j|X))$
(4) $P(C_i|X) = \frac{P(X|C_i)P(C_i)}{P(X)}$
(5) //maximum posterior probability conditioned on $\boldsymbol{X}$
(6) ***for*** each train data // predicted class label is the class C_i
(7) $P(X|C_i) = \prod_{k=1}^{n} P(x_k|C_i) = P(x_1|C_i) \times P(x_2|C_i) \times \ldots \times P(x_n|C_i)$
(8) $\arg\max P(X|C_i)$
(9) //Predict each test sample's class

Figure 7.2: General naive Bayesian classifier algorithm.

Figure 7.2 provides the steps of Bayesian classifier algorithm in detail.

The procedural steps regarding the naive Bayes classifier algorithm in Figure 7.2 are explained in order given below:

Step (1) Maximum posterior probability conditioned on $\boldsymbol{X}$. m classes exist as follows: $C_1, C_2, \ldots, C_m$. There is dataset denoted as $\boldsymbol{X}$. D being a training set of datasets and their associated class labels, each dataset is denoted by an n-dimensional attribute vector, $\boldsymbol{X} = (x_1, x_2, \ldots, x_n)$. Second, let us suppose that there exist m classes, $C = [c_1, c_2, \ldots, c_m]$. With the dataset, $\boldsymbol{X}$, the classifier will make the prediction that $\boldsymbol{X}$ belongs to the class that has the maximum posterior probability conditioned on $\boldsymbol{X}$.

Step (2–4) Classifier estimates that dataset $\boldsymbol{X}$ belongs to the C_i class and if $(P(C_i|X) > P(C_j|X))$

for $1 \leq j \leq m, j \neq i$. This is a requirement for the $\boldsymbol{X}$ to belong to C_i class.

Step (5–9) In this case, $P(C_i|X)$ is maximized and called the maximum posterior hypothesis. Through Bayes' theorem (eq. 7.6) the relevant calculation is done as follows:

It is required only to maximize $P(X|C_i)P(C_i)$ because $P(X)$ is constant for all of the classes, if you do not know the class prior probabilities, it is common to assume that the classes are likely to be equal with $P(C_1) = P(C_2) = \ldots = P(C_m)$, hence, one would maximize $P(X|C_i)$. If not maximized, then one may maximize $P(X|C_i)P(C_i)$. Estimation of class prior probabilities can be done by $P(C_i) = |C_{i,D}|/D$, in which $|C_{i,D}|$ is the number of training datasets for class C_i *in D*.

Example 7.1 Sample dataset (see Table 2.19) has been chosen from WAIS-R dataset, that is, by using Naive Bayes classifier algorithm, let us calculate the probability of which class (*patient* or *healthy*) an individual will belong to. Let us assume that the individual has following attributes (Table 7.1):

x_1 : *School education* = Bachelor, x_2 : *Age* = Middle, x_3 : *Gender* = Female

Table 7.1: Sample WAIS-R Dataset (School education, age and gender) on Table 2.19.

Individual(ID)	School education	Age	Gender	Class
1	Junior high School	Old	Male	Patient
2	Primary School	Young	Male	Healthy
3	Bachelor	Middle	Female	Healthy
4	Junior high School	Middle	Male	Patient
5	Primary School	Middle	Male	Patient
6	Bachelor	Old	Female	Patient
7	Primary School	Young	Female	Healthy
8	Junior high School	Middle	Female	Patient

Solution 7.1 In order to calculate the probability as to which class (whether patient or healthy) an individual (with the following attributes = *School Education*: Bachelor, x_2 = *Age*: Middle, x_3 = *Gender*: Female) will belong to, let us apply the Naive Bayes algorithm in Figure 7.2.

Step (1) Let us calculate the probability of which class (*patient* or *healthy*) the individuals with *School education, Age and Gender* attributes in WAIS-R sample dataset in Table 7.2 will belong to.

The probability of class to belong to for each attribute is calculated in Table 7.2. Classification of the individuals is labeled as C_1 = *Patient* and C_2 = *Healthy*. The probability for each attribute (*School education, Age, Gender*) belonging to which class (*Patient, Healthy*) has been calculated through $P(X|C_1)P(C_1)$ ve $P(X|C_2)P(C_2)$ (see eq. 7.2). The steps for this probability can be found as follows:

Step (2–7) (a) $P(X|C_1)P(C_1)$ and (b) $P(X|C_2)P(C_2)$ calculation for the probability;

(a) $P(X|C_1)P(C_1)$ calculation for the conditional probability:
For the calculation of the conditional probability of $(X|Class = Patient)$, the probability calculations are done distinctively for the $X, D = [x_1, x_2, \ldots, x_5]$ values.

$$P(SchoolEducation = Bachelor|Class = Patient) = \frac{1}{5}$$

Table 7.2: Sample WAIS-R Data Set conditional probabilities.

Attributes	Value	Class			
		Patient		Healthy	
		Number	Probability	Number	Probability
School Education	Primary School	1	1/5	2	2/3
	Junior high School	3	3/5	0	0
	Bachelor	1	1/5	1	1/3
Age	Young	0	0	2	2/3
	Middle	3	3/5	1	1/3
	Old	2	2/5	0	0
Gender	Male	3	3/5	1	1/3
	Female	2	2/5	2	2/3

$$P(Age = Middle|Class = Patient) = \frac{3}{5}$$

$$P(Gender = Female|Class = Patient) = \frac{2}{5}$$

Thus, using eq. (7.3), following result is obtained:

$$P(X|Class = Patient) = \left(\frac{1}{5}\right)\left(\frac{3}{5}\right)\left(\frac{2}{5}\right) = \frac{6}{125}$$

Now, let us also calculate the probability of $P(Class = Patient)$

$$P(Class = Patient) = \frac{5}{8} dir.$$

Thus,

$$P(X|Class = Patient)P(Class = Patient) = \left(\frac{6}{125}\right)\left(\frac{5}{8}\right) \cong 0.03$$

The probability of an individual with the $x_1 = School\ education$: Bachelor, $x_2 = Age$: Middle, $x_3 = Gender$: Female, attributes to belong to *Patient* class has been obtained as 0.03 approximately.

(b) $P(X|C_2)P(C_2)$ calculation for the probability
In order to calculate the conditional probability of $(X|Class = Healthy)$, the probability calculations are done distinctively for the $X, D = [x_1, x_2, x_3]$ values.

$$P(SchoolEducation = Bachelor|Class = Healthy) = \frac{1}{3}$$

$$P(Age = Middle|Class = Healthy) = \frac{1}{3}$$

$$P(Gender = Female|Class = Healthy) = \frac{2}{3}$$

Thus, using eq. (7.3), following result is obtained:

$$P(X|Class = Healthy) = \left(\frac{1}{3}\right)\left(\frac{1}{3}\right)\left(\frac{2}{3}\right) = \frac{2}{27}$$

Now, let us also calculate the probability of $P(Class = Healthy)$.

$$P(Class = Healthy) = \frac{3}{8}$$

Thus,

$$P(X|Class = Healthy)P(Class = Healthy) = \left(\frac{2}{27}\right)\left(\frac{3}{8}\right) \cong 0.027$$

The probability of an individual with the $x_1 = School\ education$: Bachelor, $x_2 = Age$: Middle, $x_3 = Gender$: Female, attributes to belong to *Healthy* class has been obtained as 0.027 approximately.

Step (8–9) Maximum posterior probability is conditioned on $\boldsymbol{X}$. According to maximum posterior probability method, $\arg\max_{C_i}\{P(X|C_i)P(C_i)\}$ value is calculated.

Maximum probability value is taken from the results obtained in **Step (2–7)**.

$$\arg\max_{C_i}\{P(X|C_i)P(C_i)\} = \max\{0.03, 0.027\} = 0.03$$

$$x_1 = School\ education: \textbf{Bachelor}$$

$$x_2 = Age: \textbf{Middle}$$

$$x_3 = Gender: \textbf{Female}$$

It has been identified that an individual (with the attributes given above) with an obtained probability result of *0.03* belongs to *Patient* class.

7.1 Naive Bayesian classifier algorithm (and its types) for the analysis of various data

Bayesian network algorithm has been applied on economy (U.N.I.S.) dataset, multiple sclerosis (MS) dataset and WAIS-R dataset in Sections 7.1.1, 7.1.2 and 7.1.3, respectively.

7.1.1 Naive Bayesian classifier algorithm for the analysis of economy (U.N.I.S.)

As the second set of data, we will use in the following sections, some data related to economies for U.N.I.S. countries. The attributes of these countries' economies are data regarding *years, gross value added at factor cost ($), tax revenue, net domestic credit and gross domestic product (GDP) growth (annual %)*. Economy (U.N.I.S.) Dataset is made up of a total of 18 attributes. Data belonging to economies of USA, New Zealand, Italy and Sweden from 1960 to 2015 are defined on the basis of attributes provided in Table 2.8 Economy (U.N.I.S.) Dataset (http://data.worldbank.org) [15], to be used in the following sections. For the classification of D matrix through Naive Bayesian algorithm the first-step training procedure is to be employed. For the training procedure, 66.66% of the D matrix can be split for the training dataset (151×18), and 33.33% as the test dataset (77×18).

Following the classification of the training dataset being trained with Naive Bayesian algorithm, we can perform the classification of the test dataset. It is normal that the steps for the procedure in the algorithm may seem complex at first look but if you focus on the steps and develop an insight into them, it would be sufficient to go ahead. At this point, we can take a close look at the steps in Figure 7.4.

Let us do the step by step analysis of the *Economy Dataset* according to Bayesian classifier in Example 7.2 based on Figure 7.3.

Example 7.2 The sample Economy Dataset (see Table 7.3) has been chosen from Economy (U.N.I.S.) Dataset (see Table 2.8). Let us calculate the probability as to which class (U.N.I.S.) a country with

x_1 : *Net Domestic Credit* = 49, 000 (*Current LCU*),
x_2 : *Tax Revenue* = 2, 934(% *of GDP*), x_3 : *Year* = 2003 attributes (Table 7.3) will fall under Naive Bayes classifier algorithm.

Solution 7.2 In order to calculate the probability as to which class USA, New Zealand, Italy and Sweden, a country with the following

x_1 : *Net Domestic Credit* = 49, 000 (*Current LCU*),
x_2 : *Tax Revenue* = 2, 934(% *of GDP*), x_3 : *Year* = 2003) attributes will belong to, let us apply the Naive Bayes algorithm in Figure 7.3.

Table 7.3: Selected sample from the Economy Dataset.

ID	Net domestic credit (% of GDP)	Tax revenue (% of GDP)	Year	Class
1	5,323	2,934	2003	USA
2	4,924	2,622	2004	USA
3	3,140	9,385	2003	New Zealand
4	3,715	5,642	2003	Italy
5	6,421	1,085	2003	Sweden
6	13,860	1,066	2006	Sweden

Algorithm: Naive Bayesian Classifier Algorithm for Economy (U.N.I.S.) Data Set

Input: MS Training set including 151 number of samples $D = [x_1, x_2, \ldots, x_{151}](x_i \in R^d)$
From *18* attributes, correspondingly, $A_1, A_2, \ldots, A_{18}$.
Test data including 77 number of samples $T = [x_1, x_2, \ldots, x_{77}](x_i \in R^d)$

Result: Label data that corresponds to *285* samples $C =$ [USA, New Zealand, Italy, Sweden]

Method:

(1) //***for*** each attribute $j \neq i$ {
(2) //$P(C_i|X)$ value is calculated for all the classes
(3) ***if*** $(P(C_i|X) > P(C_j|X))$
(4) $P(C_i|X) = \frac{P(X|C_i)P(C_i)}{P(X)}$
(5) //maximum posterior probability conditioned on ***Economy (U.N.I.S.) Data Set***
(6) ***for*** each train data // predicted class label is the class C_i ($C =$ [USA, New Zealand, Italy, Sweden])
(7) $P(X|C_i) = \prod_{k=1}^{151} P(x_k|C_i) = P(x_1|C_i) \times P(x_2|C_i) \times \ldots \times P(x_{188}|C_i)$
(8) arg max $P(X|C_i)$
(9) //Predict each test sample's class

Figure 7.3: Naive Bayesian classifier algorithm for the analysis of the economy (U.N.I.S.) Dataset.

Step (1) Let us calculate the probability of which class (U.N.I.S.) the economies with *net domestic credit, tax revenue and year* attributes in economics sample dataset in Table 7.4 will belong to.

The probability of class belonging to each attribute is calculated in Table 7.4. Classification of the countries *ID* is labeled as $C_1 = UnitedStates$, $C_2 = NewZealand$, $C_3 = Italy$, $C_4 = Sweden$. The probability for each attributes (*net domestic credit, tax revenue and year*) belonging to which class: USA, New Zealand, Italy, Sweden has been calculated through $P(X|C_1)P(C_1)$, $P(X|C_2)P(C_2)$, $P(X|C_3)P(C_3)$, $P(X|C_4)P(C_4)$ (see eq. 7.2). The steps for this probability can be found as follows:

Step (2–7) (a) $P(X|USA)P(USA)$, (b) P(X|NewZealand)P(NewZealand), (c) P(X|Italy) P(Italy)*and*(d)P(X|Sweden) P(Sweden)*calculationfortheprobability*;

(a) $P(X|USA)P(USA)$ *calculation of the conditional probability*
Here, for calculating the conditional probability of $(X|Class = USA)$, the probability calculations are done distinctively for the $X = \{x_1, x_2, \ldots, x_n\}$ values.

$$P(Net\ Domestic\ Credit = (\%\ of\ GDP) = 49000|Class = USA) = \frac{1}{3}$$

$$P(Tax\ \text{Revenue} = (\%\ of\ GDP)[1000 - 3000]Class = USA) = 1$$

$$P(Year = [2003 - 2005]|Class = USA) = \frac{2}{5}$$

Using eq. (7.3), following equation is obtained:

$$P(X|Class = USA) = \left(\frac{2}{3}\right)(1)\left(\frac{2}{5}\right) = \frac{4}{12}$$

Now, let us also calculate the probability of $P(Class = USA) = \frac{2}{6}$
Thus, the following equation is calculated:

$$P(X|Class = USA)P(Class = USA) = \left(\frac{4}{12}\right)\left(\frac{2}{6}\right) \cong 0.111$$

It has been identified that an economy with the
x_1 : *Net Domestic Credit* = 49, 000 (*Current LCU*),
x_2 : *Tax Revenue* = 2, 934(% *of GDP*), x_3 :*Year* = 2003 attributes has a probability result of 0.111 about belonging to *USA* class.

(b) The conditional probability for *New Zealand* class $P(X|NewZealand)$ $P(NewZealand)$ and has been obtained as 0.
(c) The conditional probability for *Italy* class $P(X|Italy)P(Italy)$ and has been obtained as 0.
(d) $P(X|Sweden)P(Sweden)$ conditional probability calculation.

Here, in order to calculate the conditional probability of $P(X|Class)P(Sweden)$ calculations are done distinctively for the $X = \{x_1, x_2, \ldots, x_n\}$ values.

$$P(Net\ Domestic\ Credit = (\%\ of\ GDP) = 49,000|Class = Sweden) = 1$$

$$P(Tax\ \text{Revenue} = (\%\ of\ GDP)[1,000 - 3,000]Class = Sweden) = 0$$

$$P(Year = [2003 - 2005]|Class = Sweden) = \frac{1}{5}$$

Table 7.4: Sample Economy (U.N.I.S.) Dataset conditional probabilities.

Attributes	Range of attributes	Class							
		USA		New Zealand		Italy		Sweden	
		Number	Probability	Number	Probability	Number	Probability	Number	Probability
Net domestic credit	[3,140–3,715]	0	0	1	1	1	1	0	0
	[3716–6,500]	2	2/3	0	0	0	0	1	1/3
	[6,501–15,000]	0	0	0	0	0	0	1	1
Tax revenue	[1,000–3,000]	2	1	0	0	0	0	0	0
	[3,000–5,000]	0	0	0	0	0	0	0	0
	[5,000–10,000]	0	0	1	1/4	1	1/4	2	1/2
Year	[2003–2005]	2	2/5	1	1/5	1	1/5	1	1/5
	[2006–2012]	0	0	0	0	0	0	1	1

Thus, using eq. (7.3), the following equation is obtained:

$$P(X|Class = Sweden) = \left(\frac{2}{3}\right)(0)\left(\frac{1}{5}\right) = 0$$

On the other hand, we have to calculate the probability of $P(Class = Sweden) = \frac{2}{6}$ as well.

Thus,

$$P(X|Class = Sweden)P(Class = Sweden) = (0)\left(\frac{2}{6}\right) = 0$$

It has been identified that an economy with the x_1 :*Net Domestic Credit* = 49,000 (*Current LCU*), x_2 :*Tax Revenue* = 2,934(% *of GDP*), x_3 :*Year* = 2003 attributes has a probability result of 0 about belonging to *Sweden* class.

Step (8–9) Maximum posterior probability conditioned on ***X***. According to maximum posterior probability method, $\arg\max_{C_i} \{P(X|C_i)P(C_i)\}$ value is calculated.
Maximum probability value is taken from the results obtained in **Step (2–7)**.

$$\arg\max_{C_i}\{P(X|C_i)P(C_i)\} = \max\{0.111, 0, 0, 0\} = 0.111$$

x_1 : *Net Domestic Credit (Current LCU)*= **49,000**
x_2 : *Tax Revenue (% of GDP)* = **2,934**

x_3 : *Year* = **2003**

It has been identified that an economy has a probability of 0.111. It has been identified that an economy with the attributes given above has a probability result of 0.111 about belonging to *USA* class.

As a result, the data in the Economy (U.N.I.S.) with 33.33% portion allocated for the test procedure and classified as USA, New Zealand, Italy and Sweden have been classified with an accuracy rate of *41.18%* based on *Bayesian classifier algorithm.*

7.1.2 Algorithms for the analysis of multiple sclerosis

As presented in Table 2.12, multiple sclerosis dataset has data from the following groups: 76 samples belonging to RRMS, 76 samples to SPMS, 76 samples to PPMS, 76 samples belonging to healthy subjects of control group. The attributes of the control group are data regarding *brain stem (MRI 1), corpus collasum periventricular*

(MRI 2), upper cervical (MRI 3) lesion diameter size (milimetric [mm]) in the MR image and EDSS score. Data are made up of a total of 112 attributes. It is known that using these attributes of 304 individuals the data if they belong to the MS subgroup or healthy group is known. How can we make the classification as to which MS patient belongs to which subgroup of MS, including healthy individuals and those diagnosed with MS (based on the lesion diameters (*MRI 1, MRI 2, MRI 3*), number of lesion size for (*MRI 1, MRI 2, MRI 3*) as obtained from MRI images and EDSS scores)? D matrix has a dimension of (304×112). This means D matrix includes the MS dataset of 304 individuals along with their 112 attributes (see Table 2.12 for the MS dataset. For the classification of D matrix through Naive Bayesian algorithm the first-step training procedure is to be employed. For the training procedure, 66.66% of the D matrix can be split for the training dataset (203×112) and 33.33% as the test dataset (101×112). For MS Dataset, we would like to develop a linear model for MS patients and healthy individuals whose data are available. Following the classification of the training dataset being trained with Naive Bayesian algorithm, we can do the classification of the test dataset.

The procedural steps of the algorithm may seem complicated initially. Yet, the only thing one will have to do is to focus on the steps and comprehend them. For this purpose, we can inquire into the steps presented in Figure 7.4.

In order to be able to do classification through Bayesian classifier algorithm, the test data has been selected randomly from the MS Dataset.

Let us now think for a while about how to classify the MS dataset provided in Table 2.12 in accordance with the steps of Bayesian classifier algorithm (Figure 7.3.)

Based Figure 7.4, let us calculate the analysis of MS Dataset according to Bayesian classifier step by step in Example 7.4.

Example 7.3 Our sample dataset (see Table 7.3 below) has been chosen from MS dataset (see Table 2.12). Let us calculate the probability as to which class (*RRMS, SPMS, PPMS, Healthy*) an individual (with $x_1 = EDSS: 6,\ x_2 = MRI\ 2 2: 4,\ x_3 = Gender$: *Female* on Table 7.5) will fall under Naive Bayesian classifier algorithm.

Solution 7.3 In order to calculate the probability as to which class (*RRMS, SPMS, PPMS, Healthy*) an individual (with the following attributes [with $x_1 = EDSS: 6$, $x_2 = MRI\ 2: 4,\ x_3 = Gender: Female$]) attributes will belong to, let us apply the Naive Bayes algorithm in Figure 7.4.

Step (1) Let us calculate the probability of which class (*RRMS, SPMS, PPMS and Healthy*) the individuals with *EDSS, MRI 2 and Gender* attributes in MS sample dataset in Table 7.6 will belong to.

The probability of class to belong to for each attribute is calculated in Table 7.6. Classification of the individuals is labeled as $C_1 =$ *RRMS*, $C_2 =$ *SPMS*, $C_3 =$ *PPMS*, $C_4 =$ *(Healthy on Number, Gender)* belonging to which class (*RRMS, SPMS, PPMS,*

Algorithm: Naive Bayesian Classifier Algorithm for MS Data Set

Input: MS Training set including *203* number of samples $D = [x_1, x_2, \ldots, x_{203}](x_i \in R^d)$
From *120* attributes, correspondingly, $A_1, A_2, \ldots, A_{120}$.
Test data including *101* number of samples $T = [x_1, x_2, \ldots, x_{101}](x_i \in R^d)$

Result: Label data that corresponds to *304* samples C = [PPMS, RRMS, SPMS, Healthy]

Method:

(1) //***for*** each attribute $j \neq i${
(2) //$P(C_i|X)$ value is calculated for all the classes
(3) ***if*** $((P(C_i|X) > P(C_j|X))$
(4) $P(C_i|X) = \frac{P(X|C_i)P(C_i)}{P(X)}$
(5) //maximum posterior probability conditioned on ***MS Data Set***
(6) ***for*** each train data // predicted class label is the class C_i (*C = [PPMS, RRMS, SPMS, Healthy]*)
(7) $P(X|C_i) = \prod_{k=1}^{203} P(x_k|C_i) = P(x_1|C_i) \times P(x_2|C_i) \times \ldots \times P(x_{203}|C_i)$
(8) $\arg\max P(X|C_i)$
(9) //Predict each test sample's class

Figure 7.4: Naive Bayesian classifier algorithm for the analysis of the MS Dataset.

Table 7.5: Selected sample from the MS Dataset.

Individual (ID)	EDSS	MRI 2	Gender	Class
1	6	10	*Male*	SPMS
2	7	11	*Male*	PPMS
3	3.5	15	*Female*	RRMS
4	5.5	7	*Male*	SPMS
5	6	7	*Male*	PPMS
6	6	10	*Female*	RRMS
7	0	0	*Female*	Healthy
8	3	22	*Female*	SPMS

Healthy has been calculated through $P(X|C_1)P(C_1)$, $P(X|C_2)P(C_2)$, $P(X|C_3)P(C_3)$, $P(X|C_4)P(C_4)$ (see eq. 7.2). The steps for this probability can be found as follows:

Step (2–7) (a) $P(X|RRMS)P(RRMS)$, (b) $P(X|SPMS)P(SPMS)$, (c) $P(X|PPMS)P(PPMS)$ and (d) $P(X|Healthy)P(Healthy)$ calculation for the probability;

(a) $P(X|RRMS)P(RRMS)$ *calculation of the conditional probability*
Here, in order to calculate the conditional probability of $(X|Class = RRMS)$, the probability calculations are done distinctively for the $X = \{x_1, x_2, \ldots, x_n\}$ values.

$$P(EDSS = (Score)7|Class = RRMS) = \frac{1}{2}$$

$$P(MRI2 = (Number)[3-10]|Class = RRMS) = \frac{1}{2}$$

$$P(Gender = Female|Class = RRMS) = \frac{1}{2}$$

Thus, using eq. (7.3) the following result is obtained:

$$P(X|Class = RRMS) = \left(\frac{1}{2}\right)\left(\frac{1}{2}\right)\left(\frac{2}{4}\right) = \frac{1}{8}$$

Now let us also calculate $(Class = RRMS) = \frac{2}{8}$ probability.

Thus, following equation is obtained:

$$P(X|Class = RRMS)P(Class = RRMS) = \left(\frac{1}{8}\right)\left(\frac{2}{8}\right) \cong 0.031$$

The probability of an individual with $x_1 = EDSS$: 6, $x_2 = MRI$ 2: 4, $x_3 = Gender$: *Female* attributes to belong to *SPMS* class has been obtained as approximately 0.031.

(b) *$P(X|SPMS)P(SPMS)$ calculation of conditional probability*:
Here, in order to calculate the conditional probability of $(X|Class = RRMS)$, the probability calculations are done distinctively for the $X = \{x_1, x_2, \ldots, x_n\}$ values.

$$P(EDSS = 7|Class = SPMS) = \frac{1}{3}$$

$$P(MRI2 = (Range)[3-10]|Class = SPMS) = \frac{1}{4}$$

$$P(Gender = Female|Class = SPMS) = \frac{1}{4}$$

Thus, using eq. (7.3) following result is obtained:

$$P(X|Class = SPMS) = \left(\frac{1}{3}\right)\left(\frac{1}{4}\right)\left(\frac{1}{4}\right) = \frac{1}{48}$$

Now let us also calculate $(Class = SPMS) = \frac{3}{8}$ probability.

Thus, following equation is obtained:

Table 7.6: Sample MS Dataset conditional probabilities.

Attributes	Range of Attributes	Class							
		RRMS		SPMS		PPMS		Healthy	
		Number	Probability	Number	Probability	Number	Probability	Number	Probability
EDSS	6	1	1/2	1	1/2	1	1/2	0	0
	7	0	0	0	0	1	1/2	0	0
	3.5	1	1/2	0	0	0	0	0	0
	5.5	0	0	1	1/2	0	0	0	0
	8	0	0	0	0	0	0	0	0
	0	0	0	0	0	0	0	0	0
	3	0	0	1	1/2	0	0	0	0
	0	0	0	0	0	0	0	1	1
MRI 2	[3–10]	1	1/2	1	1/2	1	1/4	0	0
	[10–22]	2	1/2	2	1/2	1	1/4	0	0
Gender	Male	0	0	2	1/2	2	2/4	0	0
	Female	2	1/2	1	1/2	0	0	1	1/4

$$P(X|Class = SPMS)P(Class = SPMS) = \left(\frac{1}{48}\right)\left(\frac{3}{8}\right) \cong 0.002$$

The probability of an individual with the
$x_1 = EDSS$: 6, $x_2 =$ *MRI* 2: 4, $x_3 =$ *Gender*: *Female* attributes to belong to *SPMS* class has been obtained as approximately 0.002.

(c) The conditional probability for *PPMS* class $P(X|PPMS)P(PPMS)$ has been obtained as 0.
(d) The conditional probability for *Healthy* class $P(X|Healthy)P(Healthy)$ has been obtained as 0.

Step (8–9) Maximum posterior probability conditioned on ***X***. According to maximum posterior probability method, $\arg\max_{C_i} \{P(X|C_i)P(C_i)\}$ value is calculated. Maximum probability value is taken from the results obtained in **Step (2–7)**.

$$\arg\max_{C_i}\{P(X|C_i)P(C_i)\} = \max\{0.031, 0.002, 0, 0\} = 0.031$$

$$x_1 = EDSS\text{: } \mathbf{6}$$

$$x_2 = MRI2\text{: } \mathbf{4}$$

$$x_3 = Gender\text{: } \mathbf{Female}$$

It has been identified that an individual (with the attributes given above) with an obtained probability result of *0.031* belongs to *RRMS* class.

As a result, the data in the *MS Dataset* with 33.33% portion allocated for the test procedure and classified as *RRMS, SPMS, PPMS* and *Healthy* have been classified with an accuracy rate of *52.95%* based on Bayesian classifier algorithm.

7.1.3 Naive Bayesian classifier algorithm for the analysis of mental functions

As presented in Table 2.19, the WAIS-R dataset has 200 data belonging to patient and 200 samples belonging to healthy control group. The attributes of the control group are data on *School Education, Age, Gender and D.M.* Data comprised a total of 21 attributes. The data on the group they belong to (patient or healthy group) is known using these attributes of 400 individuals. Given this, how would it be possible for us to make the classification as to which individual belongs to which patient or healthy individuals and ones diagnosed with WAIS-R test (based on the *School Education, Age, Gender and D.M*)? *D* matrix dimension is 400 × 21. This means *D* matrix consists of the WAIS-R dataset with 400 individuals as well as their 21 attributes (see Table 2.19) for the WAIS-R dataset. For the classification of *D* matrix through

Bayesian algorithm, the first step in the training procedure is made use of 66.66% of the D matrix can be split for the training dataset (267 × 21), and 33.33% as the test dataset (133 × 21) for the training procedure.

Following the classification of the training dataset being trained with Bayesian algorithm, we classify the test dataset. As mentioned above, the steps about the procedures of the algorithm may seem complicated at first hand; however, it would be of help to concentrate on the steps and develop an understanding of them. Let us look into the steps elicited in Figure 7.5.

Although the procedural steps of the algorithm may seem complicated at first glance. The only thing you have to do is to concentrate on the steps and grasp them. For this, let us have a close look at the steps provided in Figure 7.5.

Let us do the *WAIS-R Dataset* analysis step by step analysis according to Bayesian classifier in Example 7.4 as based on Figure 7.5.

Example 7.4 Our sample dataset (see Table 7.7 below) has been chosen from WAIS-R Dataset (see Table 2.19). Let us calculate the probability as to which class (*Patient, Healthy*) an individual with

$x_1: School\ education = Primary\ School,\ x_2: Age = Old, x_3: Gender = Male$ attributes (Table 7.7) will fall under Naive Bayes classifier algorithm.

Solution 7.4 In order to calculate the probability as to which class (*Patient, Healthy*) an individual with the following attributes

$x_1: School\ Education = Primary\ School,\ x_2: Age = Old,\ x_3: Gender: Male$ will belong to, let us apply the Naive Bayes algorithm in Figure 7.5.

Step (1) Let us calculate the probability of which class (*Patient, Healthy*) the individuals with *School education, Age, Gender* attributes in WAIS-R sample dataset in Table 7.2 will belong to.

The probability of class to belong to for each attribute is calculated in Table 7.2. Classification of the individuals is labeled as C_1 = *Patient*, C_2 = *Healthy*. The probability for each attribute (*School education, Age, Gender*) belonging to which class (*Patient, Healthy*) has been calculated through $P(X|C_1)P(C_1)$ and $P(X|C_2)P(C_2)$, (see eq. 7.2). The steps for this probability can be found as follows:

Step (2–7) (a) $P(X|Patient)P(Patient)$, (b) $P\ (X|Healthy)P(Healthy)$ calculation for the probability:

(a) $P(X|Patient)P(Patient)$ calculation of conditional probability:
Here, in order to calculate the conditional probability of $(X|Class = Patient)$, the probability calculations are done distinctively for the $X = \{x_1, x_2, \ldots, x_n\}$ values.

$$P(School\,education = Primary\,School|Class = Patient) = \frac{1}{5}$$

Table 7.7: Selected sample from the Economy Dataset.

Individual (ID)	School education	Age	Gender	Class
1	Junior high School	Old	Male	Patient
2	Primary School	Young	Male	Healthy
3	Bachelor	Middle	Female	Healthy
4	Junior high School	Middle	Male	Patient
5	Primary School	Middle	Male	Patient
6	Bachelor	Old	Female	Patient
7	Primary School	Young	Female	Healthy
8	Junior high School	Middle	Female	Patient

Algorithm: Naive Bayesian Classifier Algorithm for WAIS-R Data Set

Input: MS Training set including *267* number of samples $D = [x_1, x_2, \ldots, x_{267}](x_i \in R^d)$
From *21* attributes, correspondingly, $A_1, A_2, \ldots, A_{21}$.
Test data including *133* number of samples $T = [x_1, x_2, \ldots, x_{133}](x_i \in R^d)$

Result: Label data that corresponds to *400* samples $C =$ [Patient, Healthy]

Method:

//***for*** each attribute $j \neq i$ {
    //$P(C_i|X)$value is calculated for all the classes
    if $((P(C_i|X) > P(C_j|X))$
        $P(C_i|X) = \frac{P(X|C_i)P(C_i)}{P(X)}$
    //maximum posterior probability conditioned on ***WAIS-R Data Set***
    for each train data // predicted class label is the class C_i
    ($C =$ [Patient, Healthy])
        $P(X|C_i) = \prod_{k=1}^{267} P(x_k|C_i) = P(x_1|C_i) \times P(x_2|C_i) \times \ldots \times P(x_{188}|C_i)$
    arg max $P(X|C_i)$
    //Predict each test sample's class

Figure 7.5: Naïve Bayesian classifier algorithm for the WAIS-R Dataset analysis.

$$P(Age = Old|Class = Patient) = \frac{2}{5}$$

$$P(Gender = Male|Class = Patient) = \frac{3}{5}$$

Thus, using eq. (7.3) following result is obtained:

$$P(X|Class = Patient) = \left(\frac{1}{5}\right)\left(\frac{2}{5}\right)\left(\frac{3}{5}\right) = \frac{6}{125}$$

Now, let us calculate $P(Class = Patient) = \frac{5}{8}$ probability as well.

Thus, the following result is obtained:

$$P(X|Class = Patient)P(Class = Patient) = \left(\frac{6}{125}\right)\left(\frac{5}{8}\right) \cong 0.03$$

It has been identified that an individual with the

x_1 : *School education* = *Primary School*, x_2 : *Age* = *Old*, x_3 : Gender: Male attributes has a probability result of about 0.03 belonging to *Patient* class.

(b) $P(X|Healthy)P(Healthy)$ calculation of conditional probability:
Here, in order to calculate the conditional probability of $(X|Class = Healthy)$, the probability calculations are done distinctively for the $X = \{x_1, x_2, \ldots, x_n\}$ values.

$$P(School\ Education = Primary\ School|Class = Healthy) = \frac{2}{3}$$

$$P(Age = Old|Class = Healthy) = 0$$

$$P(Gender = Male|Class = Healthy) = \frac{1}{3}$$

Thus, using eq. (7.3) the following result is obtained:

$$P(X < Class = Healthy) = \left(\frac{2}{3}\right)(0)\left(\frac{1}{3}\right) = 0$$

Now, let us calculate $P(Class = Healthy) = \frac{3}{8}$ probability as well.

Thus, the following result is obtained:

$$P(X|Class = Healthy)P(Class = Healthy) = (0)\left(\frac{3}{8}\right) = 0$$

It has been identified that an individual with the

x_1 : *School education* = *Primary School*, x_2 : *Age* = *Old*, x_3 : *Gender* : *Male* attributes has a probability result of about 0 belonging to *Healthy* class.

Step (8–9) Maximum posterior probability conditioned on ***X***. According to maximum posterior probability method, $\arg\max_{C_i}\{P(X|C_i)P(C_i)\}$ value is calculated.

Maximum probability value is taken from the results obtained in **Step (2–7).**

$$\arg\max_{C_i}\{P(X|C_i)P(C_i)\} = \max\{0.03, 0\} = 0.03$$

x_1 : *School education* = **Primary School**

$$x_2 : Age \ = \textbf{Old}$$

$$x_3 : Gender \ = \textbf{Male}$$

It has been identified that an individual (with the attributes given above) with an obtained probability result of *0.03* belongs to *Patient* class.

How can we classify the WAIS-R Dataset in Table 2.19 in accordance with the steps of Bayesian classifier algorithm (Figure 7.5).

As a result, the data in the WAIS-R dataset with 33.33% portion allocated for the test procedure and classified as patient and healthy have been classified with an accuracy rate of *56.93%* based on *Bayesian classifier algorithm.*

References

[1] Kelleher JD, Brian MN, D'Arcy A. Fundamentals of machine learning for predictive data analytics: Algorithms, worked examples, and case studies. USA: MIT Press, 2015.

[2] Hristea, FT. The Naïve Bayes Model for unsupervised word sense disambiguation: Aspects concerning feature selection. Berlin Heidelberg: Springer Science&Business Media, 2012.

[3] Huan L, Motoda H. Computational methods of feature selection. USA: CRC Press, 2007.

[4] Aggarwal CC, Chandan KR. Data clustering: Algorithms and applications. USA: CRC Press, 2013.

[5] Nolan D, Duncan TL. Data science in R: A case studies approach to computational reasoning and problem solving. USA: CRC Press, 2015.

[6] Han J, Kamber M, Pei J, Data mining Concepts and Techniques. USA: The Morgan Kaufmann Series in Data Management Systems, Elsevier, 2012.

[7] Kubat M. An Introduction to Machine Learning. Switzerland: Springer International Publishing, 2015.

[8] Stone LD, Streit RL, Corwin TL, Bell KL. Bayesian multiple target tracking. USA: Artech House, 2013.

[9] Tenenbaum J. B., Thomas LG. Generalization, similarity, and Bayesian inference. *Behavioral and Brain Sciences*, USA, 2001, 24(4), 629–640.

[10] Spiegelhalter D, Thomas A, Best N, Gilks W, BUGS 0.5: Bayesian inference using Gibbs sampling. *MRC Biostatistics Unit, Institute of Public Health, Cambridge,* UK, 1996, 1–59.

[11] Simovici D, Djeraba C. Mathematical Tools for Data Mining. London: Springer, 2008.

[12] Carlin BP, Thomas AL. Bayesian empirical Bayes methods for data analysis. USA: CRC Press, 2000.

[13] Hamada MS, Wilson A, Reese SC, Martz H. Bayesian Reliability. New York: Springer Science & Business Media, 2008.

[14] Carlin BP, Thomas AL. Bayesian methods for data analysis. USA: CRC Press, 2008.

[15] http://www.worldbank.org/

8 Support vector machines algorithms

This chapter covers the following items:
- Algorithm for support vectors
- Algorithm for classifying linear and nonlinear
- Examples and applications

Support vector machines (SVMs) are used as a method for classifying linear and nonlinear data. This chapter discusses on how SVM works. It basically utilizes a nonlinear mapping for converting the original training data into a higher dimension in which it finds the linear optimal separating hyperplane (a "decision boundary" splitting the datasets of a class from another one). If an appropriate nonlinear mapping is used with an adequately high dimension, data from two classes can at all times be split by a hyperplane [1–5]. Through the use of *support vectors* ("essential" training datasets), SVM finds these hyperplane and *margins* (these are defined by the support vectors).

Vladimir Vapnik and his colleagues Bernhard Boser and Isabelle Guyon worked on a study regarding SVMs in 1992. This proves to be the first paper in this subject. In fact, SVMs have been existing since 1960s but the groundwork was made by Vapnik and his colleagues. SVMs have become an appealing option recently. The training time pertaining to even the fastest SVMs can be slow to a great extent; this may be counted as a drawback. However, the accuracy is their superior feature. They have high level of accuracy because of their capability in modeling complex nonlinear decision boundaries. In addition, they are less inclined to overfitting compared to other methods. They also present a compact description of the learned model [5]. You may use SVMs both for numeric prediction and for classification. It is possible to apply SVMs to certain fields such as speaker identification, object recognition as well as handwritten digit recognition. They are also used for the benchmark time-series prediction tests.

8.1 The case with data being linearly separable

SVMs pose a mystery. We can start solving the mystery with a simple case: this is a problem with two classes and the classes are separable in a linear way [6–10]. Dataset D is given as (X_1, y_1), (X_2, y_1), ... $(X_{|D|}, y_{|D|})$: $\boldsymbol{X}_i$ is the set of training datasets with associated class labels, y_i. Each y_i can take one of the two values: they can take +1 or −1 (i.e., $y_i \in \{+1, -1\}$).

https://doi.org/10.1515/9783110496369-008

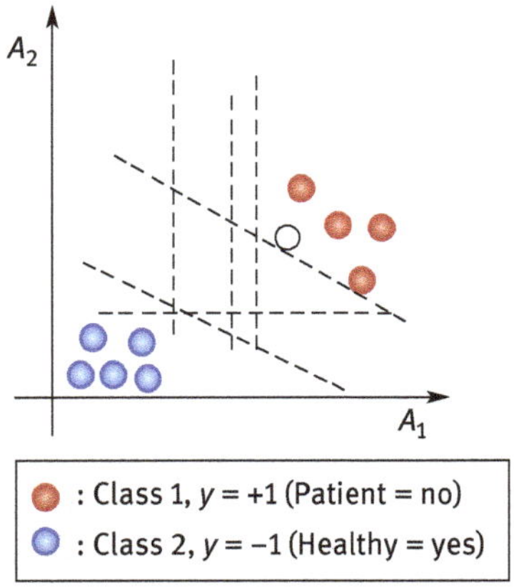

Figure 8.1: 2D training data are separable in a linear fashion. There are an infininte number of possible separating hyperplanes (decision boundaries). Some of these hyperplanes are presented in Figure 8.1 with dashed lines. Which one would be the best one? [11].

This matches with the classes *patient = no* and *healthy = yes as shown in Table 2.19* respectively. Two input attributes: A_1 and A_2, as given in Figure 8.1, the 2D data are *separable in a linear fashion.* The reason is that you can draw a straight line to separate all the class datasets C_1 from all the datasets of class −1.

Drawing the number of separating lines is infinite. The "best" one refers to the one being aspired to have the minimum classification error on earlier unseen datasets. To find the best line, one would like to find the best possible separating *plane* (for data that are 3D attributes). For n dimensions, researchers would wish to find the best *hyperplane.* In this book, *hyperplane* refers to the decision boundary being sought irrespective of the number of input attributes. So, in other words, how can we find the best hyperplane? For the aim of finding the best hyperplane, an approach related to SVM can deal with this concern by looking for the *maximum marginal hyperplane* (MMH) [11]. Figure 8.2 presents the two possible separating hyperplanes and the associated margins thereof. Before getting into the definition of margins, we can have an intuitive glance at this figure. In this figure, it can be seen that both the hyperplanes can classify all the given datasets in a correct way. It is intituitively anticipated that hyperplane that has the larger margin is more accurate for the classification procedure of the future datasets compared to the hyperplane that has a smaller margin. Because of this difference, SVM performs a search for the hyperplane that has the largest margin or the MMH during the learning or training phase. The largest separation between classes is given by the associated margin.

Figure 8.3 presents the flowchart steps for SVM algorithm.

Acccording to the binary (linear) support vector machine algorithm in Figure 8.4, the following can be stated:

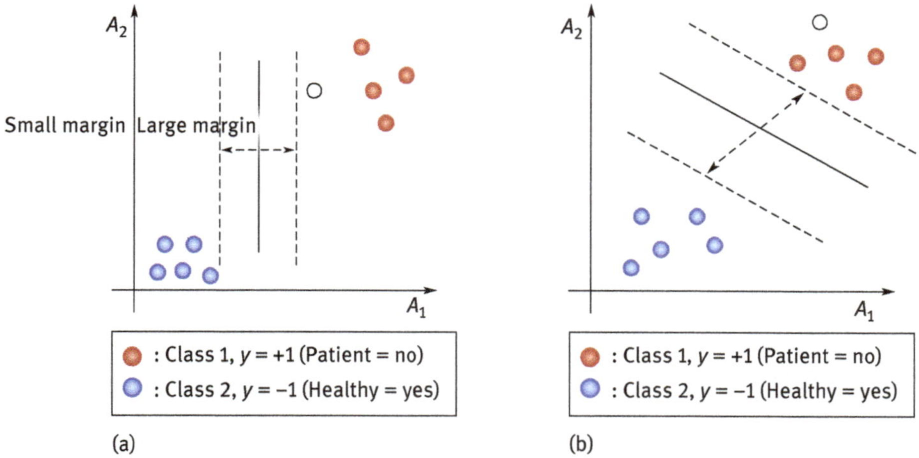

Figure 8.2: In these figures, we can see the two possible separating hyperplanes along with their associated margins. Which one would be a better one? Is it the one with the larger margin(b) that should have a greater generalization accuracy? [11].

Algorithm: Support Vector Machine (Binary SVM or Linear SVM)

Input: Training set including m number of samples $D = [x_1, x_2, ..., x_m](x_i \in R^d)$
Test data including t number of samples $T = [x_1, x_2, ..., x_t](x_k \in R^d)$

Result: label data that corresponds to n number of samples $y_i \in \{+1, -1\}$.

Method:

(1) ***while*** each sample i of belonging to the training set ***do*** {
(2) ***while*** each class j of belonging to the training set ***do*** {
(3) // Calculate H matrix
(4) $H_{ij} = y_i y_j x_i x_j$
(5) $a_i \geq 0 \forall_i$ and $\sum_{i=1}^{l} a_i y_i = 0$
(6) // provided that the conditions are fulfilled a value is found via quadratic programming aid that would maximize the equation in the 6th step.
(7) $\sum_{i=1}^{l} a_i y - \frac{1}{2} a^T H_a$
(8) $w = \sum_{i=1}^{l} a_i y_i x_i$ {
(9) // Calculation of support vectors
(10) ***if***$(a_i > 0)$ {
(11) //Support vectors that are obtained are assigned to the S variable in the 11th step
(12) $b = \frac{1}{N_S} = \sum_{s \in S}(y_s - \sum_{m \in S} a_m y_m x_m . x_s)$// w and b values are obtained.
(13) // class label of the test data is identified y'
(14) $y' = \text{sgn}(w.X' + b)$
(15) }

Figure 8.3: Linear SVM algorithm flowchart.

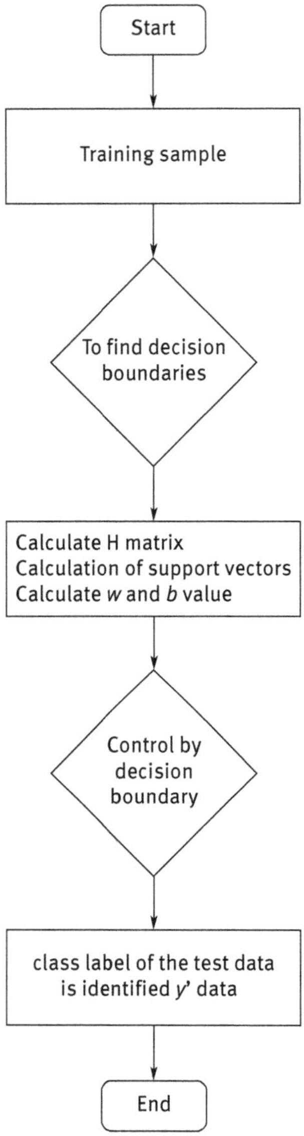

Figure 8.4: General binary (linear) support vector machine algorithm.

Step (1–6) The shortest distance from a hyperplane to one side of the margin of the hyperplane has the same value as the shortest distance from the hyperplane to the other side of its margin, the "sides" of the margin being parallel to the hyperplane. As for MMH, this distance is the shortest distance from the MMH to the closest training dataset of one class or the other one.

The denotation of separation of hyperplane can be done as follows:

$$W.X + b = 0 \tag{8.1}$$

According to this denotation, W is a weight vector, which means $W = \{w_1, w_2, \ldots, w_n\}$, n the number of attributes and b is a scalar (frequently called bias) [11–13]. If you consider input attributes A_1 and A_2, in Figure 8.2(b), training datasets are 2D (e.g., $X = (x_1, x_2)$), x_1 and x_2 are the values of attributes A_1 and A_2, respectively for X. If b is considered an additional weight, w_0, eq. (8.1) can be written as follows:

$$w_0 + w_1x_1 + w_2x_2 = 0 \tag{8.2}$$

Based on this, any point above the separating hyperplane fulfills the following criteria:

$$w_0 + w_1x_1 + w_2x_2 < 0 \tag{8.3}$$

Correspondingly, any point below the separating hyperplane fulfills the following criteria:

$$w_0 + w_1x_1 + w_2x_2 > 0 \tag{8.4}$$

It is possible to attune the weights to make the hyperplanes that define the sides of the margin to be written as follows:

$$H_1 = w_0 + w_1x_1 + w_2x_2 \geq 1 \, for \, y_i = +1 \tag{8.5}$$

$$H_2 = w_0 + w_1x_1 + w_2x_2 \leq 1 \, for \, y_i = -1 \tag{8.6}$$

Any dataset that lies on or above H_1 belongs to class +1, and any dataset that lies on or below H_2 belongs to class −1. Two inequalities of eqs. (8.5) and (8.6) are combined and the following is obtained:

$$y_i(w_0 + w_1x_1 + w_2x_2) \geq 1, \ \forall_i \tag{8.7}$$

All training datasets on hyperplanes H_1 or H_2, the "sides" that define the margin fulfill the conditions of eq. (8.2). These are termed as *support vectors*, and they are close to the separating MMH equally [14]. Figure 8.4 presents the support vectors in a way that is bordered with a bolder (thicker) border. Support vectors are basically regraded as the datasets that prove to be hardest to be classified and yield the most information for the classification process.

It is possible to obtain a formula for the maximal marginal size. The distance from the separating hyperplane to any point on H_1 is $\frac{1}{\|W\|}$:$\|w\|$, which is the Euclidean norm of W, and $\sqrt{WW}$. If $W = \{w_1, w_2, \ldots, w_n\}$, then $\sqrt{W.W} = \sqrt{w_1{}^2 + w_2{}^2 + \ldots + w_n{}^2}$. This is equal to the distance from any point on H_2 to the separating hyperplane, thereby $\frac{2}{\|W\|}$ is the maximal margin (Figure 8.5).

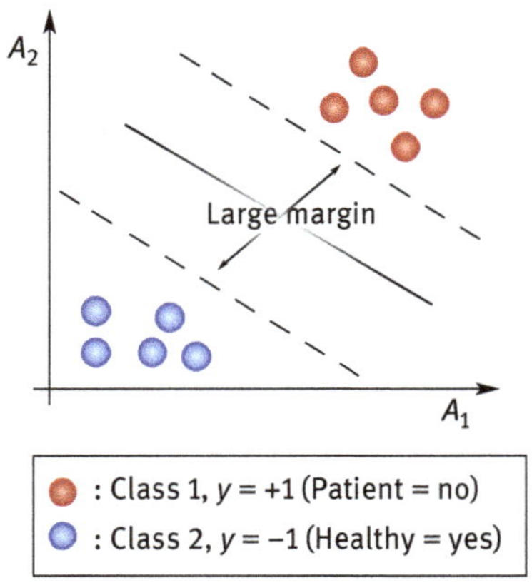

Figure 8.5: Support vectors: the SVM finds out the maximum separating hyperplane, which means the one with the maximum distance between the nearest training dataset. The support vectors are shown with a thicker border [11].

Step (7–15) Following these details, you may ask how an SVM finds the MMH and the support vectors. For this purpose, eq. (8.7) can be written again using certain tricks in maths. This turns into a concept known as constrained (convex) quadratic optimization problem in the field. Readers who would like to research further can resort to some tricks (which are not the scope of this book) using a Lagrangian formulation as well as solution that employs Karush–Kuhn–Tucker (KKT) conditions [11–14]. If your data are small with less than 2,000 training datasets, you can use any of the optimization software package to be able to find the support vectors and MMH. If you have larger data, then you will have to resort to more efficient and specific algorithms for the training of SVMs. Once you find the support vectors and MMH, you get a trained SVM. The MMH is a linear class boundary. Therefore, the corresponding SVM could be used to categorize data that are separable in a linear fashion. This is trained as *linear SVM*. The MMH can be written again as the decision boundary in accordance with the Lagrangian formulation:

$$d(X^T) = \sum_{i=1}^{l} y_i a_i X_i X^T + b_0 \tag{8.8}$$

Here, y_i is the class label of support vector X_i; X^T is a test dataset; $a_i X_i X^T$ and b_0 are numeric parameters automatically determined through SVM algorithm or optimization. Finally, l is the number of support vectors.

For the test dataset, X^T, you can insert it into eq. (8.8), and do a checking for seeing the sign of the result, which shows you the side of the hyperplane that the test dataset falls. If you get a positive sign, this is indicative of the fact that X^T falls on or above the MMH with the SVM predicting that X^T belongs to class +1 (representing *patient=no*, in our current case). For a negative side, X^T falls on or below the MMH with the class prediction being −1 (representing *healthy= yes*).

We should make note of two significant issues at this point. First of all, the complexity of the learned classifier is specified by the number of support vectors and not by the data dimensionality. It is due to this fact that SVMs are less inclined to overfitting compared to other methods. We see support vectors as the critical training datasets lying at the position closest to the decision boundary (MMH). If you remove all the other training datasets and repeat the training, you could find the same separating hyperplane. You can also use the number of support vectors found so that you can make the computation of an upper bound on the expected error rate of the SVM classifier. This classifier is not dependent of the data dimensionality. An SVM that has a small number of support vectors gives decent generalization despite the high data dimensionality.

8.2 The case when the data are linearly inseparable

The previous section has dealt with linear SVMs for the classification of data that are separable linearly. In some instances, we face with data that are not linearly separable (as can be seen from Figure 8.6). When the data cannot be separated in a linear way, no straight line can be found to separate the classes. As a result, linear SVMs would not be able to offer an applicable solution. However, there is a solution for this situation because linear SVMs can be extended to create *nonlinear SVMs* for the classification of *linearly inseparable data*, which is also named as *nonlinearly separable data*, or *nonlinear data*). These SVMs are thus able to find boundaries of nonlinear decision (nonlinear hypersurfaces) in the input space. How is then the linear

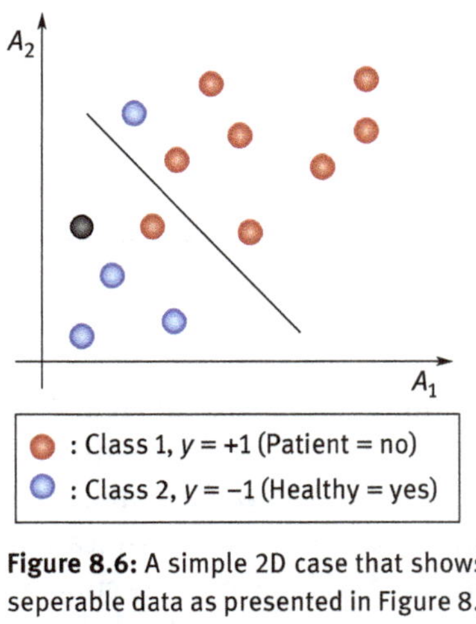

Figure 8.6: A simple 2D case that shows data that are inseperable linearly. Unlike the linear seperable data as presented in Figure 8.1, in this case, drawing a straight line seperating the classes is not a possible action. On the contrary, the decision boundary is nonlinear [11].

approach extended? When the approach for linear SVMs is extended, a nonlinear SVM is to be obtained [13–15]. For this, there are two basic steps. As the initial step, by using a nonlinear mapping, the original input data are transformed into a higher dimensional space. The second step follows after the transformation into new higher space. In this subsequent step, a search for a linear separating hyperplane starts in the new space. As a result, there is again the quadratic optimization problem. The solution of this problem is reached when you use the linear SVM formulation. The maximal marginal hyperplane that is found in the new space matches up with a nonlinear separating hypersurface in the original space.

The algorithmic flow in nonlinear SVM algorithm is presented in Figure 8.7.

Algorithm: Support Vector Machine (Nonlinear SVM)

Input: Training set including m number of samples $D = [x_1, x_2, ..., x_m](x_i \in R^d)$
Test data including t number of samples $T = [x_1, x_2, ..., x_t](x_k \in R^d)$

Result: label data that corresponds to n number of samples $C = [c_1, c_2, ..., c_m]$

Method:

(1) ***while*** each sample i of belonging to the training set ***do*** {
(2) ***while*** each class j of belonging to the training set ***do*** {
(3) // Calculate H matrix
(4) $H_{ij} = y_i y_j \phi(x_i)\phi(x_j)$
(5) $a_i \geq 0 \forall_i$ and $\sum_{i=1} a_i y_i = 0$
(6) // provided that the conditions are fulfilled a value is found via quadratic programming aid that would maximize the equation in the 6th step.
(7) $\sum_{i=1}^{l} a_i y - \frac{1}{2} a^T H_a$
(8) $w = \sum_{i=1}^{l} a_i y_i \phi(x_i)$ {
(9) // Calculation of support vectors
(10) ***if***$(a_i > 0)$ {
(11) // Support vectors that are obtained are assigned to the S variable in the 11th step
(12) $b = \frac{1}{N_S} = \sum_{s \in S}(y_s - \sum_{m \in S} a_m y_m \phi(x_m).\phi(x_s))$// w and b values are obtained.
(13) // class label of the test data is identified y'
(14) $y' = \mathrm{sgn}(w.\phi(X') + b)$
(15) }

Figure 8.7: General binary (*nonlinear*) support vector machine algorithm.

Step (1–5) As can be seen in Figure 8.7, this is not without problems, though. One problem is related to the fact how the nonlinear mapping to a higher dimensional space is chosen. The second one is related to the fact that the relevant computation involved in the process is a costly one. You may resort to eq. (8.8) for the classification of a test dataset X^T. The dot product should be calculated with each of the support vectors. The dot product of two vectors are as follows: $X^T = (x_1^T, x_2^T, ..., x_n^T)$and

$X_i = (x_{i1}, x_{i2}, ..., x_{in})$ is $x_1^T x_{i1} + x_2^T x_{i2} + ... + x_n^T x_{in}$. In training, a similar dot product should be calculated a number of times if you are to find the MMH, which would end up being an expensive approach. Hereafter, the dot product computation required is very heavy and costly. In order to deal with this situation, another trick would be of help, and there exists another trick in maths, fortunately!

Step (6–15) It occurs as in solving the quadratic optimization problem regarding the linear SVM (while looking for a linear SVM in the new higher dimensional space); the training datasets come out only in the dot product forms, $\phi(X_i)\phi(X_j)$ at this point $\phi(X)$is merely the nonlinear mapping function [16–18]. This function is applied for the transformation of the training datasets. Rather than computing the dot product on the transformed datasets, it happens to be mathematically similar to administering a kernel function, $K(X_i, X_j)$ to the original input data:

$$K(X_i, X_j) = \phi(X_i)\phi(X_j) \tag{8.9}$$

In each place that $\phi(X_i).\phi(X_j)$ is seen in the training algorithm, it can be replaced by $K(X_i, X_j)$. Thus, you will be able to do all of your computations in the original input space that has the potential of having a significantly lower dimensionality. Owing to this, you can avoid the mapping in a safe manner while you do not even have to know what mapping is. You can implement this trick and later move on to finding a maximal separating hyperplane. This procedure resembles the one explained in Section 8.1 but it includes the placing of a user-specified upper bound, C, on the Lagrange multipliers, a_i. Such upper bound is in the best way determined in an experimental way. Another relevant question is about some of the kernel functions that can be used. Kernel functions can allow the replacement of the dot product scenario. Three permissible kernel functions are given below [17–19]:

Polynomial kernel of degree h: $K(X_i, X_j) = (X_i.X_j + 1)^h$

Gaussian radial basis function kernel: $K(X_i, X_j) = e^{-\|X_i - X_j\|^2/2\sigma^2}$

Sigmoid kernel: $K(X_i, X_j) = \tanh(kX_i, X_j - \delta)$

All of these three kernel functions have results in a different nonlinear classifier in the original input space. Upholders of neural network would think that the resulting decision hyperplanes found for nonlinear SVMs is the same as the ones that have been found by other recognized neural network classifiers. As an example, SVM with a Gaussian radial basis function (RBF) gives the same decision hyperplane as a neural network kind, known as a RBF network. SVM with a sigmoid kernel is comparable to a simple two-layer neural network, also termed as a multilayer perceptron that does not have any hidden layers. There exist no unique rules available to determine which admissible kernel will give rise to the most accurate SVM. During the practice process, the chosen particular kernel does not usually make a substantial difference about the subsequent accuracy [18–19]. It is known that there is always a finding of a global solution by SVM training (as opposed to neural networks like back propagation).

The linear and nonlinear SVMs have been provided for binary (two-class) classification up until this section. It is also possible to combine the SVM classifiers for the multi-class case. About SVMs, an important aim of research is to enhance the speed for the training and testing procedures to render SVMs a more applicable and viable choice for very big sets of data (e.g., millions of support vectors). Some other concerns are the determination of the best kernel for a certain dataset and exploration of more efficient methods for cases that involve multi-classes.

Example 8.1 shows the nonlinear transformation of original data into higher dimensional space. Please have a look at this presented example: A *3D* input vector $X = (x_1, x_2, x_3)$ is mapped into 6D space. By using Z space for the mapping, $\varphi_1(X) = x_1$, $\varphi_2(X) = x_2$, $\varphi_3(X) = x_3$, $\varphi_4(X) = (x_1)^2$, $\varphi_5(X) = x_1x_2$, $\varphi_6(X) = x_1x_3$. A decision hyperplane in the new space is $d(Z) = WZ + b$. In this formulation, W and Z are vectors, which is supposed to be linear. W and b are solved and afterwards substituted back in order to allow the linear decision hyperplane in the new (Z) space to correspond to a nonlinear second-order polynomial in the original *3D* input space:

$$d(Z) = w_1x_1 + w_2x_2 + w_3x_3 + w_4(x_1)^2 + w_5x_1x_2 + w_6x_1x_3 + b$$

$$= w_1z_1 + w_2z_2 + w_3z_3 + w_4z_4 + w_5z_5 + w_6z_6 + b$$

8.3 SVM algorithm for the analysis of various data

SVM algorithm has been applied on economy (U.N.I.S.) dataset, multiple sclerosis (MS) dataset and WAIS-R dataset in Sections 8.3.1, 8.3.2 and 8.3.3, respectively.

8.3.1 SVM algorithm for the analysis of economy (U.N.I.S.) Data

As the second set of data, we will use in the following sections some data that are related to economies for U.N.I.S. countries. The attributes of economies of these countries are data regarding *years, unemployment youth male(% of male labor force ages 15–24) (national estimate), gross value added at factorcost ($), tax revenue and gross domestic product (GDP) growth (annual %)*. Economy (U.N.I.S.) Dataset is made up of a total of 18 attributes. Data belonging to economies of USA, New Zealand, Italy and Sweden from 1960 to 2015 are defined on the basis of attributes given in Table 2.8 Economy (U.N.I.S.) Dataset (http://data.worldbank.org) [15], to be used in the following sections for the classification of D matrix through SVM algorithm. The first-step training procedure is to be employed. For the training procedure, 66.66% of the D matrix can be split for the training dataset (151×18), and 33.33% as the test dataset (77×18).

Following the classification of the training dataset being trained with SVM algorithm, we can classify the test dataset. Although the procedural steps of the algorithm may seem complicated at first glance, the only thing you have to do is to concentrate on the steps and grasp them. For this, let us have a look at the steps provided in Figure 8.8 closely.

Algorithm: Support Vector Machine (Nonlinear SVM)

Input: Training set including 151number of samples $D = [x_1, x_2, ..., x_{151}](x_i \in R^d)$
Test data including 77 number of samples $T = [x_1, x_2, ..., x_{77}](x_k \in R^d)$

Result: label data that corresponds to 228 number of samples
$C =$ [USA, New Zealand, Italy, Sweden]

Method:

(1) ***while*** each sample i of belonging to the training set **do** { // $i = 1,...,151$
(2) ***while*** each class j of belonging to the training set **do** {// $j = 1, 2, 3, 4$
(3) // Calculate H matrix
(4) $H_{ij} = y_i y_j \phi(x_i)\phi(x_j)$
(5) $a_i \geq 0 \forall_i$ and $\sum_{i=1}^{l} a_i y_i = 0$
(6) // provided that the conditions are fulfilled a value is found via quadratic programming aid that would maximize the equation in the 6th step.
(7) $\sum_{i=1}^{l} a_i y - \frac{1}{2} a^T H_a$
(8) $w = \sum_{i=1}^{l} a_i y_i \phi(x_i)$ {
(9) // Calculation of support vectors
(10) ***if***$(a_i > 0)$ {
(11) // Support vectors that are obtained are assigned to the S variable in the 11th step
(12) $b = \frac{1}{N_S} = \sum_{s \in S}(y_s - \sum_{m \in S} a_m y_m \phi(x_m).\phi(x_s))$//$w$ and b values are obtained.
(13) // class label of the test data is identifiedy′
(14) $y' = \text{sgn}(w.\phi(X') + b)$
(15) }

Figure 8.8: SVM algorithm for the analysis of the economy (U.N.I.S.) Dataset.

In order to do classification with multi-class SVM algorithm, the test data is randomly chosen as 33.3% from the Economy (U.N.I.S) Dataset.

Economy (U.N.I.S.) Dataset has four classes. Let us analyze the example given in Example 8.2 in order to understand how the classifiers are found for the Economy dataset being trained by multi-class SVM algorithm.

Example 8.2

USA: $X_1 = \begin{vmatrix} -1 \\ 1 \end{vmatrix}; X_2 = \begin{vmatrix} 1 \\ -1 \end{vmatrix}$

New Zealand: B: $X_3 = \begin{vmatrix} -1 \\ -1 \end{vmatrix};\ X_4 = \begin{vmatrix} 1 \\ 1 \end{vmatrix}$

Since y = $\begin{vmatrix} 1 \\ 1 \\ -1 \\ -1 \end{vmatrix}$, let us find the classifier of this sample problem.

The aspects regarding the vectors presented in Example 8.2 are defined as in Figure 8.8.

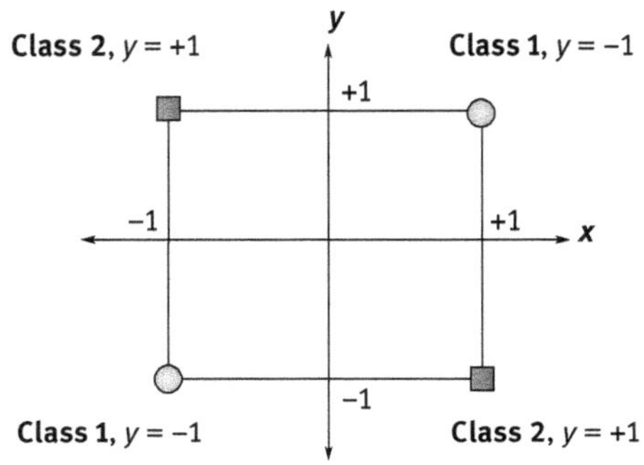

Let us rewrite the $n = 4$ relation in eq. (8.8):

Step (1–5)

$L(a) = \sum_{i=1}^{n} a_i - \frac{1}{2} a^T H_a$, here we see $H_{ij} = y_i y_j (K(X_i, X_j)$. Since the second-degree polynomial function is $K(X_i, X_j) = (X_i.X_j + 1)^2$, the matrix can be calculated as follows:

$$H_{11} = y_1 y_2 (1 + X_1^T X_2)^2 = (1)(1)\left(1 + [1 - 1]\begin{bmatrix} -1 \\ 1 \end{bmatrix}\right)^2 = 1$$

$$H_{12} = y_1 y_2 (1 + X_1^T X_1)^2 = (1)(1)\left(1 + [1 - 1]\begin{bmatrix} 1 \\ -1 \end{bmatrix}\right)^2 = 9$$

If all the H_{ij} values are calculated in this way, the following matrix is obtained:

$$H = \begin{bmatrix} 1 & 9 & -1 & -1 \\ 9 & 1 & -1 & -1 \\ -1 & -1 & 1 & 9 \\ -1 & -1 & 9 & 1 \end{bmatrix}$$

Step (8–17) In this stage, in order to obtain the α value, it is essential to solve the $1 - H_a = 0$ and $\begin{bmatrix} 1 \\ 1 \\ 1 \\ 1 \end{bmatrix} - \begin{bmatrix} 1 & 9 & -1 & -1 \\ 9 & 1 & -1 & -1 \\ -1 & -1 & 1 & 9 \\ -1 & -1 & 9 & 1 \end{bmatrix} a = 0$. As a result of the solution, $a_1 = a_2 = a_3 = a_4 = 0.125$ is obtained. This means all the samples are accepted as support vectors. These results obtained fulfill the condition of $\sum_{i=1}^{4} a_i y_i = a_1 + a_2 - a_3 - a_4 = 0$ in eq. (8.8).

The $\phi(x)$ corresponding for the second-degree $K(X_i, X_j) = (1 + X_i^T X_j)$ is calculated as $Z = \begin{bmatrix} 1 & \sqrt{2}x_1\sqrt{2}x & \sqrt{2}x_1x & x_1 & x \end{bmatrix}$ based on eq. (8.8).

In this case, in order to find the w weight vector, the calculation is performed as follows: $w = \sum_{i=1}^{4} a_i y_i(\phi(x_i))$

For economies of USA and New Zealand, $X_1 = \begin{vmatrix} 1 \\ -1 \end{vmatrix}$; $X_2 = \begin{vmatrix} -1 \\ 1 \end{vmatrix}$; $X_3 = \begin{vmatrix} 1 \\ 1 \end{vmatrix}$; $X_4 = \begin{vmatrix} -1 \\ -1 \end{vmatrix}$ and $y = \begin{vmatrix} 1 \\ 1 \\ -1 \\ -1 \end{vmatrix}$, w vector is calculated as follows:

$$\begin{aligned} w &= (0.125)\{(1)\,[1\ -\sqrt{2}\sqrt{2} - \sqrt{2}\ 1\ 1] + (1)\,[1\ \sqrt{2} - \sqrt{2} - \sqrt{2}\ 1\ 1] \\ &\quad + (-1)[1 - \sqrt{2} - \sqrt{2}\sqrt{2}\ 1\ 1] + (-1)\,[1\ \sqrt{2}\sqrt{2}\sqrt{2}\ 1\ 1]\} \\ &= (0.125)\begin{bmatrix} 0 & 0 & 0 & -4\sqrt{2} & 0 & 0 \end{bmatrix} \\ &= \begin{bmatrix} 0 & 0 & 0 & \frac{-\sqrt{2}}{2} & 0 & 0 \end{bmatrix} \end{aligned}$$

Step (6)
Such being the case, the classifier is obtained as follows:

$$= \begin{bmatrix} 0 \\ 0 \\ 0 \\ \frac{-\sqrt{2}}{2} \\ 0 \\ 0 \end{bmatrix} \begin{bmatrix} 1 & \sqrt{2}x_1 & \sqrt{2}x_2 & \sqrt{2}x_1x_2 & x_1^2 & x_2^2 \end{bmatrix} = -x_1x_2$$

The test accuracy rate has been obtained as *95.2%* in the classification procedure of Economy (U.N.I.S.) dataset with four classes through *SVM polynomial kernel* algorithm.

8.3.2 SVM algorithm for the analysis of multiple sclerosis

As presented in Table 2.12, multiple sclerosis (MS) dataset has data from the following groups: 76 samples belonging to RRMS, 76 samples to SPMS, 76 samples to PPMS and 76 samples (see chapter 2, Table 2.11) belonging to healthy subjects of control group.

The attributes of the control group are data regarding *brain stem (MRI 1), corpus collasum periventricular (MRI 2), upper cervical (MRI 3) lesion diameter size (milimetric [mm]) in the MR image and EDSS* score. Data are made up of a total of 112 attributes. It is known that using these attributes of 304 individuals as the data if they belong to the MS subgroup or healthy group is known. How can we make the classification as to which MS patient belongs to which subgroup of MS, including healthy individuals and those diagnosed with MS (based on the lesion diameters [*MRI 1, MRI 2, MRI 3*]), the number of lesion size for (*MRI 1, MRI 2, MRI 3*) as obtained from MRI images and EDSS scores)? D matrix has a dimension of (*304 × 112*). This means D matrix includes the MS dataset of 304 individuals along with their 112 attributes (see Table 2.12 for the MS dataset). For the classification of D matrix through SVM algorithm, the first-step training procedure is to be employed. For the training procedure, 66.66% of the D matrix can be split for the training dataset (203 × 112) and 33.33% as the test dataset (*101 × 112*).

Following the classification of the training dataset being trained with SVM algorithm, we can do the classification of the test dataset.

33.3% portion has been selected from the MS dataset randomly in order to do the classification with multi-class SVM algorithm. Test data was randomly selected from the MS dataset as 33.3%.

The MS dataset has four classes. Let us study the example given in Example 8.2 in order to understand how the classifiers are found through the training of the MS dataset with multi-class SVM algorithm.

Example 8.3

SPMS: $X_1 = \begin{vmatrix} 1 \\ -1 \end{vmatrix}; X_2 = \begin{vmatrix} -1 \\ 1 \end{vmatrix}$

RRMS: $X_3 = \begin{vmatrix} 1 \\ 1 \end{vmatrix};\ X_4 = \begin{vmatrix} -1 \\ -1 \end{vmatrix}$

As we see that y= $\begin{vmatrix} 1 \\ 1 \\ -1 \\ -1 \end{vmatrix}$, let us find the classifier of the sample problem.

The aspects regarding the vectors presented in Example 8.3 can be seen in Figure 8.9.

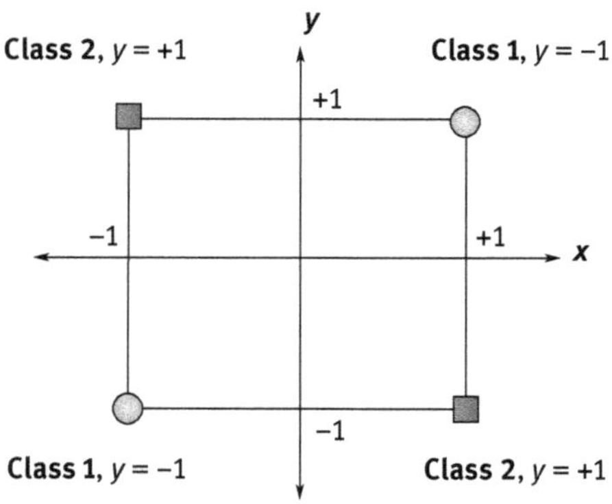

Algorithm: Support Vector Machine (Nonlinear SVM)

Input: Training set including *203* number of samples $D = [x_1, x_2, ..., x_{203}](x_i \in R^d)$

Test data including *101* number of samples $T = [x_1, x_2, ..., x_{101}](x_k \in R^d)$

Result: label data that corresponds to *304* number of samples C = [PPMS, RRMS, SPMS, Healthy]

Method:

(1) ***while*** each sample i of belonging to the training set ***do*** { // $i = 1,...,203$
(2) ***while*** each class j of belonging to the training set ***do*** {//$j = 1, 2, 3, 4$
(3) // Calculate H matrix
(4) $H_{ij} = y_i y_j \phi(x_i)\phi(x_j)$
(5) $a_i \geq 0 \forall_i$ and $\sum_{i=1}^{l} a_i y_i = 0$
(6) // provided that the conditions are fulfilled a value is found via quadratic programming aid that would maximize the equation in the 6th step.
(7) $\sum_{i=1}^{l} a_i y - \frac{1}{2} a^T H_a$
(8) $w = \sum_{i=1}^{l} a_i y_i \phi(x_i)$ {
(9) // Calculation of support vectors
(10) ***if***$(a_i > 0)$ {
(11) // Support vectors that are obtained are assigned to the S variable in the 11th step (12)
(12) $b = \frac{1}{N_S} = \sum_{s \in S}(y_s - \sum_{m \in S} a_m y_m \phi(x_m).\phi(x_s))$// w and b values are obtained.
(13) // class label of the test data is identified y'
(14) $y' = \text{sgn}(w.\phi(X') + b)$
(15) }

Figure 8.9: SVM algorithm for the analysis of the MS Dataset.

Let us rewrite $n = 4$ relation in eq. (8.8):

$L(a) = \sum_{i=1}^{n} a_i - \frac{1}{2} a^T H_a$, here we see $H_{ij} = y_i y_j (K(X_i, X_j)$. The second-degree polynomial function is $K(X_i, X_j) = (X_i.X_j + 1)^2$, the matrix can be calculated as follows:

Step (1–7)

$$H_{11} = y_1 y_2 (1 + X_1^T X_1)^2 = (1)(1)\left(1 + [1 - 1]\begin{bmatrix} 1 \\ -1 \end{bmatrix}\right)^2 = 9$$

$$H_{12} = y_1 y_2 (1 + X_1^T X_2)^2 = (1)(1)\left(1 + [1 - 1]\begin{bmatrix} -1 \\ 1 \end{bmatrix}\right)^2 = 1$$

If all the H_{ij} values are calculated in this way, the matrix provided below will be obtained:

$$H = \begin{bmatrix} 9 & 1 & -1 & -1 \\ 1 & 9 & -1 & -1 \\ -1 & -1 & 9 & 1 \\ -1 & -1 & 1 & 9 \end{bmatrix}$$

Step (8–15) In order to get the a value at this stage, it is essential to solve these systems: $1 - H_a = 0$ and $\begin{bmatrix} 1 \\ 1 \\ 1 \\ 1 \end{bmatrix} - \begin{bmatrix} 9 & 1 & -1 & -1 \\ 1 & 9 & -1 & -1 \\ -1 & -1 & 9 & 1 \\ -1 & -1 & 1 & 9 \end{bmatrix} a = 0$. As a result of the solution, we obtain $a_1 = a_2 = a_3 = a_4 = 0.125$. Such being the case, all the samples are accepted as the support vectors. These results obtained fulfill the condition presented in eq. (8.8), which is the condition of $\sum_{i=1}^{4} a_i y_i = a_1 + a_2 - a_3 - a_4 = 0$.

In order to find the weight vector w, $w = \sum_{i=1}^{4} a_i y_i(\phi(x_i))$ is to be calculated in this case.

$$w = (0.125)\Big\{(1)\ \left[1\,\sqrt{2} - \sqrt{2} - \sqrt{2}\ 1\ 1\right] + (1)\ \left[1 - \sqrt{2}\sqrt{2} - \sqrt{2}\ 1\ 1\right]$$

$$+(-1)\left[1\,\sqrt{2}\sqrt{2}\sqrt{2}\ 1\ 1\right] + ((-1)\left[1 - \sqrt{2} - \sqrt{2}\sqrt{2}\ 1\ 1\right]\Big\}$$

$$= (0.125)\begin{bmatrix} 0 & 0 & 0 & -4\sqrt{2} & 0 & 0 \end{bmatrix}$$

$$= \begin{bmatrix} 0 & 0 & 0 & \frac{-\sqrt{2}}{2} & 0 & 0 \end{bmatrix}$$

In this case, the classifier is obtained as follows:

$$= \begin{bmatrix} 0 \\ 0 \\ 0 \\ \frac{-\sqrt{2}}{2} \\ 0 \\ 0 \end{bmatrix} \begin{bmatrix} 1 & \sqrt{2}x_1 & \sqrt{2}x_2 & \sqrt{2}x_1x_2 & x_1^2 & x_2^2 \end{bmatrix} = -x_1x_2$$

The test accuracy rate has been obtained as *88.9%* in the classification procedure of the MS dataset with four classes through *SVM polynomial kernel* algorithm.

8.3.3 SVM algorithm for the analysis of mental functions

As presented in Table 2.19, the WAIS-R dataset has data 200 samples belonging to *patient* and 200 samples to *healthy* control group. The attributes of the control group are data regarding *school education, gender and D.M* (see Chapter 2, Table 2.18). Data are made up of a total of 21 attributes. It is known that using these attributes of 400 individuals, the data if they belong to patient or healthy group are known. How can we make the classification as to which individual belongs to which patient or healthy individuals and those diagnosed with WAIS-R test(based on the *school education, gender and D.M*)?D matrix has a dimension of 400 × 21. This means D matrix includes the WAIS-R dataset of 400 individuals along with their 21 attributes (see Table 2.19) for the WAIS-R dataset. For the classification of D matrix through SVM the first-step training procedure is to be employed. For the training procedure, 66.66% of the D matrix can be split for the training dataset (*267 × 21*), and 33.33% as the test dataset (*133 × 21*).

Following the classification of the training dataset being trained with SVM algorithm, we can do the classification of the test dataset (Figure 8.10)

Algorithm: Support Vector Machine (Binary SVM or Linear SVM))

Input: Training set including *267* number of samples $D = [x_1, x_2, ..., x_{267}](x_i \in R^d)$

Test data including *133* number of samples $T = [x_1, x_2, ..., x_{133}](x_k \in R^d)$

Result: label data that corresponds to *400* number of samples$y_i \in \{+1, -1\}$//*Patient, Healthy*

Method:

(1) ***while*** each sample i of belonging to the training set ***do*** { //$i = 1, ..., 267$
(2) ***while*** each class j of belonging to the training set ***do*** {// $j = 1,2,3,4$
(3) // Calculate H matrix
(4) $H_{ij} = y_i y_j x_i x_j$
(5) $a_i \geq 0 \forall_i$ and $\sum_{i=1}^{l} a_i y_i = 0$
(6) // provided that the conditions are fulfilled a value is found via quadratic programming aid that would maximize the equation in the 6th step.
(7) $\sum_{i=1}^{l} a_i y - \frac{1}{2} a^T H_a$
(8) $w = \sum_{i=1}^{l} a_i y_i x_i$ {
(9) // Calculation of support vectors
(10) ***if***$(a_i > 0)$ {
(11) // Support vectors that are obtained are assigned to the S variable in the 11th step
(12) $b = \frac{1}{N_S} = \sum_{s \in S}(y_s - \sum_{m \in S} a_m y_m x_m . x_s)$// w and b values are obtained.
(13) // class label of the test data is identified y'
(14) $y' = \text{sgn}(w.X' + b)$
(15) }

Figure 8.10: Binary (linear) support vector machine algorithm for the analysis of WAIS-R.

In order to do classification with linear SVM algorithm, the test data is randomly chosen as 33.3% from the WAIS-R dataset.

WAIS-R dataset has two classes. Let us analyze the example given in Example 8.4 in order to understand how the classifiers are found for WAIS-R dataset being trained by linear (binary) SVM algorithm.

Example 8.4 For the WAIS-R dataset that has *patient* and *healthy* classes, for *patient* class with **(0,0)(0,1)** values and *healthy* class with (1,1) values, let us have a function that can separate these two classes from each other in a linear fashion:

The vectors relevant to this can be defined in the following way:

$$x_1 = \begin{bmatrix} 0 \\ 0 \end{bmatrix}, \; x_2 = \begin{bmatrix} 0 \\ 1 \end{bmatrix}, \; x_3 = \begin{bmatrix} 1 \\ 1 \end{bmatrix}, \; y = \begin{bmatrix} 1 \\ 1 \\ -1 \end{bmatrix}$$

Lagrange function was in the pattern as presented in eq. (8.8). If the given values are written in their place, the calculation of Lagrange function can be done as follows:

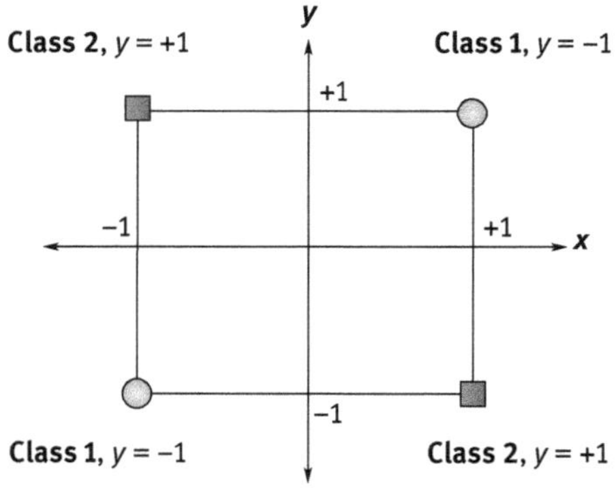

Step (1–5)

$$d(X^T) = a_1 + a_2 + a_3 - \tfrac{1}{2}(a_1 a_1 y_1 y_1 {x_1}^T x_1 + a_1 a_2 y_1 y_2 {x_1}^T x_2 + a_1 a_3 y_1 y_3 {x_1}^T x_3$$

$$+ a_2 a_1 y_2 y_1 {x_2}^T x_1 + a_2 a_2 y_2 y_2 {x_2}^T x_2 + a_1 a_3 y_2 y_3 {x_2}^T x_3$$

$$+ a_3 a_1 y_3 y_1 {x_3}^T x_1 + a_3 a_2 y_3 y_2 {x_3}^T x_2 + a_3 a_3 y_3 y_3 {x_3}^T x_3)$$

Since $x_1 = \begin{bmatrix} 0 \\ 0 \end{bmatrix}$ we can write the denotation mentioned above in the way presented as follows:

$$d(X^T) = a_1 + a_2 + a_3 - \frac{1}{2}(a_2^2 y_2^2 x_2^T x_2 + a_2 a_3 y_2 y_3 x_2^T x_3 + a_3 a_2 y_3 y_2 x_3^T x_2 + a_3^2 y_3^2 x_3^T x_3)$$

$$d(X^T) = a_1 + a_2 + a_3 - \frac{1}{2}\left(a_2^2(1)^2[0 \quad 1]\begin{bmatrix}0\\1\end{bmatrix} + a_2a_3(1)(-1)[0 \quad 1]\begin{bmatrix}0\\1\end{bmatrix}\right.$$

$$\left. + a_3a_2(-1)(1)[1 \quad 1]\begin{bmatrix}0\\1\end{bmatrix} + a_3^2(-1)^2[1 \quad 1]\begin{bmatrix}1\\1\end{bmatrix}\right)$$

$$d(X^t) = a_1 + a_2 + a_3 - \frac{1}{2}\{a_2^2 - 2a_2a_3 + 2a_3^2\}$$

Step (6–11) However, since $k\sum_{i=1}^{3} y_i a_i = 0$ from $a_1 + a_2 - a_3 = 0$ relation, $a_1 + a_2 = a_3$ is obtained. If this value is written in its place in $d(X^T)$, in order to find the values of $d(X^T) = 2a_3 - \frac{1}{2}\{a_2^2 - 2a_2a_3 + 2a_3^2\}a_2$ and a_3 the derivatives of the functions are taken and equaled to zero.

$$\frac{\partial d(X^T)}{\partial a_2} = -2a_2 + 2a_3 = 0$$

$$\frac{\partial d(X^T)}{\partial a_3} = 2 + a_2 - 2a_3 = 0$$

If these two last equations are solved, $a_2 = 2, a_3 = 2$ is obtained. $a_1 = 0$ is obtained as well. This means, it is possible to do the denotation as $a = \begin{bmatrix}0\\2\\2\end{bmatrix}$. Now, we can find the w and b values.

$$w = a_2y_2x_2 + a_3y_3x_3$$

$$= 2(1)\begin{bmatrix}0\\1\end{bmatrix} + 2(-1)\begin{bmatrix}1\\1\end{bmatrix} = \begin{bmatrix}0\\2\end{bmatrix} - \begin{bmatrix}2\\2\end{bmatrix} = \begin{bmatrix}-2\\0\end{bmatrix}$$

$$b = \frac{1}{2}\left(\frac{1}{y_2}x_2^T\begin{bmatrix}-2\\0\end{bmatrix} + \frac{1}{y_3}x_3^T\begin{bmatrix}-2\\0\end{bmatrix}\right)$$

$$= \frac{1}{2}\left(\frac{1}{1} - [0 \quad 1]\begin{bmatrix}-2\\0\end{bmatrix} + \frac{1}{(-1)} - [1 \quad 1]\begin{bmatrix}-2\\0\end{bmatrix}\right) = 1$$

Step (12–15) As a result, $\{x_i\}$ classification for the observation values to be given as new will be as follows:

$$f(x) = \mathrm{sgn}(wx + b)$$

$$= \mathrm{sgn}\left(\begin{bmatrix}-2\\0\end{bmatrix}[x_1 \quad x_2] + 1\right)$$

$$= \mathrm{sgn}(-2x_1 + 1)$$

In this case, if we would want to classify the new $x_4 = \begin{bmatrix} 0 \\ 4 \end{bmatrix}$ observation value, we see that $n(-2x_1+1) > 0$. Therefore, it is understood that the observation in question is in the positive zone (*patient*). For $x_i > 0$, all the observations will be in the negative zone, namely in *healthy* area, which is seen clearly.

In this way, the WAIS-R datasets with two classes (samples) are linearly separable. In the classification procedure of WAIS-R dataset through polynomial SVM algorithm yielded a test accuracy rate of 98.8%.

We can use SVM kernel functions in our datasets (MS Dataset, Economy (U.N.I.S.) Dataset and WAIS-R Dataset) and do the application accordingly. The accuracy rates for the classification obtained can be seen in Table 8.1.

Table 8.1: The classification accuracy rates of SVM kernel functions.

Data Sets	SVM kernel functions		
	Polynominal kernel (%)	**Gaussian kernel (%)**	**Sigmoid kernel (%)**
Economy (U.N.I.S.)	95.2	96.1	91.2
MS	88.9	88.5	89.8
WAIS-R	98.8	99	98.5

As presented in Table 8.1, the classification of datasets such as the WAIS-R data that can be separable linearly through SVM kernel functions can be said to be more accurate compared to the classification of multi-class datasets such as those of MS Dataset and Economy Dataset.

References

[1] Larose DT. Discovering knowledge in data: An introduction to data mining, USA: John Wiley & Sons, Inc., 90–106, 2005.

[2] Schölkopf B, Smola AJ. Learning with kernels: Support vector machines, regularization, optimization, and beyond. USA: MIT press, 2001.

[3] Ma Y, Guo G. Support vector machines applications. New York: Springer, 2014.

[4] Han J, Kamber M, Pei J, Data mining Concepts and Techniques. USA: The Morgan Kaufmann Series in Data Management Systems, Elsevier, 2012.

[5] Stoean, C, Stoean, R. Support vector machines and evolutionary algorithms for classification. Single or Together. Switzerland: Springer International Publishing, 2014.

[6] Suykens JAK, Signoretto M, Argyriou A, Regularization, optimization, kernels, and support vector machines. USA: CRC Press, 2014.

[7] Wang SH, Zhang YD, Dong Z, Phillips P. Pathological Brain Detection. Singapore: Springer, 2018.

[8] Kung SY. Kernel methods and machine learning. United Kingdom: Cambridge University Press, 2014.

[9] Kubat M. An Introduction to Machine Learning. Switzerland: Springer International Publishing, 2015.

[10] Olson DL, Delen D. Advanced data mining techniques. Berlin Heidelberg: Springer Science & Business Media, 2008.

[11] Han J, Kamber M, Pei J, Data mining Concepts and Techniques. USA: The Morgan Kaufmann Series in Data Management Systems, Elsevier, 2012.

[12] Witten IH, Frank E, Hall MA, Pal CJ. Data Mining: Practical machine learning tools and techniques. USA: Morgan Kaufmann Series in Data Management Systems, Elsevier, 2016

[13] Karaca Y, Zhang YD, Cattani C, Ayan U. The differential diagnosis of multiple sclerosis using convex combination of infinite Kernels. *CNS & Neurological Disorders-Drug Targets (Formerly Current Drug Targets-CNS & Neurological Disorders)*, 2017, 16(1), 36–43.

[14] Tang J, Tian Y. A multi-kernel framework with nonparallel support vector machine. *Neurocomputing, Elsevier*, 2017, 266, 226–238.

[15] Abe S. Support vector machines for pattern classification, London: Springer, 2010.

[16] Burges CJ. A tutorial on support vector machines for pattern recognition. *Data Mining and Knowledge Discovery*. 1998, 2 (2), 121–167.

[17] Lipo Wang. Support Vector Machines: Theory and Applications. Berlin Heidelberg: Springer Science & Business Media, 2005.

[18] Smola AJ, Schölkopf B. A tutorial on support vector regression. Statistics and Computing, Kluwer Academic Publishers Hingham, USA, 2004, 14 (3), 199–222.

[19] Hsu CW, Lin CJ. A comparison of methods for multiclass support vector machines. *IEEE Transactions on Neural Networks*, 2002, 13 (2), 415–425.

9 *k*-Nearest neighbor algorithm

This chapter covers the following items:

- Algorithm for computing the distance between the neighboring data
- Examples and applications

The methods used for classification (Bayesian classification, decision tree induction, rule-based classification, support vector machines, classification by back propagation and classification based on association rule mining) mentioned up until this point in our book have been the examples of *eager learners*. By *eager learners*, we mean that there will be a construction of a generalization based on a set of training datasets prior to obtaining new datasets to be classified. The new datasets are also in the category of test. It is called eager because the learned model is ready and willing to do the classification of the datasets that have not been seen previously [1–4]. As opposed to a lazy approach where the learner waits until the last minute before it does any model of construction for the classification of a test dataset, *lazy learner or learning from your neighbors* just carries out a simple storing or very minor processing, and waits until given to a test dataset. Upon seeing the test dataset, it carries out generalization for the classification of the dataset depending on its similarity to the training datasets stored. When we make a contrast with eager learning methods, lazy learners as its name suggests do less work when a training dataset is presented but it is supposed to do more work when it does a numeric estimation or a classification. Since lazy learners store the training datasets, or "instances", we also call them. *instance-based learners*. Lazy learners prove to be expensive in terms of computational issues while they are engaged in numeric estimation of a classification since they need efficient techniques for storage. They also present little explanation for the structure of the data. Another aspect of lazy learners is that they support incremental learning, capable of modeling complex decision spaces that have hyperpolygonal shapes that are not likely to be described easily through other learning algorithms [5].

One of the examples of lazy learners is *k-nearest neighbor classifiers* [6]. The first description of *k-nearest* neighbor method was made in the early 1950s. The method requires intensive work with large training sets. It gained popularity after the 1960s (this was the period when increased power of computation became accessible). As in the 1960s, they have been used extensively in pattern recognition field. Based on learning by analogy, nearest neighbor classifiers include a comparison of a particular test dataset with training datasets that are similar to it. The training datasets are depicted by *n* attributes. Each dataset represents a point in an *n*-dimensional space. Thus, all the training datasets are stored in *n*-dimensional pattern space. If there is a

https://doi.org/10.1515/9783110496369-009

case of an unknown dataset, then a *k-nearest neighbor classifier* will look for a pattern space for the *K training* datasets, which will be nearest/closest to the unknown dataset [7–8]. These *K training* datasets are therefore the *k* “nearest neighbors” of the unknown dataset. We define closeness based on a distance metric similar to the Euclidean distance, which is between two points or datasets: $X_1 = (x_{11}, x_{12}, \ldots, x_{1n})$ and $X_2 = (x_{21}, x_{22}, \ldots, x_{2n})$ that gives the following formula:

$$dist(X_1, X_2) = \sqrt{\sum_{i=1}^{n} (x_{1i} - x_{2i})^2} \tag{9.1}$$

We can explain this procedure as follows: the difference between the corresponding values of the attribute in dataset X_1 and in dataset X_2 is taken for each numeric attribute. The square of this difference is taken and accumulated. From the total accumulated distance count, the square root is taken. The values of each attribute is normalized before the use of eq. (9.1). Thus, this process allows the prevention of attributes that have initially large ranges (e.g., net domestic credit) from balancing attributes with initially smaller ranges (e.g., binary attributes). Min-max normalization can be utilized for transforming a value v of a numeric attribute A to v' in the range [0, 1]:

$$v' = \frac{v - \min_A}{\max_A - \min_A} \tag{9.2}$$

$\min_A$ and $\max_A$ are the minimum and maximum values of attribute A. In *k-nearest* neighbor classification, the most common class is allotted to the unknown dataset among its *k-nearest* neighbors. Nearest neighbor classifiers are also used for numerical prediction to return a real-valued prediction for a particular dataset that is unknown [9–10]. What about the distance to be calculated for attributes that are not numeric but categorical or nominal in nature? For such nominal attributes, comparing the corresponding value of the attribute in dataset with the one in dataset X_2 is used as a simple method. If two of them are identical (e.g., datasets X_1 and X_2 have the EDSS score of 4), the difference between the two is considered and taken as 0. If the two are different (e.g., dataset X_1 is 4. Yet, dataset X_2 is 2), the difference is regarded and taken as 1. As for the missing values, generally the following is to be applied: the maximum possible difference is assumed if the value of a given attribute A is missing in dataset X_1 and/or dataset X_2. The difference value is taken as 1 for nominal attributes if the corresponding values of A is missing or both of them are missing. In an alternative scenario, let us assume that A is numeric and missing from both X_1 and X_2 datasets, then the difference is also taken to be 1. If only one value is missing and the other one that is called v' is existent and normalized, it is possible to take the difference as $|1 - v'|$ or $|0 - v'|(|1 - v'|\text{or } v')$, depending on which one is higher.

Another concern at this point is about determining a good value for k *(the number of neighbors)*. It is possible to do the determination experimentally. You start with $k = 1$, and use a test set in order to predict the classifier's error rate. This is a process which you can repeat each time by incrementing k to permit one more neighbor. You may therefore select the k value that yields the minimum error rate. Generally, when the number of training datasets is large, then the value of k will also be large. Thereby, you base the prediction decisions pertaining to classification and numeric on a larger slice of the stored datasets. Distance-based comparisons are used by nearest neighbor classifiers. Distance-based comparisons essentially give equal weight to each of the attributes. For this reason, it is probable that they may have poor accuracy with irrelevant or noisy attributes. This method has been subjected to certain modification so that one can merge attribute weighting and the pruning of datasets with noisy data. Choosing a distance metric is a crucial decision to make. It is also possible that you use the Manhattan (city block) distance or other distance measurements. It is known that nearest neighbor classifiers are tremendously slow while they are carrying out the classification of test datasets. Let us say D is a training dataset of $|D|$datasets and $k = 1$, and $O(|D|)$ comparisons are needed for the classification of a given test dataset. You can reduce the number of comparisons to $O(\log|D|)$ by arranging the stored datasets into search trees. Doing the application in parallel fashion may allow the reduction of the running time to a constant: $O(1)$ independent of $|D|$. There are some other techniques that can be employed for speeding the classification time. For example, you can make use of *partial distance* calculations and *editing* the stored datasets. In the former one, namely in *partial distance* method, you calculate the distance depending on a subset of the n attributes. In case this is beyond a threshold, then you halt the further computation for the prearranged stored dataset. This process goes on with the next stored dataset. As for the *editing method*, training datasets that end up being useless are removed. In a way, a clearing is made for the unworkable ones, which is also known as *pruning* or *condensing* because of the method reducing the total number of stored datasets [11–13].

The flowchart of k-nearest neighbor algorithm is provided in Figure 9.1.

Below you can find the steps of k-nearest neighbor algorithm (Figure 9.2).

According to this set of steps, each dataset signifies a point in an n-dimensional space. Thereby, all the training datasets are stored in an n-dimensional pattern space. For an unknown dataset, a k-nearest neighbor classifier explores the pattern space for the *K training* datasets that are closest to the unknown dataset with distance as shown in the second step of the flow. The *K training* datasets give us the k "nearest neighbors" of the unknown dataset.

In algorithm Figure 9.2, the step by step progress stages have been provided for the k-nearest neighbor algorithm. This concise summary will show the readers that it is simple to follow this algorithm as the steps will show.

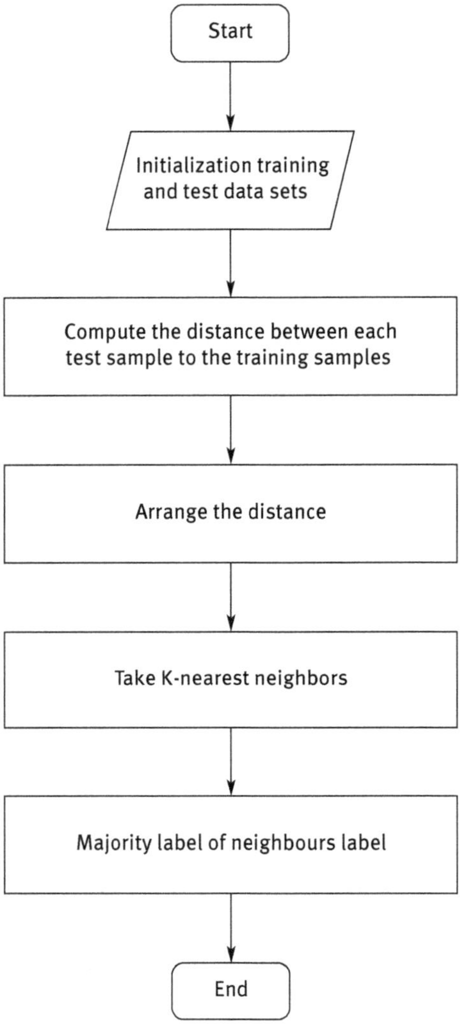

Figure 9.1: *k*-Nearest neighbor algorithm flowchart. From Zhang and Zhou [3].

Identifying the training and dataset to be classified, *D* training dataset that contains *n* number of samples is applied to *k*-nearest neighbor algorithm as the input. The attributes of each sample in *D* training set are represented as $x_1, x_2, \ldots, x_n$.

Compute distance $dist(x_i, x_j)$: Each sample of the test dataset to be classified, the distance to all the samples in the training dataset is calculated according to eq. (9.1). Which training samples the smallest *K unit* belongs to is calculated among the distance vector obtained (*argsort* function denotes the ranking of the test sample according to the distance value regarding the samples) [3]. Take a look at the class label data of the *k* number of neighbors that are closest. Among the label data of the neighbors, the most recurring value is identified as the class of the *x* test data.

Algorithm: K-nearest Neighbor

Input: Training set including n number of samples $D = [x_1, x_2, ..., x_n](x_i \in R^d)$
Test data including k number of samples $T = [x_1, x_2, ..., x_k](x_t \in R^d)$
Result: label data that corresponds to m number of samples $C = [c_1, c_2, ..., c_m]$

Method:

(1) // Identification of training and test data
(2) ***while*** each sample i of belonging to the training set ***do*** {
(3) ***while*** each class j of belonging to the training set ***do*** {
(4) Compute distance $dist(x_i, x_j)$
(5) }
(6) // It is calculated which training samples the smallest K *unit* belongs to among the distance vector obtained
(7) $index = argsort(dist(x_i, x_j))$ the first k one
(8) **return** majority label for $\{x_i \text{ where } i \in index\}$
(9) }

Figure 9.2 General *k*-nearest neighbor algorithm.

9.1 *k*-Nearest algorithm for the analysis of various data

k nearest algorithm has been applied on economy (U.N.I.S.) dataset, multiple sclerosis (MS) dataset and WAIS-R dataset in Sections 9.1.1, 9.1.2 and 9.1.3, respectively.

9.1.1 *k*-Nearest algorithm for the analysis of Economy (U.N.I.S.)

As the second set of data, we will use in the following sections, some data related to U.N.I.S. countries' economies. The attributes of these countries' economies are data regarding *years, unemployed youth male(% of male labor force ages 15–24) (national estimate), gross value added at factorcost ($), tax revenue, gross domestic product (GDP) growth (annual %)*. Economy (U.N.I.S.) Dataset is made up of a total of 18 attributes. Data belonging to economies of countries such as USA, New Zealand, Italy and Sweden from 1960 to 2015 are defined on the basis of attributes given in Table 2.8 Economy (U.N.I.S.) Dataset, to be used in the following sections. For the classification of D matrix through *k*-nearest neighbor algorithm, the first-step training procedure is to be employed. For the training procedure, 66.66% of the D matrix can be split for the training dataset (151×18) and 33.33% as the test dataset (77×18).

How can we find the class of a test data belonging to Economy (U.N.I.S.) dataset provided in Table 2.8 according to *k*-nearest neighbor algorithm? We can a find the class label of a sample data belonging to Economy (U.N.I.S.) dataset through the following steps (Figure 9.3) specified below:

Algorithm: K-nearest Neighbor

Input: Training set including 151 number of samples $D = [x_1, x_2, ..., x_{151}](x_i \in R^d)$
K=3 nearest neighbor to be examined
Test data including 77 number of samples $T = [x_1, x_2, ..., x_{77}](x_k \in R^d)$

Output:
label data that corresponds to *228* number of samples
C = [USA, New Zealand, Italy, Sweden]

Method:
(1) // Identification of training and test data
(2) **while** each sample *151(i)* of belonging to the training set **do** {
(3) **while** each class *77(j)*of belonging to the test set **do** {
(4) Compute distance $dist(x_i, x_j)$
(5) }
(6) // It is calculated which training samples the smallest *K unit* belongs to among the distance vector obtained
(7) $index = argsort(dist(x_i, x_j))$ the first k one
(8) **return** majority label for $\{x_i \text{ where } i \in index\}$
(9) }

Figure 9.3: *k*-Ne arest neighbor algorithm for the analysis of the Economy (U.N.I.S.) dataset.

Step (1) Let us split the bank dataset as follows: 66.66% as the training data ($D = 151 \times 18$) and 33.33% as the test data ($T = 77 \times 18$).

Step (2–3) Let us calculate the distance of the data to be classified to all the samples in accordance with the Euclid distance (eq. 9.1)

$D = (12000, 45000, ..., 54510)$ and $T = (2000, 21500, ..., 49857)$, and the distance between the data is computed according to eq. (9.1).

$$dist(D, T) = \sqrt{\sum_{i=1}^{18} (12000 - 20000)^2 + (45000 - 21500)^2 + ... + (54510 - 49857)^2}$$

Please bear in mind that our Economy dataset comprised 18 individual attributes and we see that it is $(1 \le i \le 18)$

Step (5–9) Let us find which class the smallest *k* number of neighbors belong to with regard to training samples among the distance vector obtained $dist(D, T) =$ $dist(D, T) = [dist(D, T)_1, dist(D, T)_2, ..., dist(D, T)_{18}]$

How can we find the class of a test data belonging to Economy data according to *k*-nearest neighbor algorithm?

We can find the class label of a sample data belonging to Economy dataset through the following steps specified below:

Step (1) The Economy dataset samples are positioned to a 2D coordinate system as provided below: the blue dots represent *USA economy*, the red ones represent *New Zealand economy*, the orange dots represent *Italy economy*, and finally the purple dots show *Sweden economy* classes.

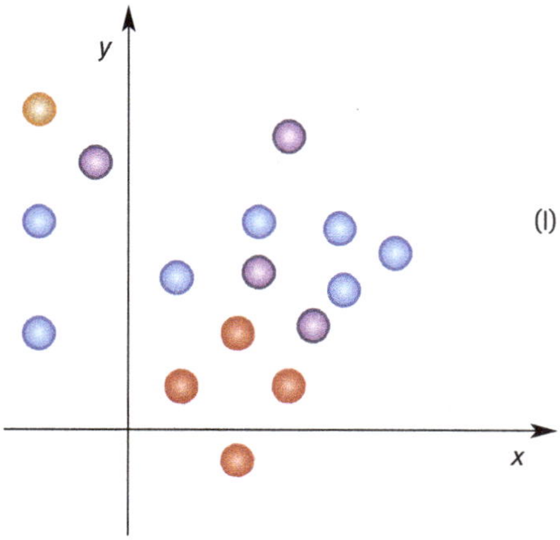

(I)

Step (2) Assume that the test data that we have classified on the coordinate plane where Economy dataset exists corresponds to the green dot. The green dot selected from the dataset of Economy of *USA*, *NewZealand*, *Italy*, *Sweden* and data (whose class is not identified) has been obtained randomly from the dataset, which is the green dot.

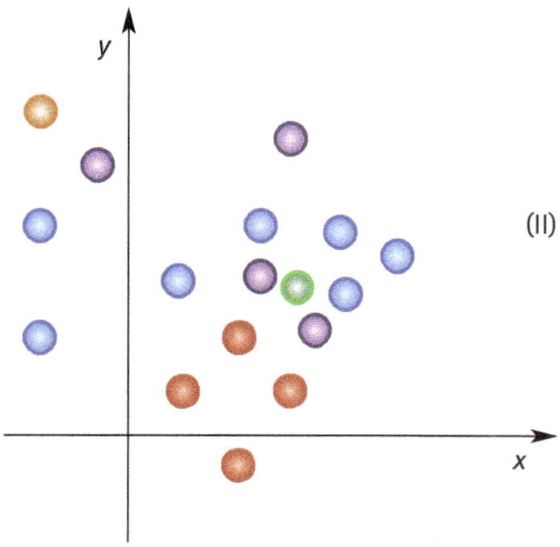

(II)

Step (3) Let us calculate the distances of the new test data to the training data according to Euclid distance formula (eq. 9.1). By ranking the distances calculated, find the three nearest neighbors to the test data.

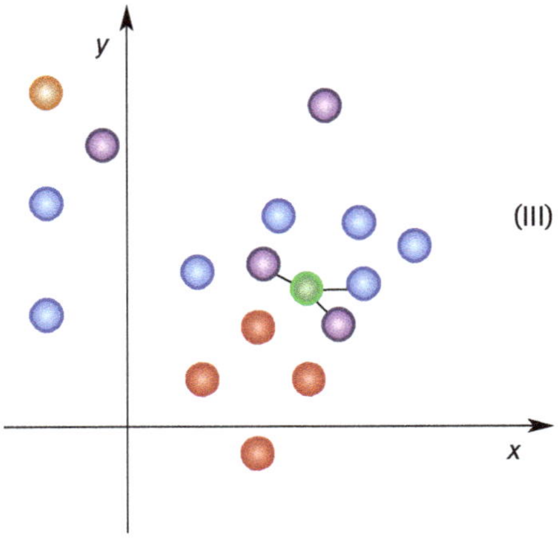

Step (4) Two of the three nearest members belong to purple dot class. Therefore, we can classify the class of our test data as purple (Sweden economy).

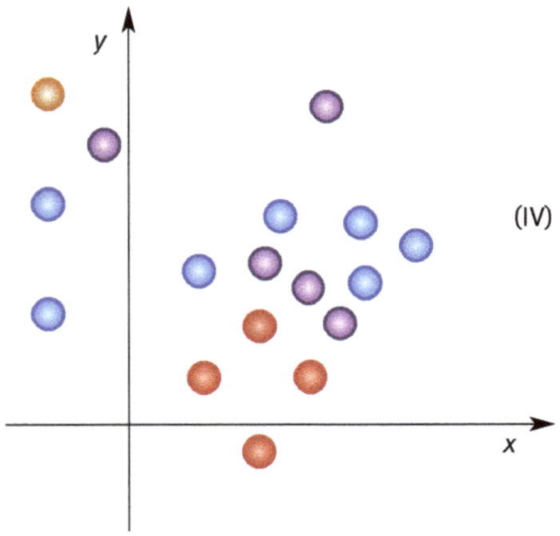

As a result, the data in the Economy dataset with 33.33% portion allocated for the test procedure and classified as USA (U.), New Zealand(N.), Italy(I.) and Sweden(S.) have been classified with an accuracy rate of *84.6%* based on three *nearest neighbor algorithm*.

9.1.2 *k*-Nearest algorithm for the analysis of multiple sclerosis

As presented in Table 2.12, multiple sclerosis dataset has data from the following groups: 76 samples belonging to RRMS, 76 samples to SPMS, 76 samples to PPMS and 76 samples belonging to healthy subjects of control group. The attributes of the control group are data regarding *brain stem (MRI 1), corpus callosum periventricular*

(MRI 2), upper cervical (MRI 3) lesion diameter size (milimetric [mm]) in the MR image and EDSS score. Data are made up of a total of 112 attributes. It is known that using these attributes of 304 individuals the data if they belong to the MS subgroup or healthy group is known. How can we make the classification as to which MS patient belongs to which subgroup of MS (including healthy individuals and those diagnosed with MS (based on the lesion diameters (*MRI 1, MRI 2, MRI 3*), number of lesion size for *MRI 1, MRI 2, MRI 3* as obtained from MRI images and EDSS scores?) D matrix has a dimension of (304×112). This means D matrix includes the MS dataset of 304 individuals along with their 112 attributes (see Table 2.12 for the MS dataset). For the classification of D matrix through Naive Bayesian algorithm the first-step training procedure is to be employed. For the training procedure, 66.66% of the D matrix can be split for the training dataset (203×112) and 33.33% as the test dataset (101×112).

Following the classification of the training dataset being trained with k-nearest algorithm, we can do the classification of the test dataset (Figure 9.4).

Algorithm: K-nearest Neighbor

Input: Training set including *203* number of samples $D = [x_1, x_2, \ldots, x_{203}](x_i \in R^d)$
K=3 nearest neighbor to be examined
Test data including *101* number of samples $T = [x_1, x_2, \ldots, x_{101}](x_t \in R^d)$

Output:
label data that corresponds to *304* number of samples
C = [PPMS, RRMS, SPMS, Healthy]

Method:
(1) // Identification of training and test data
(2) ***while*** each sample *203(i)* of belonging to the training set ***do*** {
(3) ***while*** each class *101(j)*of belonging to the test set ***do*** {
(4) Compute distance $dist(x_i, x_j)$
(5) }
(6) // It is calculated which training samples the smallest *K unit* belongs to among the distance vector obtained
(7) $index = argsort(dist(x_i, x_j))$the first k one
(8) **return** majority label for $\{x_i \text{ where } i \in index\}$
(9) }

Figure 9.4: *k*-Nearest neighbor algorithm for the analysis of the MS dataset.

To be able to classify through k-nearest neighbor algorithm, the test data has been selected randomly from the MS dataset as 33.33%.

Let us now think for a while about how to classify the MS dataset provided in Table 2.12 in accordance with the steps of k-nearest neighbor algorithm (Figure 9.3).

Step (1) Let us identify 66.66% of the MS dataset as the training data ($D = 203 \times 112$) and 33.33% as the test data ($t = 101 \times 112$).

Step (2–3) Let us calculate the distance of the data to be classified to all the samples in accordance with the Euclid distance (eq. 9.1.). $D = (1, 4, ..., 5.5)$ and $T = (2, 0, ..., 7)$, and let us calculate the distance between the data according to eq. (9.1).

$$dist(D, T) = \sqrt{\sum_{i=1}^{120} (1-2)^2 + (4-0)^2 + ... + (5.5-7)^2}$$

Please bear in mind that the MS dataset comprised 112 attributes (attribute) $(1 \le i \le 120)$

Step (5–9) Let us now find which class the smallest k number of neighbors belong to among the distance vector obtained $dist(D, T) = [dist(D, T)_1, dist(D, T)_2, ..., dist(D, T)_{120}]$.

How can we find the class of a test data belonging to the MS dataset according to *k*-nearest neighbor algorithm?

We can find the class label of a sample data belonging to MS dataset through the following steps specified below:

Step (1) The MS dataset samples are positioned to a 2D coordinate system as provided below: the blue dots represent *PPMS*, the red ones represent *SPMS*, the orange dots represent *RRMS*, and purple dots represent the healthy subjects class.

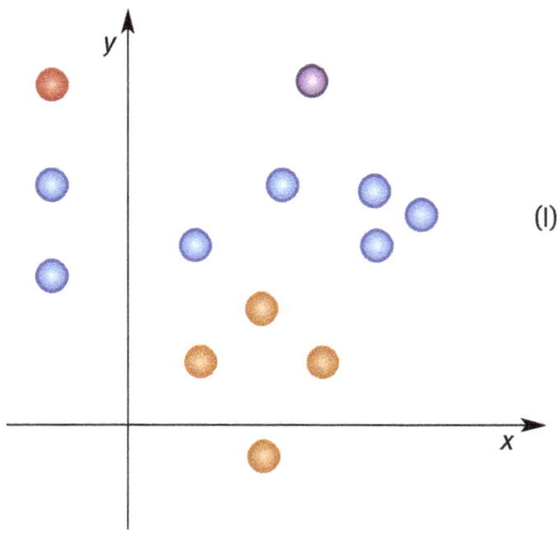

(I)

Step (2) Assume that the test data that we have classified on the coordinate plane where the MS dataset exists corresponds to the green dot. The green dot selected from the dataset of the MS data (*PPMS, RRMS, SPMS*) and data (whose class is not identified) has been gained randomly from the dataset.

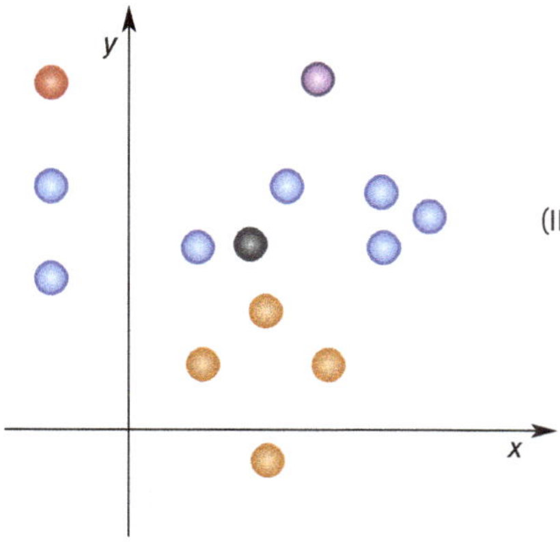

(II)

Step (3) The distances of the subgroups in the other *MS dataset* whose selected class is not identified is calculated in accordance with Euclid distances. The Euclid distances of the test data to the training data are calculated. Let us find the three nearest neighbors by ranking the distances calculated from the smallest to the greatest (unclassified).

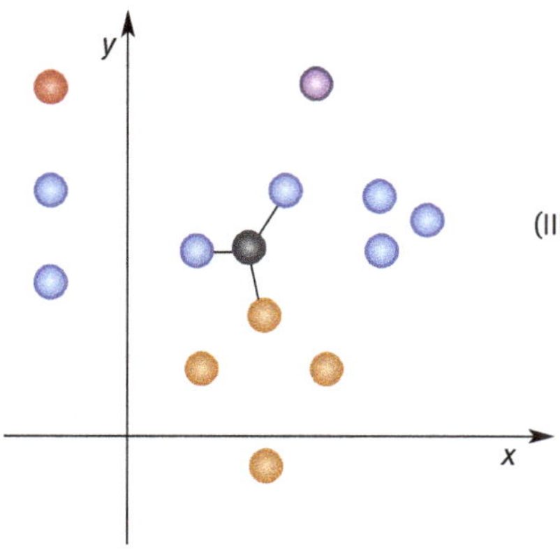

(III)

Step (4) The two of the three nearest neighbors belonging to the blue dot class have been identified according to Euclid distance formula. Based on this result, we identify the class of the test data as blue. The dot denoted in blue color can be classified as *PPMS* subgroup of MS.

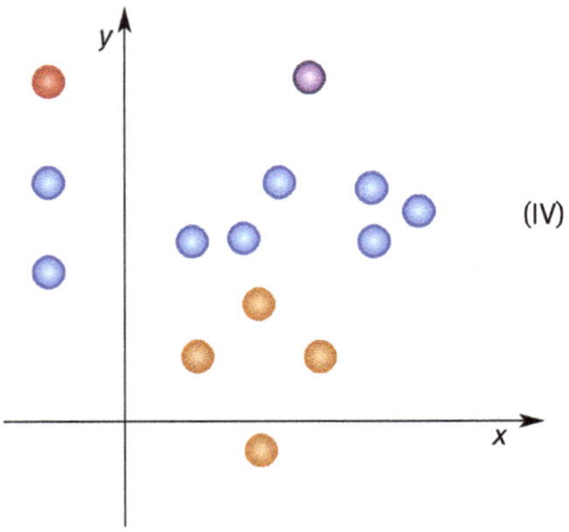

As a result, the data in the MS dataset with 33.33% portion allocated for the test procedure and classified as *RRMS, SPMS, PPMS* and *Healthy* have been classified with an accuracy rate of *68.31%* based on *three nearest neighbor algorithm.*

9.1.3 *k*-Nearest algorithm for the analysis of mental functions

As presented in Table 2.19, the WAIS-R dataset has the following data: 200 belonging to patient and 200 samples belonging to healthy control group. The attributes of the control group are data regarding *School Eeducation, Gender and D.M.* Data are made up of a total of 21 attributes. It is known that using these attributes of 400 individuals, the data if they belong to patient or healthy group are known. How can we make the classification as to which individual belongs to which patient or healthy individuals and those diagnosed with WAIS-R test *(based on the School Education, Gender and D.M)?D* matrix has a dimension of 400×21. This means *D* matrix includes the WAIS-R dataset of 400 individuals along with their 21 attributes (see Table 2.19) for the WAIS-R dataset. For the classification of *D* matrix through *k*-nearest neighbor algorithm, the first-step training procedure is to be employed. For the training procedure, 66.66% of the *D* matrix can be split for the training dataset (267×21) and 33.33% as the test dataset (133×21). Following the classification of the training dataset being trained with *k*-nearest algorithm, we can classify the test dataset. Following the classification of the training dataset being trained with *k*-nearest algorithm, we can classify the test dataset.

How can we classify the WAIS-R dataset in Table 2.19 in accordance with the steps of *k*-nearest neighbor algorithm (Figure 9.5)?

Step (1) Let us identify the 66.66% of WAIS-R dataset as the training data ($D = 267 \times 21$) and 33.33% as the test data ($k = 133 \times 21$).

Algorithm: K-nearest Neighbor

Input: Training set including *267* number of samples $D = [x_1, x_2, ..., x_{267}](x_i \in R^d)$
k=3 nearest neighbor to be examined
Test data including *133*number of samples $T = [x_1, x_2, ..., x_{133}](x_k \in R^d)$

Output:
label data that corresponds to *400* number of samples C = [Healthy, Patient]

Method:
(1) // Identification of training and test data
(2) ***while*** each sample *267(i)* of belonging to the training set ***do*** {
(3) ***while*** each class *133(j)*of belonging to the test set ***do*** {
(4) Compute distance $dist(x_i, x_j)$
(5) }
(6) // It is calculated which training samples the smallest *K unit* belongs to among the distance vector obtained
(7) $index = argsort(dist(x_i, x_j))$ the first k one
(8) **return** majority label for $\{x_i \text{ where } i \in index\}$
(9) }

Figure 9.5: *k*-Nearest neighbor algorithm for the analysis of WAIS-R dataset.

Step (2–3) Let us calculate the distance of the data to be classified to all the samples in accordance with the Euclid distance (eq. 9.1). $D = (1, 1, ..., 1.4)$ and $T = (3, 0, ..., 2.7)$, and let us calculate the distance between the data according to eq. (9.1).

$$dist(D, T) = \sqrt{\sum_{i=1}^{21} (1-3)^2 + (1-0)^2 + ... + (1.4-2.7)^2}$$

Please bear in mind that the WAIS-R dataset comprised 21 attributes as administered to the individuals and we see that $(1 \le i \le 21)$

Step (5–9) Let us now find which class the smallest k number of neighbors belong to among the distance vector obtained $dist(D, T) = [dist(D, T)_1, dist(D, T)_2, ..., dist(D, T)_{21}]$

How can we find the class of a test data belonging to WAIS-R test data according to *k*-nearest neighbor algorithm?

We can find the class label of a sample data belonging to WAIS-R dataset through the following steps specified below:

Step (1) The WAIS-R dataset samples are positioned to a 2D coordinate system as provided below: the blue dots represent *Patient* and the red ones represent *Healthy* class.

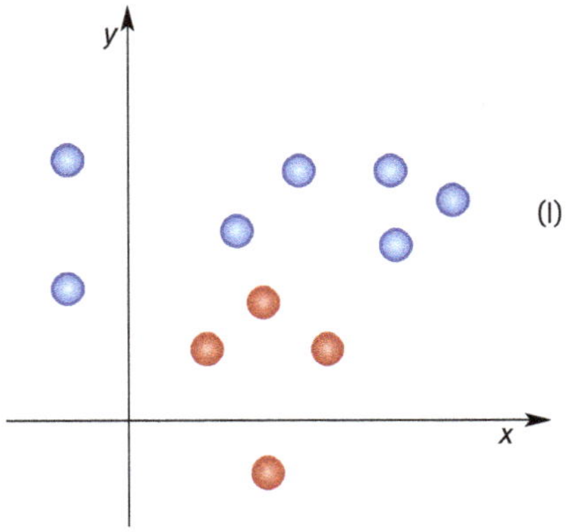

Step (2) Assume that the test data that we have classified on the coordinate plane where WAIS-R dataset exists corresponds to the green dot. The green dot selected from the dataset of WAIS-R dataset (*Healthy*, *Patient*) and data (whose class is not identified) has been obtained randomly from the dataset.

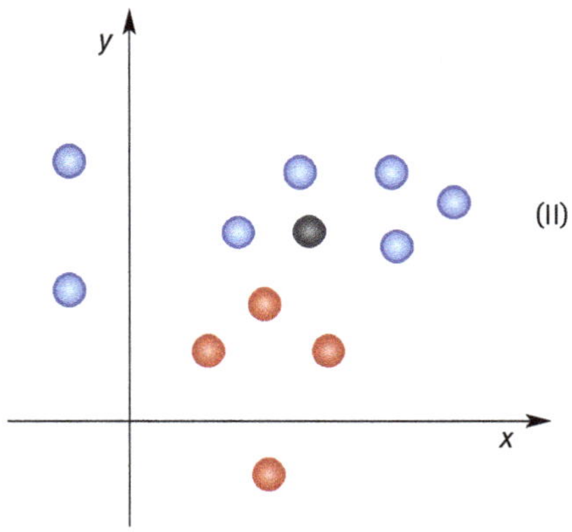

Step (3) Let us calculate the distances of the new test data to the training data according to Euclid distance formula (eq. 9.1). By ranking the distances calculated, find the two nearest neighbors to the test data.

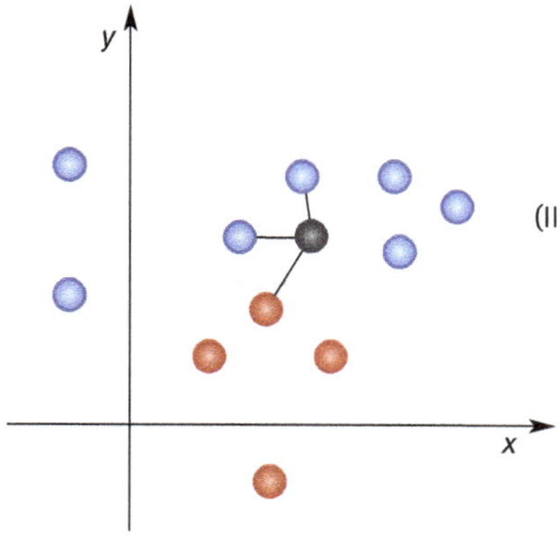

(III)

Step (4) As the two nearest neighbors belong to the blue dot group, we can identify the class of our test data as blue that refers to *Patient*.

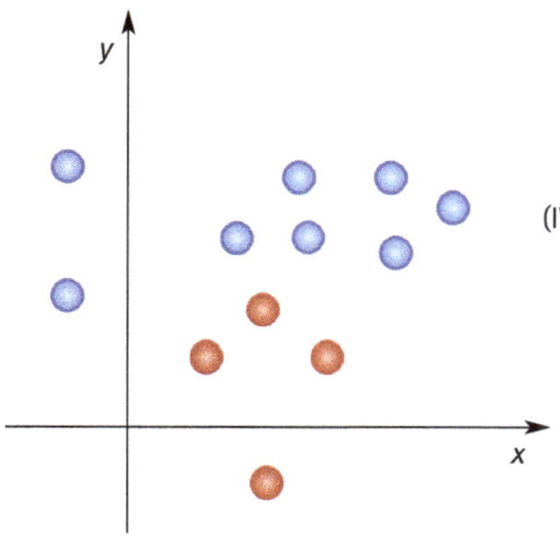

(IV)

As a result, the data in the WAIS-R dataset with 33.33% portion allocated for the test procedure and classified as Patient and Healthy have been classified with an accuracy rate of *96.24%* based on *two nearest neighbor algorithm*.

References

[1] Papadopoulos AN, Manolopoulos Y. Nearest neighbor search: A database perspective. USA: Springer Science & Business Media, 2006.

[2] Altman NS. An introduction to kernel and nearest-neighbor nonparametric regression. *The American Statistician*, 1992, 46(3), 175–185.

[3] Han J, Kamber M, Pei J, Data mining Concepts and Techniques. USA: The Morgan Kaufmann Series in Data Management Systems, Elsevier, 2012.

[4] Kramer O. Dimensionality reduction with unsupervised nearest neighbors. Berlin Heidelberg: Springer, 2013.

[5] Larose DT. Discovering knowledge in data: An introduction to data mining, USA: John Wiley & Sons, Inc., 90–106, 2005.

[6] Beyer K, Goldstein J, Ramakrishnan R, Shaft U. When is "nearest neighbor" meaningful?. In International conference on database theory. Springer: Berlin Heidelberg, 1999, January, 217–235.

[7] Simovici D, Djeraba C. Mathematical Tools for Data Mining. London: Springer, 2008.

[8] Khodabakhshi M, Fartash M. Fraud detection in banking using KNN (K-Nearest Neighbor) algorithm. In Intenational Conf. on Research in Science and Technology. London, United Kingdom, 2016.

[9] Wang SH, Zhang YD, Dong Z, Phillips P. Pathological Brain Detection. Singapore: Springer, 2018.

[10] Witten IH, Frank E, Hall MA, Pal CJ. Data Mining: Practical machine learning tools and techniques. USA: Morgan Kaufmann Series in Data Management Systems, Elsevier, 2016.

[11] Shmueli G, Bruce PC, Yahav I, Patel NR, Lichtendahl Jr, K. C. Data mining for business analytics: concepts, techniques, and applications in R. USA: John Wiley & Sons, Inc., 2017.

[12] Karaca Y, Cattani C, Moonis M, Bayrak Ş, Stroke Subtype Clustering by Multifractal Bayesian Denoising with Fuzzy C Means and K-Means Algorithms, *Complexity,Hindawi*, 2018.

[13] Rogers S, Girolami M. A first course in machine learning. USA: CRC Press, 2016.

10 Artificial neural networks algorithm

This chapter covers the following items:
- Algorithm for neural network
- Algorithm for classifying with multilayer perceptron
- Examples and applications

Artificial neural networks (ANN) serve as a computational tool modeled on the interconnection of neurons in the nervous systems of the human brain as well as the brain of other living things and organisms; these are also named and known as "artificial neural nets" and "neural nets." Biological neural nets (BNN) are known as the naturally occurring equivalent of the ANN. Both ANN and BNN are network systems that are composed of atomic components. Such components are known as the "neurons." ANN differ significantly from biological networks but many of the characteristics and concepts pertaining to biological systems are reproduced authentically in the artificial systems. ANN are sort of nonlinear processing systems. Such systems are appropriate for an extensive spectrum of tasks. These tasks are particularly those in which there are not any existing algorithms for the completion of the task. It is possible to train ANN for solving particular problems by using sample data as well as a teaching method. In this way, it would be possible for the identically constructed ANN to be capable of conducting various tasks in line with the training it has received. ANN can perform generalizations and acquire the ability of similarity recognition among different input patterns, particularly patterns that have been corrupted by noise.

Section 10.1 discusses an ANN topology/structure definition. The description of backpropagation algorithm is given in Section 10.2. The description of learning vector quantization (LVQ) algorithm is in Section 10.3.

10.1 Classification by backpropagation

First let us start with the definition of *backpropagation*. In fact, it is a neural network learning algorithm. The foundations of the neural networks field were laid by neurobiologists along with psychologists. Such people were trying to find a way to develop and test computational analogs of neurons. General definition of a *neural network* is as follows: neural network is regarded as a set of connected input/output units with each connection having a weight related to it [1–6]. There are certain phases involved. For example, during the learning phase, the network learns by adjusting the weights so that it will be capable of predicting the correct class label pertaining to the input datasets. Following a short definition, we can have a brief discussion on the advantages and disadvantages of neural networks. A drawback is that neural networks have long

https://doi.org/10.1515/9783110496369-010

training times; for this reason, they are more appropriate for the connectivity of applications between the units. In the process, a number of parameters is required; these parameters are usually determined best empirically (to illustrate, we can give the examples of network structure). Due to their poor interpretability level, neural networks have been subject to criticism. For humans, it is a difficult task to construe the symbolic meaning that is behind the learned weights and *hidden units* in the network. For such reasons, neural networks have proven to be less appropriate for data mining purposes. As for the advantages, we know that neural networks include high tolerance for noisy data. In addition, they have the ability of classifying patterns where they have not been trained. Even if you do not have a lot of knowledge of the relationships between the attributes and classes, you may use neural networks. In addition, neural networks are considered to be well suited for continuous-valued inputs and outputs [7–11]. This quality is missing in most decision tree algorithms, though. As for applicability to real life, which is also the emphasis of our book, neural networks are said to be successful in terms of a wide range of data related to real world and fieldwork. They are used in pathology field, handwritten character recognition as well as in laboratory medicine. Neural network algorithms also have parallel characteristics with the parallelization techniques being able to be used for accelerating the process of computation. Another application for neural network algorithms is seen in rule extraction from trained neural networks, thanks to certain techniques developed recently. All of these developments and uses indicate the practicality and usefulness of neural networks for data mining particularly while doing numeric prediction as well as classification.

Varied kinds of neural network algorithms and neural networks exist. Among them we see backpropagation as the most popular one. Section 10.1.1 will provide insight into multilayer feed-forward networks (it is the kind of neural network on which the backpropagation algorithm performs).

10.1.1 A multilayer feed-forward neural network

The backpropagation algorithm carries out learning on a multilayer feed-forward neural network. It learns a set of weights for predicting the class label of datasets repetitively or iteratively [5–11]. A multilayer feed-forward neural network is composed of the following components: an input layer, one or more hidden layers and an output layer. Figure 10.1 presents an illustration of a multilayer feed-forward network.

As you can see from Figure 10.1, units exist on each of the layers. The inputs to the network signify the attributes that are measured for each training pertaining to the dataset. The inputs are fed concurrently into the units that constitute the *input layer* [11]. Subsequently, such inputs pass through the input layer. They are subsequently weighted and fed concurrently to a second layer of a unit that resembles a neuron. This neuronlike unit is also known as the *hidden layer*. The outputs of the hidden layer units may well be input to another hidden layer; this process is carried

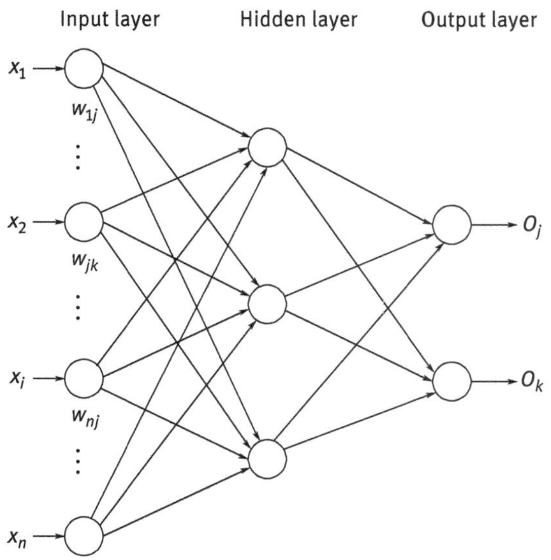

Figure 10.1: Multilayer feed-forward neural network.

out this way. We know that the number of hidden layers is arbitrary. However, thinking in terms of practice, only one of them is used as a general practice. The weighted outputs of the last hidden layer are input to units that constitute the *output layer*, emitting the prediction of the network for the relevant datasets.

Following the layers, here are some definitions for the units:

Input units are the units in the input layer. The units in the hidden layers and output layer are called *output units*. In Figure 10.1, a multilayer neural network with two layers of output units can be seen. This is why it is also called a *two-layer* neural network. It is *fully connected* because each unit supplies input to each unit in the next forward layer. A weighted sum of the outputs from units in the former layer is taken as input by each of the output unit (see Figure 10.4). A nonlinear function is applied onto the weighted input. Multilayer feed-forward neural networks are capable of modeling the class prediction as a nonlinear combination of the inputs. In terms of statistics, they perform a nonlinear regression [12–14].

10.2 Feed-forward backpropagation (FFBP) algorithm

Backpropagation involves processing a dataset of training datasets iteratively. For the prediction of the network, each dataset is compared with the actual known *target* value. For classification problems, the target value could be the known class label of the training dataset. For numeric predictions and continuous values, you may have to modify the weights if you want to minimize the mean-squared error between the prediction of the network and the actual target value. And such modifications are carried out in the backward direction, that is, from the output layer through each

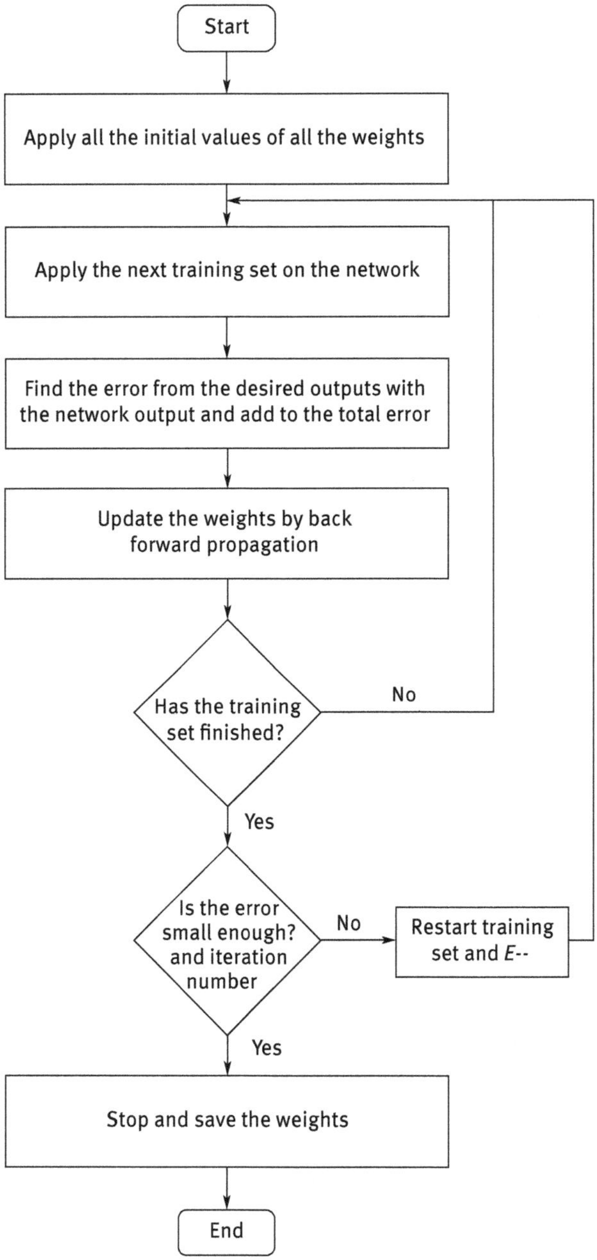

Figure 10.2: Feed-forward backpropagation algorithm flowchart.

hidden layer down to the first hidden layer. This is why the name backpropagation is appropriate. The weights usually converge at the end; it is not always the case though. As a result, you come to an end of the learning process as it stops. The flowchart of the algorithm is shown in Figure 10.2. The summary of the algorithm can be found in Figure 10.3. The relevant steps are defined with respect to inputs, outputs and errors. Despite appearing a bit difficult at the beginning, each step is in fact simple. So in time, it is possible that you become familiar with them [9–15].

The general configuration of FFBP algorithm is given as follows:

Algorithm: Feed Forward Backpropagation

Input: Training set including m number of samples $D = [x_1, x_2, \ldots, x_m] (x_i \in R^d)$

Test data including t number of samples $T = [x_1, x_2, \ldots, x_t] (x_t \in R^d)$

γ, the learning rate

The number of hidden layers J

The number of neuron on each layer $L = \{L_1, L_2, \ldots, L_j\}$

The number of iteration E

Result:

label data that corresponds to m number of samples ($Y = [y_1, y_2, \ldots, y_n]$)

Method:

```
//Initialize all weights and biases in network;
    W = { W_L1, W_L2, ..., W_LJ } values are assigned randomly.
while the condition not satisfied {
        for each training dataset X in D {
        // Propagate the inputs forward:
                for each input layer unit j {
                        O_j = I_j; // the output of an input unit refers to its actual input value
                for each hidden or output layer unit j {
                    calculate the net input (I_j)of unit j regarding the previous layer, i
                    compute the output (O_j) of each unit j}
        // Back propagate the errors:
        for each unit j in the output layer
                calculate the error Err_j
        for each unit j in the hidden layers, from the last to the first hidden layer
                calculate the error regarding the next higher layer, Err_j
        for each weight w_ij in network {
                weight increment (Δw_ij)
                    weight update (Δw_ij)}
        for each bias θ_j in network {
                bias increase (Δθ_j)
                bias update (θ_j)}
}}}
for each training dataset X in T {
// Propagate the inputs forward:
        for each input layer unit j {
                O_j = I_j; // the output of an input unit refers to its actual input value
        for each hidden or output layer unit j {
                calculate the net input (I_j) of unit j regarding the previous layer, i
                compute the output (O_j) of each unit j}
}}}
```

Figure 10.3: General feedforward backpropagation algorithm.

In these steps, you *initialize the weights* to small random numbers that can vary from −1.0 to 1.0, or −0.5 to 0.5. There is a *bias* in each unit linked with it. You also initialize the biases corresponding to small random numbers.

Each training dataset $\boldsymbol{X}$ is processed by the following steps:

Initially, the training dataset is fed to the input layer of the network. The step is called the *propagation of the inputs forward*. As a result, the inputs pass through the input units but they remain unchanged. Figure 10.4 shows the hidden output layer unit: The inputs to unit j are outputs from the previous layer, multiplied by their matching weights to form a weighted sum. This sum is added to the bias that is associated with unit j. Nonlinear activation function is applied to the net input. Due to practical reasons, the inputs to unit j are labeled $y_1, y_2, ..., y_n$, where unit j in the first hidden layer; these inputs would match the input dataset $x_1, x_2, ..., x_n$ [16].

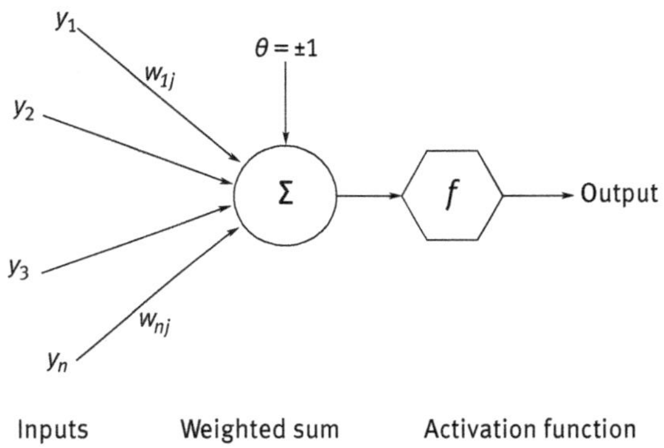

Figure 10.4: Hidden or output layer unit j: The inputs to unit j happen to be the outputs from the previous layer. So as to form a weighted sum, added to the bias associated with unit j, these are multiplied by their corresponding weights. In addition, a nonlinear activation function is administered to the net input. In simple words, the inputs to unit j are marked as $y_1, y_2, ..., y_n$. If unit j in the first hidden layer, then these inputs would correspond to the input dataset $x_1, x_2, ..., x_n$. From Jiawel and Kamber [11].

Its output, O_j, equals to its input value, I_j. Later, the calculation of the net input and output of each unit in the hidden as well as output layers is done. The net input in the hidden or output layers is calculated as a linear combination of its inputs. Figure 10.1 shows an illustration to make the understanding easier. Figure 10.4 presents a hidden layer or output layer unit. Each of these units have inputs. The units' outputs are connected to it in the former layer [17–19]. It is known that each connection has a weight. Each input that is connected to the unit is multiplied by the weight corresponding to it and the sum is performed for the calculation of the net input, j, in a hidden or an output layer; the net input, I_j, to unit j is in line with the following denotation:

$$I_j = \sum_i w_{ij} O_i + \theta_j \tag{10.1}$$

where w_{ij} is the weight of the connection from unit i in the former layer to unit j; O_i is the output of unit I from the former layer and θ_j represents the *bias* of the unit. The bias functions as a threshold because it is for the alteration of the unit activity. Each unit found in the hidden and output layers takes its net input. As a next step, an *activation* function is applied to it; the illustration is shown in Figure 10.4. The function indicates the neuron regarding the activation of characterized by the unit. Here, the *sigmoid* or *logistic* function is used. The net input I_j to unit j, then θ_j for the output of unit j, is calculated in line with the following formula:

$$O_j = \frac{1}{1+e^{-I_j}} \tag{10.2}$$

This function also has the ability of mapping a large input area onto the smaller range of 0 to 1; it is also referred to as a *squashing function*. The logistic function is differentiable; it is nonlinear as well. It permits the backpropagation algorithm for the modeling of the classification problems that are inseparable linearly. The output values, O_j, for each hidden layer, together with the output layer, are calculated. This computation gives the prediction pertaining to the network [20–22]. For real-life and practical areas, saving the intermediate output values at each unit is a good idea since you will need them once again at later stages when you are doing the back-propagation of the error. This hint can reduce the amount of computation required, thus saving you a lot of energy and time as well.

In the step of backpropagating the error, as the name suggests, the error is propagated backward. How is it done? It is done by updating the weights and biases to reflect the error of the prediction of the network. For a unit j in the output layer, the error Err_j is given as follows:

$$\mathrm{Err}_j = O_j(1-O_j)(T_j - O_j) \tag{10.3}$$

Here, O_j is the actual output of unit j, and T_j is the target value that is known regarding the relevant training dataset. $O_j(1-O_j)$ is the derivative of the logistic function. If you want to calculate the error of a hidden layer unit j, you have to consider the weighted sum of the errors of the units that are connected to unit j in the subsequent layer. We can find the error of a hidden layer unit j as follows:

$$\mathrm{Err}_j = O_j(1-O_j)\sum_k Err_k w_{jk} \tag{10.4}$$

where w_{jk} is the weight of the connection from unit j to a unit k in the next higher layer, and Err_k is the error of unit k.

The weights and biases are updated for the reflection of the propagated errors. The following equations are used for the updating of the weights:

$$\Delta w_{ij} = (l)\text{Err}_j O_j \tag{10.5}$$

$$w_{ij} = w_{ij} + \Delta w_{ij} \tag{10.6}$$

In these equations Δw_{ij} refers to the change in weight w_{ij}.

In eq. (10.5), / is the learning rate, a constant that has a value between 0.0 and 1.0. In backpropagation, there is a learning involved, that is, to use a gradient descent method for performing the search of a set of weights. It is important that this will fit the training data so that the mean-squared distance between the class prediction of the network and the known target value of the datasets can be minimized [21]. The learning rate is helpful because it allows to prevent being getting caught at a local minimum in decision space. It also prompts to find the minimum global. If you have a learning rate that is too small, learning pace will be really slow. On the other hand, if the learning rate is extremely high, then it is possible to see oscillation between inadequate solutions occurring. An important rule to bear in mind is that you have to set the learning rate to $1/t$, with t being the iteration number through the training set up until then.

The following equations provided are used for updating the biases:

$$\Delta\theta_j = (l)\text{Err}_j \tag{10.7}$$

$$\theta_j = \theta_j + \Delta\theta_j \tag{10.8}$$

In these equations, $\Delta\theta_j$ is the change in bias θ_j. The biases and weights are updated following the presentation of each dataset, which also known as *case updating*. In the strategy called *iteration* (*E*) *updating*, the weight and bias increments/increases may well be accumulated in variables alternatively, and so this will allow updating the biases and weights after all the datasets in the training set are presented. Here, *iteration* (*E*) is a common term, which is defined as one iteration through the training set. Theoretically, the mathematical derivation of backpropagation uses iteration updating. As we have mentioned in the beginning of the book, real-life practices may be different than theoretical knowledge. As in this case, in practice, we see that case updating is more common since it is likely to give outcomes that are more accurate. As we know, accuracy is vital for scientific studies. Especially for medicine, accuracy affects a lot of vital decisions to be made for the life of patients [20–22].

Let us consider the *terminating condition*: Training stops in the process when the following are observed:

- all Δw_{ij} in the previous iteration are too small and below the identified threshold, or
- the percentage of datasets incorrectly classified in the previous iteration is below the threshold, or
- a prespecified number of iterations have terminated.

In practice, we require hundreds of thousands of iterations before the weights reach a convergence.

The efficiency of *backpropagation* is often questionable. The efficiency of the calculation is reliant on the time allocated for the training of the network. For $|D|$ datasets and w weights, each iteration needs $O(|D| \times w)$ time. In the worst-case scenario, one may imagine the number of iterations could be exponential in n, the number of inputs. In real life, the time required for the networks to reach convergence is changeable. It is possible to accelerate the training time by means of certain techniques. *Simulated annealing* is one such technique that guarantees convergence to a global optimum.

10.2.1 FFBP algorithm for the analysis of various data

FFBP algorithm has been applied on economy (U.N.I.S.) dataset, multiple sclerosis (MS) dataset and WAIS-R dataset in Sections 10.2.1.1, 10.2.1.2 and 10.2.1.3, respectively.

10.2.1.1 FFBP algorithm for the analysis of economy (U.N.I.S.)

As the second set of data, in the following sections, we use some data related to U. N.I.S. countries. The attributes of these countries' economies are data regarding *years, unemployment, GDP per capita (current international \$), youth male (% of male labor force ages 15–24) (national estimate), ..., GDP growth (annual %).* (Data is composed of a total of 18 attributes.) Data belonging to USA, New Zealand, Italy, Sweden economies from 1960 to 2015 are defined based on the attributes given in Table 2.8. Economy dataset (http://data.worldbank.org) [15] is used in the following sections. For the classification of D matrix through FFBP algorithm, the first step training procedure is to be employed. For the training procedure, 66.66% of the D matrix can be split for the training dataset (151 ;× ;18), and 33.33% as the test dataset (77 × 18).

Following the classification of the training dataset being trained with FFBP algorithm, we can do the classification of the test dataset. Although the procedural steps of the algorithm may seem complicated at first, the only thing you have to do is to concentrate on the steps and grasp them. We can have a close look at the steps provided in Figure 10.7:

Algorithm: Feed Forward Backpropagation for Economy (U.N.I.S.) Data Set

Input: Training set including 151 number of samples $D = [x_1, x_2, \ldots, x_{151}] (x_i \in R^d)$
Test data including 77 number of samples $T = [x_1, x_2, \ldots, x_{77}] (x_t \in R^d)$
γ, the learning rate; 0.7
The number of hidden layers J
The number of neuron on each layer $L = \{L_1, L_2, \ldots, L_j\}$
The number of iteration E = 1000

Result:
label data that corresponds to *228* number of samples
(C = [USA, New Zealand, Italy, Sweden])

Method:

//Initialize all weights and biases in network;
$W = \{W_{L_1}, W_{L_2}, \ldots, W_{L_{10}}\}$ values are assigned randomly.
while the condition not satisfied {
  for each training dataset X in D {
  // Propagate the inputs forward:
    for each input layer unit j {
      $O_j = I_j$; // the output of an input unit refers to its actual input value
    for each hidden or output layer unit j {
      calculate the net input (I_j) of unit j regarding the previous layer, i
      compute the output (O_j) of each unit j}
    // Back propagate the errors:
    for each unit j in the output layer
      calculate the error Err_j
    for each unit j in the hidden layers, from the last to the first hidden layer
      calculate the error regarding the next higher layer, Err_j
    for each weight w_{ij} in network {
      weight increment (Δw_{ij})
      weight update (Δw_{ij})}
    for each bias θ_j in network {
      bias increase ($\Delta\theta_j$)
      bias update (θ_j)}
}}}
for each training dataset X in T {
// Propagate the inputs forward:
    for each input layer unit j {
      $O_j = I_j$; // the output of an input unit refers to its actual input value
    for each hidden or output layer unit j {
      calculate the net input (I_j) of unit j regarding the previous layer, i
      compute the output (O_j) of each unit j}
}}}
//Predict test samples' class

Figure 10.5: Feed forward backpropagation algorithm for the analysis of the Economy (U.N.I.S.) dataset.

In order to do classification with FFBP algorithm, the test data from the economy dataset has been selected randomly as 33.3%. Figure 10.5 presents the multilayer feed-forward neural network of the economy dataset. The learning coefficient has been identified as 0.7 as can be seen from Figure 10.5. Table 10.5 lists the bias values and the initial values of the network. Calculation has been done by using the data attributes of the first country in the economy dataset. There are 18 attributes belonging to each subject in the economy dataset. If the economy with $X = (1, 1,..., 1)$ data has the label of 1, the relevant classification will be USA. Dataset learns from the network structure and the output value of each node is calculated, which is listed in Table 10.6. Calculations are done for each of the nodes and in this way learning procedure is realized through the computations with the network. The errors are also calculated by the learning process. If the errors are not optimum error expected, the network realizes learning by the errors it makes. By doing backpropagation, the learning process is carried on. The error values are listed in Table 10.7. The weights and bias values updated can be seen in Table 10.6.

As can be seen in Figure 10.5, for the multilayer perceptron network, the sample economy data $X = (1, 0,..., 1)$ have been chosen randomly. Now, let us find the class in the sample USA, New Zealand, Italy and Sweden from X as the dataset. The training procedure of the X through FFBP continues until the minimum error rate is achieved based on the iteration number to be determined. In Figure 10.8, the steps of FFBP algorithm have been provided as applied for each iteration.

Steps (1–3) Initialize all weights and biases in the network step (see Figure 10.6).

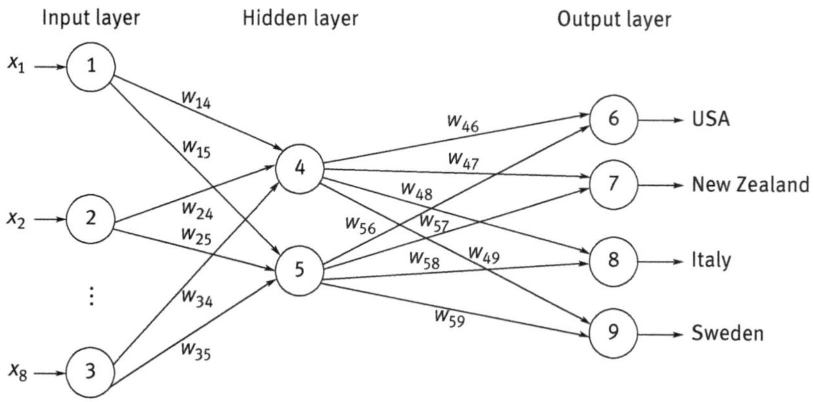

Figure 10.6: Economy dataset of multilayer feed-forward neural network.

Steps (4–28) Before starting the training procedure of X sample listed in Table 10.1 by FFBP network, the initial input values (x), weight values (w) as well as bias values (θ) are introduced to Multilayer Perceptron (MLP) network.

In Table 10.2, net input (I_j) values are calculated along with one hidden layer (w) value based on eq. (10.1). The value of the output layer (O_j) is calculated according to eq. (10.2). The calculations regarding the learning in FFBP algorithm are carried out based on the node numbers specified.

Tables 10.1 and 10.2 represent the hidden layer, *j*: 4, 5, output layer *j*: 6, 7, 8, 9 and node numbers. In addition, the calculations regarding learning in FFBP are carried out according to these node numbers.

Table 10.1: Initial input, weight and bias values for sample economy (U.N.I.S.) dataset.

x_1	x_2	...	x_{18}	w_{14}	...	w_{15}	w_{24}	w_{25}	w_{34}	w_{35}	w_{46}
1	1	...	1	0.1	...	0.6	0.5	0.4	0.1	−0.1	0.8

w_{47}	w_{48}	w_{49}	w_{56}	w_{57}	w_{58}	w_{59}	θ_4	θ_5	θ_6	θ_7	θ_8	θ_9
0.5	0.1	−0.4	0.4	0.6	0.2	0.1	0.5	0.4	0.5	−0.6	0.2	0.3

Table 10.2: Net input and output calculations for sample economy dataset.

j	Net input, I_j	Output, O_j
4	0.1+ 0.5 + 0.1+ 0.5 = 1.2	$1/1 + e^{1.2} = 0.231$
5	0.6 + 0.4 − 0.1 + 0.4 = 1.3	$1/1 + e^{1.3} = 0.214$
6	(0.8)(0.231) + (0.4)(0.214) + 0.5 = 0.707	$1/1 + e^{0.707} = 0.330$
7	(0.5)(0.231) + (0.6)(0.214) − 0.6 = −0.3561	$1/1 + e^{0.3561} = 0.588$
8	(0.1)(0.231) + (0.2)(0.214) + 0.2 = 0.2659	$1/1 + e^{0.2659} = 0.434$
9	(−0.4)(0.231) + (0.1)(0.214) + 0.3 = 0.229	$1/1 + e^{0.229} = 0.443$

Table 10.3 lists calculation of the error at each node; the error values of each node in the output layer Err_j are calculated using eq. (10.4).

The calculation procedure of *X* sample data for the first iteration is completed by backpropagating the errors calculated in Table 10.4 from the output layer to the input layer. The bias values (θ) and the initial weight values (w) are calculated as well.

Table 10.3: Calculation of the error at each node.

J	Err_j
9	(0.443)(1 − 0.443)(1− 0.443) = 0.493
8	(0.434)(1 − 0.434)(1− 0.434) = 0.491
7	(0.588)(1 − 0.588)(1 − 0.588) = 0.484
6	(0.330)(1 − 0.330)(1 − 0.330) = 0.442
5	(0.214)(1 − 0.214)[(0.442)(0.4) + (0.484)(0.6) + (0.491)(0.2) + (0.493)(0.1)] = 0.103
4	(0.231)(1 − 0.231)[(0.442)(0.8) + (0.588)(0.5) + (0.434)(0.1) + (0.493)(−0.4)] = 0.493

Table 10.4: Calculations for weight and bias updating.

Weight (bias)	New value
w_{46}	0.8 + (0.7)(0.442)(0.231) = 0.871
w_{47}	0.5+(0.7)(0.484)(0.231) = 0.578
w_{56}	0.4 + (0.7)(0.442)(0.214) = 0.466
w_{57}	0.6 + (0.7)(0.484)(0.214) = 0.672
w_{48}	0.1 + (0.7)(0.491)(0.231) = 0.179
w_{58}	0.2 + (0.7)(0.491)(0.214) = 0.273
w_{49}	0.4 + (0.7)(0.493)(0.231) = − 0.320
w_{59}	0.1 + (0.7)(0.493)(0.214) = 0.173
w_{14}	0.1 + (0.7)(0.493)(1) = 0.445
w_{15}	0.6 + (0.7)(0.103)(1) = 0.672
w_{24}	0.5 + (0.7)(0.493)(1) = 0.845
w_{25}	0.4 + (0.7)(0.103)(1) = 0.472
w_{34}	0.1 + (0.7)(0.493)(1) = 0.445
w_{35}	− 0.1+ (0.7)(0.103)(1) = − 0.0279
θ_9	0.3 + (0.7)(0.493) = 0.645
θ_8	0.2 + (0.7)(0.491) = 0.543
θ_7	− 0.6 + (0.7)(0.484) = −0.261
θ_6	0.5 + (0.7)(0.442) = 0.809
θ_5	0.4 + (0.7)(0.103) = 0.472
θ_4	0.5 + (0.7)(0.493) = 0.845

The weight values (w) and bias values (θ) as obtained from calculation procedures are applied for Iteration 2.

The steps in FFBP network for the attributes of an economy (see Table 2.8) have been applied up to this point. The steps for calculation of Iteration 1 are provided in Figure 10.7. Table 2.8 shows that the same steps will be applied for the other individuals in economy (U.N.I.S.) dataset, and the classification of countries economies can be calculated with minimum error.

In the economy dataset application, iteration number is identified as 1,000. The iteration in training procedure will end when the dataset learns max accuracy rate with [1− (min error × 100)] by the network of the dataset.

About 33.3% portion has been randomly selected from the economy training dataset and this portion is allocated as the test dataset. A question to be addressed at this point is as follows: how can we do the classification of a sample economy dataset (whose class is not known) by using the trained economy dataset?

The answer to this question is as follows: X dataset, whose class is not certain, and which is randomly selected from the training dataset is applied to the network. Afterwards, the net input and net output values of each node are calculated. (You do not have to do the computation and/or backpropagation of the error in this case.

If the output node for each class is one, then the output node with the highest value determines the predicted class label for *X*.) The accuracy rate for classification by neural network for the test dataset has been found to be *57.234%* regarding the learning of the class labels of *USA, New Zealand, Italy* and *Sweden.*

10.2.1.2 FFBP algorithm for the analysis of MS

As presented in Table 2.12, MS dataset has data from the following groups: 76 sample belonging to RRMS, 76 sample to SPMS, 76 sample to PPMS and 76 sample belonging to healthy subjects of control group. The attributes of the control group are data regarding *brain stem* (*MRI 1*), *corpus callosum periventricular* (*MRI 2*), *upper cervical* (*MRI 3*) *lesion diameter size* (*millimetric* (*mm*)) *in the MR image and EDSS* score. Data are composed of a total of 112 attributes. It is known that using these attributes of 304 individuals, we can know whether the data belong to the MS subgroup or healthy group. How can we make the classification as to which MS patient belongs to which subgroup of MS including healthy individuals and those diagnosed with MS (based on the lesion diameters (*MRI 1, MRI 2* and *MRI 3*) and number of lesion size for (*MRI 1, MRI 2* and *MRI 3*) as obtained from MRI images and EDSS scores)? *D* matrix has a dimension of (304 × 112). This means *D* matrix includes the MS dataset of 304 individuals along with their 112 attributes (see Table 2.12 for the MS dataset). For the classification of *D* matrix through FFBP, the first step training procedure needs to be employed. For the training procedure, 66.66% of the *D* matrix can be split for the training dataset (203 × 112) and 33.33% as the test dataset (101 × 112).

Following the classification of the training dataset being trained with FFBP algorithm, we can do the classification of the test dataset. Although the procedural steps of the algorithm may seem complicated at first glance, the only thing you have to do is to concentrate on the steps and grasp them. For this, let us have a close look at the steps provided in Figure 10.5.

The test dataset has been chosen randomly as 33.33% from the MS dataset in order to do the classification through FFBP algorithm. Figure 10.6 presents the multilayer feed-forward neural network of the MS dataset. The learning coefficient has been identified as 0.8 as can be seen from Figure 10.7.

The initial and bias values of the network are listed in Table 10.1. Calculation was done using the data attributes of the first individual in the MS dataset. There are 112 attributes belonging to each subject in the MS dataset. If the person with $X = (1, 0, ..., 1)$ data has the label of 1, the relevant subgroup will be RRMS. Dataset learns from the network structure and the output value of each node is calculated, which is listed in Table 10.2. Calculations are done for each of the nodes and in this way learning procedure is realized through the computations with the network. The errors are also calculated through the learning process. If the errors are not optimum error expected, the network realizes learning through the errors it makes. By doing backpropagation,

Algorithm: Feed Forward Backpropagation for MS Data Set

Input:

D training set ($D = [x_1, x_2, ..., x_{203}]$) ($x_i \in R^d$) including n number of samples
T test set $L = \{L_1, L_2, ..., L_J\}$ including *101*- number of data that is to be classified
γ, the learning rate; 0.8
The number of hidden layers J
The number of neuron on each layer $L = \{L_1, L_2, ..., L_j\}$
The number of iteration $E = 1000$

Result:

label data that corresponds to *304* number of samples
(C = [PPMS, RRMS, SPMS, Healthy])

Method:

(1) //Initialize all weights and biases in network;
$W = \{W_{L_1}, W_{L_2}, ..., W_{L_{10}}\}$values are assigned randomly.
(2) **while** the condition not satisfied {
(3) **for** each training dataset X in D {
(4) // Propagate the inputs forward:
(5) **for** each input layer unit j {
(6) $O_j = I_j$; // the output of an input unit refers to its actual input value
(7) **for** each hidden or output layer unit j {
(8) calculate the net input (I_j) of unit j regarding the previous layer, i
(9) compute the output (O_j) of each unit j}
(10) // Back propagate the errors:
(11) **for** each unit j in the output layer
(12) calculate the error Err_j
(13) **for** each unit j in the hidden layers, from the last to the first hidden layer
(14) calculate the error regarding the next higher layer, Err_j
(15) **for** each weight w_{ij} in network {
(16) weight increment (Δw_{ij})
(17) weight update (Δw_{ij})}
(18) **for** each bias θ_j in network {
(19) bias increase ($\Delta\theta_j$)
(20) bias update (θ_j)}
(21) }}}
(22) **for** each training dataset X in T {
(23) // Propagate the inputs forward:
(24) **for** each input layer unit j {
(25) $O_j = I_j$; // the output of an input unit refers to its actual input value
(26) **for** each hidden or output layer unit j {
(27) calculate the net input (I_j) of unit j regarding the previous layer, i
(28) compute the output (O_j) of each unit j}
(29) }}}
(30) //Predict test samples' class

Figure 10.7 FFBP algorithm for the analysis of the MS dataset.

the learning process is carried on. The error values are listed in Table 10.3. The weights and bias values updated are listed in Table 10.4.

As can be seen in Figure 10.8, for the multilayer perceptron network, the sample MS data $X = (1, 0, \ldots, 1)$ have been chosen randomly. Now, let us find the class in the sample MS subgroup from X as the dataset. The training procedure of the X through FFBP continues up until the minimum error rate is achieved based on the iteration number to be determined. In Figure 10.5, the steps of FFBP algorithm are provided as applied for each iteration.

Steps (1–3) Initialize all weights and biases in network.

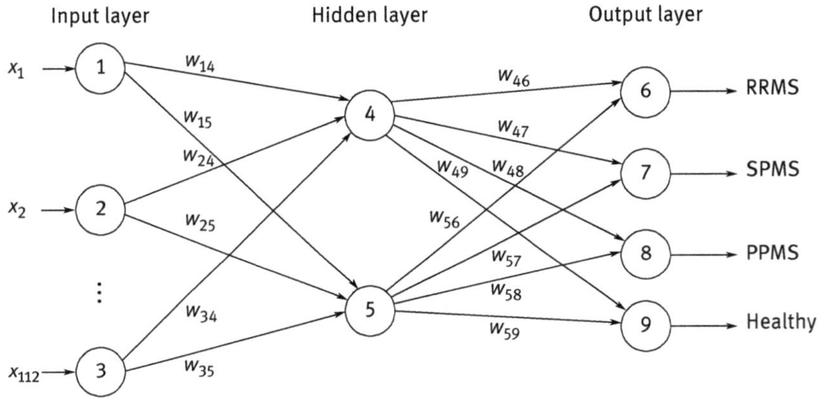

Figure 10.8: MS dataset of multilayer feed-forward neural network.

Steps (4–28) Before starting the training procedure of X sample in Table 10.1 by FFBP network, the initial input values (x), weight values (w) as well as bias values (θ) are introduced to MLP network.

In Table 10.5, net input (I_j) values are calculated along with one hidden layer (w) value in accordance with eq. (10.1). The value of the output layer (O_j) is calculated based on eq. (10.2). The calculations regarding the learning in FFBP algorithm are carried out based on the node numbers specified below.

Tables 10.5 and 10.6 provided represent the hidden layer, j: 4, 5, output layer j: 6, 7, 8, 9 and node numbers. Additionally, calculations concerning learning in FFBP are performed according to these node numbers.

Table 10.7 lists calculation of the error at each node. The error values of each node in the output layer Err_j are calculated using eq. (10.4).

The calculation procedure of X sample data for the first iteration is completed by backpropagating the errors calculated in Table 10.8 from the output layer to the input layer. The bias values (θ) and the initial weight values (w) are calculated as well. The

Table 10.5: Initial input, weight and bias values for sample MS dataset.

x_1	x_2	...	x_{112}	w_{14}	...	w_{15}	w_{24}	w_{25}	w_{34}	w_{35}	w_{46}
1	0	...	1	0.2	...	−0.3	0.1	0.1	−0.3	0.1	0.1

w_{47}	w_{48}	w_{49}	w_{56}	w_{57}	w_{58}	w_{59}	θ_4	θ_5	θ_6	θ_7	θ_8	θ_9
0.1	0.2	0.4	0.3	0.1	0.1	0.5	... 0.1	−0.1	0.4	0.1	0.2	0.3

Table 10.6: Net input and output calculations for sample MS dataset.

j	Net input, I_j	Output, O_j
4	0.1+ 0−0.3−0.1 =−0.3	$1/1 + e^{0.3} = 0.425$
5	0.2 + 0−0.2−0.2 =−0.2	$1/1 + e^{0.2} = 0.450$
6	(0.1)(0.425)+(0.3)(0.450) + 0.4 = 0.577	$1/1 + e^{0.557} = 0.359$
7	(0.1)(0.425)+(0.1)(0.450) + 0.1 = 0.187	$1/1 + e^{0.187} = 0.453$
8	(0.2)(0.425) + (0.1)(0.450) + 0.2 = 0.33	$1/1 + e^{0.33} = 0.418$
9	(0.4)(0.425)+(0.5)(0.450) + 0.3 = 0.695	$1/1 + e^{0.695} = 0.333$

Table 10.7: Calculation of the error at each node.

j	Err_j
9	(0.333)(1−0.333)(1−0.333) = 0.444
8	(0.418)(1−0.418)(1−0.418) = 0.486
7	(0.453)(1−0.453)(1−0.453) = 0.495
6	(0.359)(1−0.359)(1−0.359) = 0.460
5	(0.450)(1−0.450)[(0.460)(0.3) + (0.495)(0.1) + (0.486)(0.1) + (0.444)(0.5)] = 0.113
4	(0.425)(1−0.425) [(0.460)(0.1) + (0.495)(0.1) + (0.486)(0.1) + (0.444)(0.4)] = 0.078

weight values (w) and bias values (θ) as obtained from calculation procedures are applied for Iteration 2.

The steps in FFBP network for the attributes of an MS patient (see Table 2.12) have been applied up to this point. The way of doing calculation for Iteration 1 is provided with their steps in Figure 10.5. Table 2.12 shows that the same steps will be applied for the other individuals in MS dataset, and the classification of MS subgroups and healthy individuals can be calculated with the minimum error.

In the MS dataset application, iteration number is identified as 1,000. The iteration in training procedure will end when the dataset learns max accuracy rate with [1− (min error × 100)] by the network of the dataset.

Table 10.8: Calculations for weight and bias updating.

Weight (/bias)	New value
w_{46}	0.1 + (0.8)(0.460)(0.425) = 0.256
w_{47}	0.1 + (0.8)(0.495)(0.425) = 0.268
w_{56}	0.3 + (0.8)(0.460)(0.450) = 0.465
w_{57}	0.1 + (0.8)(0.495)(0.450) = 0.278
w_{48}	0.2 + (0.8)(0.486)(0.425) = 0.365
w_{58}	0.1 + (0.8)(0.486)(0.450) = 0.274
w_{49}	0.1 + (0.8)(0.444)(0.425) = 0.250
w_{59}	0.5 + (0.8)(0.444)(0.450) = 0.659
w_{14}	0.2 + (0.8)(0.113)(1) = 0.290
w_{15}	−0.3 + (0.8)(0.113)(1) = −0.209
w_{24}	0.1 + (0.8)(0.078)(0) = 0.1
w_{25}	0.2 + (0.8)(0.113)(0) = 0.2
w_{34}	−0.3 + (0.8)(0.078)(1) = −0.237
w_{35}	−0.2 + (0.8)(0.113)(1) = −0.109
θ_9	−0.3 + (0.8)(0.444) = 0.655
θ_8	0.2 + (0.8)(0.486) = 0.588
θ_7	0.1 + (0.8)(0.495) = 0.496
θ_6	0.4 + (0.8)(0.460) = 0.768
θ_5	−0.2 + (0.8)(0.113) = −0.109
θ_4	−0.1 + (0.8)(0.078) = −0.037

About 33.3% portion has been selected from the MS training dataset randomly and this portion is allocated as the test dataset. A question to be addressed at this point would be this: what about the case when you know the data but you do not know the subgroup details of MS or it is not known whether the group is a healthy one? How could the identification be performed for such a sample data class?

The answer to this question is as follows: *X* data, whose class is not certain, and which is randomly selected from the training dataset, are applied to the neural network. Subsequently, the net input and net output values of each node are calculated. (You do not have to do the computation and/or backpropagation of the error in this case. If, for each class, there exists one output node, then the output node that has the highest value will determine the predicted class label for *X*.) The accuracy rate for classification by neural network has been found to be *84.1%* regarding sample *test data* whose class is not known.

10.2.1.3 FFBP algorithm for the analysis of mental functions

As presented in Table 2.19, the WAIS-R dataset has data where 200 belong to *patient* and 200 sample to *healthy* control group. The attributes of the control group are data regarding *school education, gender, ..., DM.* MS data is composed of a total of 21 attributes. It is known that using these attributes of 400 individuals, we can know whether the data belong to *patient* or *Healthy* group. How can we make the classification as to which individual belongs to which patient or healthy individuals and those diagnosed with WAIS-R test (based on the school, education, gender, the DM, vocabulary, QIV, VIV ..., DM)? D matrix has a dimension of 400 × 21. This means that D matrix includes the WAIS-R dataset of 400 individuals along with their 21 attributes (see Table 2.19 for the WAIS-R dataset). For the classification of D matrix through FFBP algorithm, the first step training procedure is to be employed. For the training procedure, 66.66% of the D matrix can be split for the training dataset (267 × 21) and 33.33% as the test dataset (133 × 21). Following the classification of the training dataset being trained with FFBP algorithm.

We can classify the test dataset. Although the procedural steps of the algorithm may seem complicated at first glance, the only thing you have to do is to concentrate on the steps and grasp them. We can have a look at the steps as presented in Figure 10.9.

In order to classify with FFBP algorithm, the test data from the WAIS-R dataset have been selected randomly as 33.3%. Figure 10.9 presents the multilayer feed-forward neural network of the WAIS-R dataset. The learning coefficient has been identified as 0.5 as can be seen from Figure 10.9. The initial and bias values of the network are listed in Table 10.9. Calculation was done using the data attributes of the first individual in the WAIS-R dataset. There are 21 attributes belonging to each subject in the WAIS-R dataset. If the person with $X = (1, 0, \dots, 0)$ data has the label of 1, the relevant classification will be *patient*. Dataset learns from the network structure and the output value of each node is calculated, which is listed in Table 10.9. Calculations are done for each of the nodes, and in this way learning procedure is realized through the computations with the network. The errors are also calculated through the learning process. If the errors are not optimum error expected, the network realizes learning through the errors it makes. By doing backpropagation the learning process is carried on. The error values are listed in Table 10.10. The weights and bias values updated are listed in Table 10.11.

As shown in Figure 10.9, for the multilayer perceptron network, the sample WAIS-R data $X = (1, 0, \dots, 0)$ have been chosen randomly. Now, let us find the class in the sample *patient* or *healthy* from X as the dataset. The training procedure of the X through FFBP continues up until the minimum error rate is achieved based on the iteration number to be determined. In Figure 10.10, the steps of FFBP algorithm have been provided as applied for each iteration.

Steps (1–3) Initialize all weights and biases in network step.

Steps (4–28) Before starting the training procedure of X sample in Table 10.9 by FFBP network, the initial input values (x), weight values (w) as well as bias values (θ) are introduced to MLP network.

Algorithm: Feed Forward Backpropagation for WAIS-R Data Set

Input:

- D training set ($D = [x_1, x_2, ..., x_{267}]$) $(x_i \in R^d)$ including 267 number of samples
- γ, the learning rate; 0.5
- The number of hidden layers J
- The number of neuron on each layer $L = \{L_1, L_2, ..., L_j\}$
- The number of iteration E
- T test set ($T = [x_1, x_2, ..., x_{133}]$) $(x_i \in R^d)$ including *133*- number of data that is to be classified

Output:

label data that corresponds to *400* number of samples (C = [Healthy, Patient]

Method:

(1) //Initialize all weights and biases in network; $W = \{W_{L_1}, W_{L_2}, ..., W_{L_{10}}\}$values are assigned randomly.
(2) **while** the condition not satisfied {
(3) **for** each training dataset X in D {
(4) // Propagate the inputs forward:
(5) **for** each input layer unit j {
(6) $O_j = I_j$; // the output of an input unit refers to its actual input value
(7) **for** each hidden or output layer unit j {
(8) calculate the net input (I_j)of unit j regarding the previous layer, i
(9) compute the output (O_j) of each unit j}
(10) // Back propagate the errors:
(11) **for** each unit j in the output layer
(12) calculate the error Err_j
(13) **for** each unit j in the hidden layers, from the last to the first hidden layer
(14) calculate the error regarding the next higher layer, Err_j
(15) **for** each weight w_{ij} in network {
(16) weight increment (Δw_{ij})
(17) weight update (Δw_{ij})}
(18) **for** each bias θ_j in network {
(19) bias increase ($\Delta\theta_j$)
(20) bias update (θ_j)}
(21) }}}
(22) **for** each training dataset X in T {
(23) // Propagate the inputs forward:
(24) **for** each input layer unit j {
(25) $O_j = I_j$; // the output of an input unit refers to its actual input value
(26) **for** each hidden or output layer unit j {
(27) calculate the net input (I_j) of unit j regarding the previous layer, i
(28) compute the output (O_j) of each unit j}
(29) }}}
(30) //Predict test samples' class

Figure 10.9 FFBP algorithm for the analysis of WAIS-R dataset.

Table 10.9: Initial input, weight and bias values for sample WAIS-R dataset.

x_1	x_2	...	x_{21}	w_{14}	w_{15}	w_{24}	w_{25}	w_{34}
1	0	...	0	0.2	−0.3	0.4	0.1	−0.5

w_{35}	w_{46}	...	w_{56}	w_{57}	θ_4	θ_5	θ_6	θ_7
−0.2	−0.3	0.3	−0.2	0.1	−0.1	0.2	0.1	−0.4

Table 10.10: Net input and output calculations.

j	Net input, I_j	Output, O_j
4	0.2 + 0 + 0−0.1 = 0.1	$1/1 + e^{0.1} = 0.475$
5	−0.3 + 0 + 0 + 0.2 = −0.1	$1/1 + e^{0.1} = 0.525$
6	(−0.3)(0.475)−(0.2)(0.525) + 0.1 = −0.1475	$1/1 + e^{0.1} = 0.525$
7	(0.3)(0.475) + (0.1)(0.525)−0.4 = −0.205	$1/1 + e^{0.205} = 0.551$

Table 10.11: Calculation of the error at each node.

j	Err_j
7	(0.551)(1−0.551)(1−0.551) = 0.3574
6	(0.536)(1−0.536)(1−0.536) = 0.1153
5	(0.525)(1−0.525)[(0.525)(−0.2) + (0.3574)(0.1)] = −0.0009
4	(0.475)(1−0.475)[(0.3574)(0.3) + (0.1153)(−0.3)] = 0.0181

In Table 10.10, net input (I_j) values are calculated along with one hidden layer (w) value based on eq. (10.1). The value of the output layer (O_j) is calculated according to eq. (10.2). The calculations regarding the learning in FFBP algorithm are carried out based on the node numbers specified below.

Tables 10.10 and 10.11 list the hidden layer, j: 4, 5, output layer j: 6, 7, 8, 9 and node numbers. In addition, the calculations regarding learning in FFBP are carried out according to these node numbers.

Table 10.11 lists calculation of the error at each node; the error values of each node in the output layer Err_j are calculated using eq. (10. 4).

The calculation procedure of X sample data for the first iteration is completed by backpropagating the errors calculated in Table 10.12 from the output layer to the input layer. The bias values (θ)and the initial weight values (w) are calculated as well.

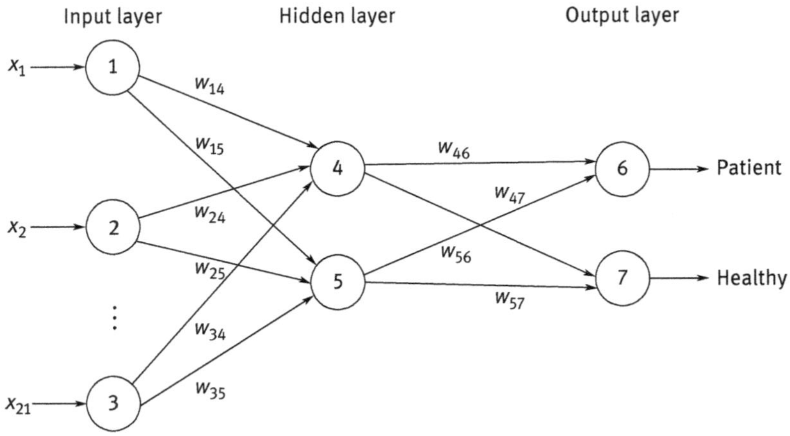

Figure 10.10: WAIS-R dataset of multilayer feed-forward neural network.

Table 10.12: Calculations for weight and bias updating.

Weight (/bias)	New value
w_{46}	−0.3 + (0.5)(0.1153)(0.475) = −0.272
w_{47}	0.3 + (0.5)(0.551)(0.475) = 0.430
w_{56}	−0.2 + (0.5)(0.1153)(0.525) = −0.169
w_{57}	0.1 + (0.5)(0.3574)(0.525) = 0.193
w_{14}	0.2 + (0.5)(0.0181)(1) = 0.209
w_{15}	− 0.3 + (0.5)(− 0.0009)(1) = 0.299
w_{24}	0.4 + (0.5)(0.0181)(0) = 0.4
w_{25}	0.1 + (0.5)(− 0.0009)(0) = 0.1
w_{34}	− 0.5 + (0.5)(0.0181)(0) = −0.5
w_{35}	0.2 + (0.5)(− 0.0009)(0) = 0.2
7	−0.4 + (0.5)(0.3574) = −0.221
6	0.1 + (0.5)(0.1153) = 0.157
5	0.2 + (0.5)(−0.0009) = 0.199
4	−0.1 + (0.5)(0.0181) = −0.090

The weight values (w) and bias values (θ) as obtained from calculation procedures are applied for Iteration 2.

The steps in FFBP network for the attributes of a WAIS-R patient (see Table 2.19) have been applied up to this point. The way of doing calculation for Iteration 1 is provided with their steps in Figure 10.9. Table 2.19 shows that the same steps will be applied for the other individuals in WAIS-R dataset, and the classification of patient and healthy individuals can be calculated with the minimum error.

In the WAIS-R dataset application, iteration number is identified as 1,000. The iteration in training procedure will end when the dataset learns max accuracy rate with [1− (min error × 100)] by the network of the dataset.

About 33.3% portion has been selected from the WAIS-R training dataset randomly and this portion is allocated as the test dataset. One question to be addressed at this point would be this: how can we do the classification of WAIS-R dataset that is unknown by using the trained WAIS-R network?

The answer to this question is as follows: X dataset, whose class is not certain, and which is randomly selected from the training dataset, is applied to the neural network. Later, the net input and net output values of each node are calculated. (You do not have to do the computation and/or backpropagation of the error in this case. If, for each class, one output node exists, in that case the output node that has the highest value determines the predicted class label for X). The accuracy rate for classification by neural network for the test dataset has been found to be *43.6%*, regarding the learning of the class labels of *patient* and *healthy*.

10.3 LVQ algorithm

LVQ [Kohonen, 1989] [23, 24] is a pattern classification method. In this method, each output unit denotes a particular category or class. Several output units should be used for each class. An output unit's weight vector is often termed as reference vector for the class represented by the unit. Throughout the training process, the output units are positioned in a way to adjust their weights by supervised training so that it is possible to approximate the decision surfaces of the theoretical Bayes classifier. We assume that a set of training patterns that know classifications are supplied, in conjunction with an initial distribution of reference vectors (each being represented as a classification that is known) [25].

Following the training process, an LVQ network makes the classification of an input vector by assigning it to the same class. Here, the output unit that has its weight vector (reference vector) is to be the closest to the input vector.

The architecture of an LVQ network presented in Figure 10.11 is the same as the architecture pertaining to a Kohonen self-organizing map without a topological structure presumed for the output units. Besides, each output unit possesses a known class for the representation [26] see Fig. 10.11.

The steps of LVQ algorithm is shown in Figure 10.12.

Figure 10.13 shows the steps of LVQ algorithm.

The impetus for the algorithm for the LVQ net is to find the output unit closest to the input vector. For this purpose, if x and w_j are of the same class, in that case, the weights are moved toward the new input vector. On the other hand, if x and w_j are from different classes, the weights are moved away from the input vector.

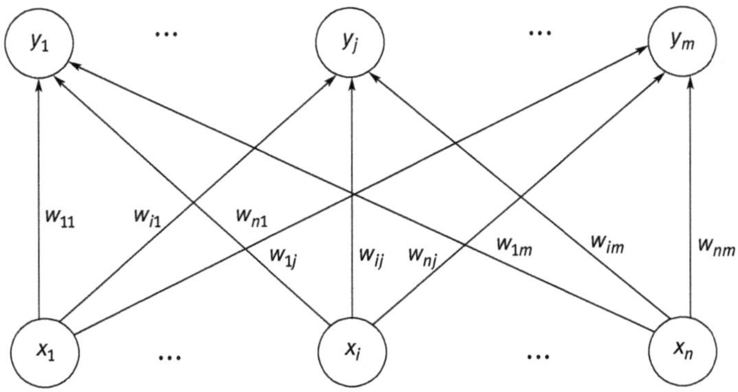

Figure 10.11: LVQ algorithm network structure.

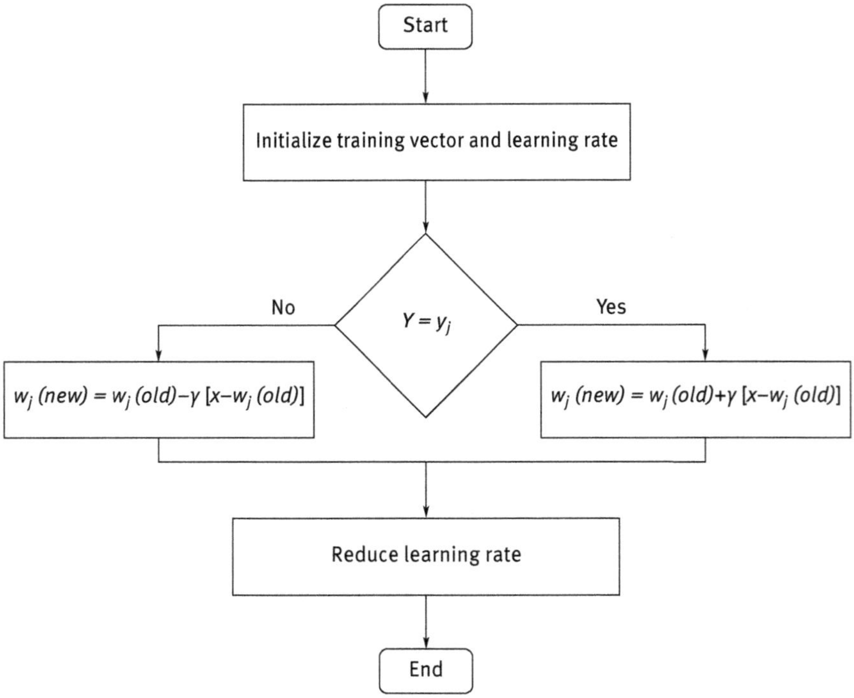

Figure 10.12: LVQ algorithm flowchart.

Figure 10.13 shows the steps of LVQ algorithm.

Step (1) Initialize reference vectors (several strategies are discussed shortly). İnitialize learning rate; y is default (0).

Step (2) While stopping condition is false, follow Steps (2–6).

Step (3) For each training input vector x, follow Steps (3) and (4).

Algorithm: Learning Vector Quantization

Input: Training set including m number of samples $D = [x_1, x_2, \ldots, x_m](x_i \in R^d)$

Test data including t number of samples $T = [x_1, x_2, \ldots, x_t](x_t \in R^d)$

Output:

label data that corresponds to m number of samples ($Y = [y_1, y_2, \ldots, y_m]$)

Method:

(1) //Initialize learning rate γ
(2) **while** (stopping condition is false){
(3) **for each** input vector $x_1, x_2, \ldots, x_n$
(4) **for** each weight w_{ij} in network {
(5) Euclidian distance between the input vector and (weight vector for) the *jth* output unit.}
//update w_j
(6) **if** $(Y = y_j)$
(7) $w_j(new) = w_j(old) + \gamma[x - w_j(old)]$
(8) **else**
(9) $w_j(new) = w_j(old) - \gamma[x - w_j(old)]$Update learning rate
(10) //Predict each test sample

Figure 10.13: General LVQ algorithm.

Step (4) Find j so that $\|x - w_j\|$ is a minimum.

Step (5) Update w_{ij}

Steps (6–9) *if* $(Y = y_j)$

$$w_j(\text{new}) = w_j(\text{old}) + \gamma[x - w_j(\text{old})].$$

else

$$w_j(\text{new}) = w_j(\text{old}) - \gamma[x - w_j(\text{old})]$$

Step (10) The condition may postulate a fixed number of iterations (i.e., executions of Step (2)) or learning rate reaching a satisfactorily small value.

A simple method for initializing the weight vectors will be to take the first n training vectors and use them as weight vectors. Each weight vector is calibrated by determining the input patterns that are closest to it. In addition, find the class the largest number of these input patterns belong to and assign that class to the weight vector.

10.3.1 LVQ algorithm for the analysis of various data

LVQ algorithm has been applied on economy (U.N.I.S.) dataset, multiple sclerosis (MS) dataset and WAIS-R dataset in Sections 10.3.1.1, 10.3.1.2 and 10.3.1.3, respectively.

10.3.1.1 LVQ algorithm for the analysis of economy (U.N.I.S.)

As the second set of data, we will use, in the following sections, some data related to economies for U.N.I.S. countries. The attributes of these countries' economies are data regarding *years, unemployment, GDP per capita (current international \$), youth male (% of male labor force ages 15–24) (national estimate), ..., GDP growth (annual %)*. (Data are composed of a total of 18 attributes.) Data belonging to USA, New Zealand, Italy or Sweden economies from 1960 to 2015 are defined based on the attributes given in Table 2.8 (economy dataset; http://data.worldbank.org) [15], to be used in the following sections. For the classification of D matrix through LVQ algorithm, the first step training procedure is to be employed. For the training procedure, 66.66% of the D matrix can be split for the training dataset (151 × 18), and 33.33% as the test dataset (77 × 18).

Following the classification of the training dataset being trained with LVQ algorithm, we can do the classification of the test dataset. Although the procedural steps of the algorithm may seem complicated at first glance, the only thing you have to do is to concentrate on the steps and grasp them. At this point, we can take a close look at the steps presented in Figure 10.14.

The sample economy data have been selected from countries economies dataset (see Table 2.8) as given in Example 10.1. Let us make the classification based on the economies' data selected using the steps for Iteration 1 pertaining to LVQ algorithm as shown in Figure 10.14.

Algorithm: Learning Vector Quantization for Economy (U.N.I.S.) Data Set
Input: Training set including 151 number of samples $D = [x_1, x_2, \ldots, x_{151}] (x_i \in R^d)$
Test data including 77 number of samples $T = [x_1, x_2, \ldots, x_{77}] (x_t \in R^d)$

Output:
label data that corresponds to m number of samples (C = [USA, New Zealand, Italy, Sweden])

Method:

//Initialize learning rate $\gamma = 0$
while (stopping condition is false){
  for each input vector $x_1, x_2, \ldots, x_{151}$
    for each weight w_{ij}in network {
Euclidian distance between the input vector and (weight vector for) the *jth* output unit.}
//update w_j
if $(Y = y_j)$
$w_j(new) = w_j(old) + \gamma[x - w_j(old)]$
else
$w_j(new) = w_j(old) + \gamma[x - w_j(old)]$Update learning rate
//Predict each test sample

Figure 10.14: LVQ algorithm for the analysis of the economy data.

Example 10.1 Four vectors are assigned to four classes for economy dataset. The following input vectors represent four classes. In Table 10.13, vector (first column) represents the net domestic credit (current LCU), GDP per capita, PPP (current international $) and GDP growth (annual %) of the subjects on which economies were administered. Class(Y) (second column) represents the USA, New Zealand, Italy and Sweden. Based on the data of the selected economies, the classification of country economy can be done for Iteration 1 of LVQ algorithm as shown in Figure 10.14. Table 10.13 lists economy data normalized with min–max normalization (see Section 3.7, Example 3.4).

Table 10.13: Selected sample from the economy dataset.

Vector	Class(*Y*)
(1, 4, 2, 0.2)	USA (labeled with 1)
(1, 3, 0.9, 0.5)	New Zealand (labeled with 2)
(1, 3, 0.5, 0.3)	Italy (labelled with 3)
(1, 6, 1, 1)	Sweden (labeled with 4)

Solution 10.1 Let us do the classification for the LVQ network, with four countries' economy chosen from economy (U.N.I.S.) dataset. The learning procedure described with vector in LVQ will go on until the minimum error is reached depending on the iteration number. Let us apply the steps of LVQ algorithm as can be seen in Figure 10.14.

Step (1) Initialize w_1, w_2 weights. Initialize the learning rate γ is 0.1.

$$w_1 = (1, 1, 0, 0), w_2 = (0, 0, 0, 1)$$

Table 10.13 lists the vector data belonging to countries' economies. LVQ algorithm was applied on them. The related procedural steps are shown in Figure 10.16 and their weight values (w) are calculated.

Iteration 1:
Step (2) Begin computation. For input vector $x = (1, 3, 0.9, 0.5)$ with $Y =$ New Zealand (labeled with 2), do Steps (3) and (4).

Step (3) $J = 2$, since x is closer to w_2 than to w_1.

Step (4) Since $Y =$ New Zealand update w_2 as follows:

$$w_2 = (0, 0, 0, 1) + 0.8[((1, 3, 0.9, 0.5) - (0, 0, 0, 1))] = (0.8, 2.4, 0.72, 0.6)$$

Iteration 2:
Step (2) For input vector $x = (1, 4, 2, 0.2)$ with $Y =$ USA (labeled with 1), do Steps (3) and (4).

Step (3) $J = 1$.

Step (4) Since $Y = 2$ update $w_1$as follows:

$$w_1 = (0.8, 2.4, 0.72, 0.6) + 0.8[((1, 4, 2, 0.2) - (0.8, 2.4, 0.72, 0.6))]$$
$$= (0.96, 3.68, 1.744, 0.28)$$

Iteration 3:
Step (2) For input vector $x = (1, 3, 0.5, 0.3)$ with Y = Italy (labeled with 3), do Steps (3) and (4).

Step (3) $J = 1$.

Step (4) Since Y = Italy update w_2 as follows:

$$w_1 = (0.96, 3.68, 1.744, 0.28) - 0.8[((1, 3, 0.5, 0.3) - (0.96, 3.68, 1.744, 0.28))]$$
$$= (0.928, 4.224, 2.896, 0.264)$$

Step (5) *This completes one iteration of training.*
Reduce the learning rate.

Step (6) *Test the stopping condition.*

As a result, the data in the economy dataset with 33.33% portion allocated for the test procedure and classified as USA, New Zealand, Italy and Sweden have been classified with an accuracy rate of *69.6%* based on LVQ algorithm.

10.3.1.2 LVQ algorithm for the analysis of MS

As presented in Table 2.12 MS dataset has data from the following groups: 76 sample belonging to RRMS, 76 sample to SPMS, 76 sample to PPMS and 76 sample belonging to healthy subjects of control group. The attributes of the control group are data regarding *brain stem* (*MRI 1*), *corpus callosum periventricular* (*MRI 2*), *upper cervical* (*MRI 3*) *lesion diameter size* (*millimetric* (*mm*)) *in the MR image and EDSS* score. Data are composed of a total of 112 attributes. It is known that using these attributes of 304 individuals, we can know that whether the data belong to the MS subgroup or healthy group. How can we make the classification as to which MS patient belongs to which subgroup of MS including healthy individuals and those diagnosed with MS (based on the lesion diameters (*MRI 1, MRI 2* and *MRI 3*) and number of lesion size for (*MRI 1, MRI 2* and *MRI 3*) as obtained from MRI images and EDSS scores)? D matrix has a dimension of (304×112). This means D matrix includes the MS dataset of 304 individuals along with their 112 attributes (see Table 2.12 for the MS dataset). For the classification of D matrix through LVQ algorithm, the first step training procedure is to be employed. For the training procedure, 66.66% of the D matrix can be split for the training dataset (203 × 112) and 33.33% as the test dataset (101 × 112). Following the classification of the training dataset being trained with LVQ algorithm, we can do the classification of the test dataset. Although the procedural steps of the algorithm may seem complicated at the beginning, the only thing you have to do is to concentrate on the steps and comprehend them. At this point, we can take a close look at the steps presented in Figure 10.15.

Algorithm: Learning Vector Quantization for MS Data Set

Input: Training set including m number of samples $D = [x_1, x_2, \ldots, x_{212}](x_i \in R^d)$

Test data including t number of samples $T = [x_1, x_2, \ldots, x_{92}](x_t \in R^d)$

Output:

label data that corresponds to m number of samples (C = [PPMS, RRMS, SPMS, Healthy])

Method:

(1) //Initialize learning rate $\gamma = 0.1$
(2) **while** (stopping condition is false){
(3) **for each** input vector $x_1, x_2, \ldots, x_{212}$
(4) **for each** weight w_{ij} in network {
(5) Euclidian distance between the input vector and (weight vector for) the *jth* output unit.} //update w_j
(6) **if** $(Y = y_j)$
(7) $w_j(new) = w_j(old) + \gamma[x - w_j(old)]$
(8) **else**
(9) $w_j(new) = w_j(old) + \gamma[x - w_j(old)]$Update learning rate
(10) //Predict each test sample

Figure 10.15 LVQ algorithm for the analysis of the MS dataset.

The sample MS data have been selected from MS dataset (see Table 2.12) in Example 10.2. Let us make the classification based on the MS patients' data selected using the steps for Iteration 1 pertaining to LVQ algorithm as shown in Figure 10.14.

Example 10.2 Five vectors are assigned to four classes for MS dataset. The following input vectors represent four classes *SPMS, PPMS, RRMS* and *Healthy*. In Table 10.14, vector (first column) represents the *EDSS, MRI 1 lesion diameter, MRI 2 lesion diameter* and *MRI 3 lesion diameter* of the MS subgroups and healthy data, whereas Class(Y) (second column) represents the MS subgroups and the healthy subjects. Based on the MS patients' data, the classification of MS subgroup and healthy can be done for Iteration 1 of LVQ algorithm as shown in Figure 10.14.

Solution 10.2 Let us do the classification for the LVQ network, with four MS patients chosen from MS dataset and one healthy individual. The learning procedure described with vector in LVQ will go on till the minimum error is reached depending on the

Table 10.14: Selected sample from the MS dataset as per Table 2.12.

Vector	Class (Y)
(6, 10, 17, 0)	SPMS (labeled with 1)
(2, 6, 8, 1)	PPMS (labeled with 2)
(6, 15, 30, 2)	RRMS (labeled with 3)
(0, 0, 0, 0)	Healthy (labeled with 4)

iteration number. Let us apply the steps of LVQ algorithm as can be seen from Figure 10.14 as well.

Step (1) Initialize w_1, w_2 weights. Initialize the learning rate γ is 0.1.

$$w_1 = (1, 1, 0, 0), w_2 = (1, 0, 0, 1)$$

In Table 10.14, the LVQ algorithm was applied on the vector data belonging to MS subgroup and healthy subjects. The related procedural steps are shown in Figure 10.15 and accordingly their weight values (w) are calculated.

Iteration 1:

Step (2) Start the calculation. For input vector x = (2, 6, 8, 1) with Y = PPMS (labeled with 2), follow Steps (3) and (4).

Step (3) $j = 2$, since x is closer to w_2 than to w_1.

Step (4) Since Y = PPMS update w_2 as follows:

$$w_2 = (1, 0, 0, 1) + 0.1[((2, 6, 8, 1) - (1, 0, 0, 1))] = (1.1, 0.6, 0.8, 1)$$

Iteration 2:

Step (2) For input vector x = (6, 10, 17, 0) with Y = SPMS (labeled with 1), follow Steps (3) and (4).

Step (3) $J = 1$

Step (4) Since Y = SPMS update w_2 as follows:

$$w_1 = (1.1, 0.6, 0.8, 1) + 0.1[((6, 10, 17, 0) - (1.1, 0.6, 0.8, 1))] = (1.59, 1.54, 2.42, 0.9)$$

Iteration 3:

Step (2) For input vector x = (0, 0, 0, 0) with Y = healthy, follow (labeled with 4) Steps (3) and (4).

Step (3) $J = 1$

Step (4) Since $Y = 2$ update w_1 as follows:

$$w_1 = (1.59, 1.54, 2.42, 0.9) - 0.1[(0, 0, 0, 0) - (1.59, 1.54, 2.42, 0.9)]$$
$$= (1.431, 1.386, 2.178, 0.891)$$

Steps (3–5) This concludes one iteration of training.
Reduce the learning rate.

Step (6) Test the stopping condition.

As a result, the data in the MS dataset with 33.33% portion allocated for the test procedure and classified as *RRMS, SPMS, PPMS* and *healthy* have been classified with an accuracy rate of *79.3%* based on LVQ algorithm.

10.3.1.3 LVQ algorithm for the analysis of mental functions

As presented in Table 2.19, the WAIS-R dataset has data where 200 belong to patient and 200 sample to healthy control group. The attributes of the control group are data regarding school education, gender, ..., DM. Data are composed of a total of 21 attributes. It is known that using these attributes of 400 individuals, we can know that whether the data belong to patient or healthy group. How can we make the classification as to which individual belongs to which patient or healthy individuals and those diagnosed with WAIS-R test (based on the school education, gender, the DM, vocabulary, QIV, VIV ..., DM)? *D* matrix has a dimension of 400 × 21. This means *D* matrix includes the WAIS-R dataset of 400 individuals along with their 21 attributes (see Table 2.19) for the WAIS-R dataset. For the classification of *D* matrix through LVQ, the first step training procedure is to be employed. For the training procedure, 66.66% of the *D* matrix can be split for the training dataset (267 × 21) and 33.33% as the test dataset (133 × 21). Following the classification of the training dataset being trained with LVQ algorithm, we classify of the test dataset.

Although the procedural steps of the algorithm may seem complicated at first glance, the only thing you have to do is to concentrate on the steps and grasp them. At this point, we can take a close look at the steps presented in Figure 10.16.

The sample WAIS-R data have been selected from WAIS-R dataset (see Table 2.19) in Example 10.3. Let us make the classification based on the patients' data selected using the steps for Iteration 1 pertaining to LVQ algorithm in Figure 10.16.

Example 10.3 Three vectors are assigned to two classes for WAIS-R dataset. The following input vectors represent two classes. In Table 10.15, vector (first column) represents the *DM, vocabulary, QIV* and *VIV* of the subjects on whom WAIS-R was administered. Class(*Y*) (second column) represents the patient and healthy subjects. Based on the data of the selected individuals, the classification of WAIS-R patient and healthy can be done for Iteration 1 of LVQ algorithm in Figure 10.16.

Solution 10.3 Let us do the classification for the LVQ network, with two patients chosen from WAIS-R dataset and one healthy individual. The learning procedure

Algorithm: Learning Vector Quantization for WAIS-R Data Set
Input: Training set including 151 number of samples $D = [x_1, x_2, \ldots, x_{267}](x_i \in R^d)$
Test data including 77 number of samples $T = [x_1, x_2, \ldots, x_{133}](x_t \in R^d)$
Output:
label data that corresponds to m number of samples C = [Healthy, Patient]
Method:
(1) //Initialize learning rate $\gamma = 0$
(2) while (stopping condition is false){
(3) **for each** input vector $x_1, x_2, \ldots, x_{267}$
(4) **for each** weight w_{ij}in network {
(5) Euclidian distance between the input vector and (weight vector for) the *jth* output unit.}
//update w_j
(6) **if** $(Y = y_j)$
(7) $w_j(new) = w_j(old) + \gamma[x - w_j(old)]$
(8) **else**
(9) $w_j(new) = w_j(old) + \gamma[x - w_j(old)]$Update learning rate
(10) //Predict each test sample

Figure 10.16: LVQ algorithm for the analysis of WAIS-R dataset.

described with vector in LVQ will go on until the minimum error is reached depending on the iteration number. Let us apply the steps of LVQ algorithm as can be seen from Figure 10.16.

Step (1) Initialize w_1, w_2 weights. Initialize the learning rate y is 0.1.

$$w_1 = (1, 1, 0, 0), w_2 = (0, 0, 0, 1)$$

In Table 10.13, the LVQ algorithm was applied on the vector data belonging to the subjects on whom WAIS-R test was administered. The related procedural steps are shown in Figure 10.16 and accordingly their weight values (w) are calculated.

Iteration 1:
Step (2) Begin computation. For input vector x = (0,0,1,1) with Y = healthy (labeled with 1), do Steps (3) and (4).

Step (3) $J = 1$, since x is closer to w_2 than to w_1.

Step (4) Since $Y = 1$ update w_2 as follows:

$$w_2 = (0, 0, 0, 1) + 0.1[((0, 0, 1, 1) - (0, 0, 0, 1))] = (0, 0, 0.1, 1)$$

Iteration 2:
Step (2) For input vector x = (2, 1, 3, 1) with Y = patient, do Steps (3) and (4).

Step (3) $J = 1$.

Step 4 Since $Y = 2$ update w_2 as follows:

$$w_2 = (1, 1, 0, 0) + 0.1[((2, 1, 3, 1) - (1, 1, 0, 0))] = (1.1, 1, 0.3, 0.1)$$

Iteration 3:

Step (2) For input vector $x = (4, 6, 4, 1)$ with Y = patient (labeled with 2), follow Steps (3) and (4).

Step (3) $J = 2$.

Step (4): Since $Y = 2$ update w_1 as follows:

$$w_1 = (1.1, 1, 0.3, 0.1) + 0.1[((4, 6, 4, 1) - (1.1, 1, 0.3, 0.1))] = (0.81, 0.5, 0.07, 0.01)$$

Steps (3–5) This finalizes one iteration of training.
Reduce the learning rate.

Step (6) Test the stopping condition.

As a result, the data in the WAIS-R dataset with 33.33% portion allocated for the test procedure and classified as patient and healthy have been classified with an accuracy rate of *62%* based on LVQ algorithm.

References

[1] Hassoun MH. Fundamentals of artificial neural networks. USA: MIT Press, 1995.

[2] Hornik K, Maxwell S, Halbert W. Multilayer feedforward networks are universal approximators. *Neural Networks*, USA, 1989, 2(5), 359–366.

[3] Annema J, Feed-Forward Neural Networks, New York, USA: Springer Science+Business Media, 1995.

[4] Da Silva IN, Spatti DH, Flauzino RA, Liboni LHB, dos Reis Alves SF. Artificial neural networks. Switzerland: Springer International Publishing, 2017.

[5] Wang SH, Zhang YD, Dong Z, Phillips P. Pathological Brain Detection. Singapore: Springer, 2018.

[6] Trippi RR, Efraim T. Neural networks in finance and investing: Using artificial intelligence to improve real world performance. New York, NY, USA: McGraw-Hill, Inc., 1992.

[7] Rabuñal, JR, Dorado J. Artificial neural networks in real-life applications. USA: IGI Global, 2005.

[8] Graupe D. Principles of artificial neural networks. Advanced series in circuits and systems (Vol. 7), USA: World Scientific, 2013.

[9] Rajasekaran S, Pai GV. Neural networks, fuzzy logic and genetic algorithm: synthesis and applications. New Delhi: PHI Learning Pvt. Ltd., 2003.

[10] Witten IH, Frank E, Hall MA, Pal CJ. Data Mining: Practical machine learning tools and techniques. USA: Morgan Kaufmann Series in Data Management Systems, Elsevier, 2016.

[11] Han J, Kamber M, Pei J, Data mining Concepts and Techniques. USA: The Morgan Kaufmann Series in Data Management Systems, Elsevier, 2012.

[12] Cross SS, Harrison RF, Kennedy RL. Introduction to neural networks. The Lancet, 1995, 346 (8982), 1075–1079.

[13] Kim P. MATLAB Deep Learning: With Machine Learning, Neural Networks and Artificial Intelligence. New York: Springer Science & Business Media, 2017.
[14] Chauvin Y, Rumelhart DE. Backpropagation: Theory, architectures, and applications. New Jersey, USA: Lawrence Erlbaum Associates, Inc., Publishers, 1995.
[15] Karaca Y, Moonis M, Zhang YD, Gezgez C. Mobile cloud computing based stroke healthcare system, *International Journal of Information Management, Elsevier*, 2018, In Press.
[16] Haykin S. Neural Networks and Learning Machines. New Jersey, USA: Pearson Education, Inc., 2009.
[17] Karaca Y, Cattani. Clustering multiple sclerosis subgroups with multifractal methods and self-organizing map algorithm. *Fractals*, 2017, 25(4). 1740001.
[18] Karaca Y, Bayrak Ş, Yetkin EF, The Classification of Turkish Economic Growth by Artificial Neural Network Algorithms, *In: Gervasi O. et al. (eds) Computational Science and Its Applications, ICCSA 2017. Springer, Lecture Notes in Computer Science,* 2017, 10405, 115–126.
[19] Patterson DW. Artificial neural networks: Theory and applications. USA: Prentice Hall PTR, 1998.
[20] Wasserman PD. Advanced methods in neural computing. New York, NY, USA: John Wiley & Sons, Inc., 1993.
[21] Zurada JM. Introduction to artificial neural systems, USA: West Publishing Company, 1992.
[22] Baxt WG. Application of artificial neural networks to clinical medicine. *The Lancet*, 1995, 346 (8983), 1135–1138.
[23] Kohonen T. Self-organization and associative memory. Berlin Heidelberg: Springer, 1989.
[24] Kohonen T. Self-Organizing Maps. Berlin Heidelberg: Springer-Verlag, 2001.
[25] Kohonen T. Learning vector quantization for pattern recognition. Technical Report TKK-F-A601, Helsinki University of Technology, Department of Technical Physics, Laboratory of Computer and Information Science, 1986.
[26] Kohonen T. Self-Organizing Maps. Berlin Heidelberg: Springer-Verlag, 2001.

11 Fractal and multifractal methods with ANN

This chapter covers the following items:
- Basic fractal explanations (definitions and descriptions)
- Multifractal methods
- LVQ algorithm application to datasets obtained from multifractal methods
- Examples and applications

11.1 Basic descriptions of fractal

The word fractal is derived from Latin *fractus*, which means "broken." Mathematician Benoit Mandelbrot coined this term in 1975 [1], in the book *The Fractal Geometry of Nature*. In this work, fractal is defined as "a fragmented or rough geometric shape of a geometric object which is possible to be split into parts, and each of such parts is (at least approximately) a reduced-size copy of the whole."

One of the most familiar fractal patterns is the Mandelbrot set, named after Benoit Mandelbrot. Producing the Mandelbrot set includes the testing of the properties of complex numbers after they are passed through an iterative function. Some questions emerge in this respect as follows: Do they stay bounded or do they tend to infinity? This "escape-time" algorithm is to a lesser amount a practical method used for producing fractals compared to the recursive techniques which shall be examined later in this chapter. An example for producing the Mandelbrot set is encompassed in the code examples.

Let us elaborate on this definition through two easy examples in Figures 11.1–11.3. Initially, we can consider a tree branching structure as in Figure 11.2.

Figure 11.1: A tree with a single root and two branches.

Pay attention to the way the tree depicted in Figure 11.2 has a single root that has two branches that join at its end. Further, these branches own two branches, maintaining a similar pattern. Let us imagine what would happen if we plucked one branch off from the tree and examine it on its own.

https://doi.org/10.1515/9783110496369-011

Figure 11.2: What would happen if we pluck one branch off from the tree and examine it on its own?

We can discover the fact that the shape of this branch we plucked is similar to the tree itself. This situation is known as *self-similarity* (as cited by Mandelbrot) and in this figure, each part is a "reduced-size copy of the whole."

The tree depicted is perfectly symmetrical in that the parts are actually the exact replicas belonging to the whole. Nevertheless, it is not always the case that fractals are perfectly self-similar. Let us have a glance at the graph of the Net Domestic Credit data (adapted from actual Italy Net Domestic Credit (US $) between 2012 and 2015).

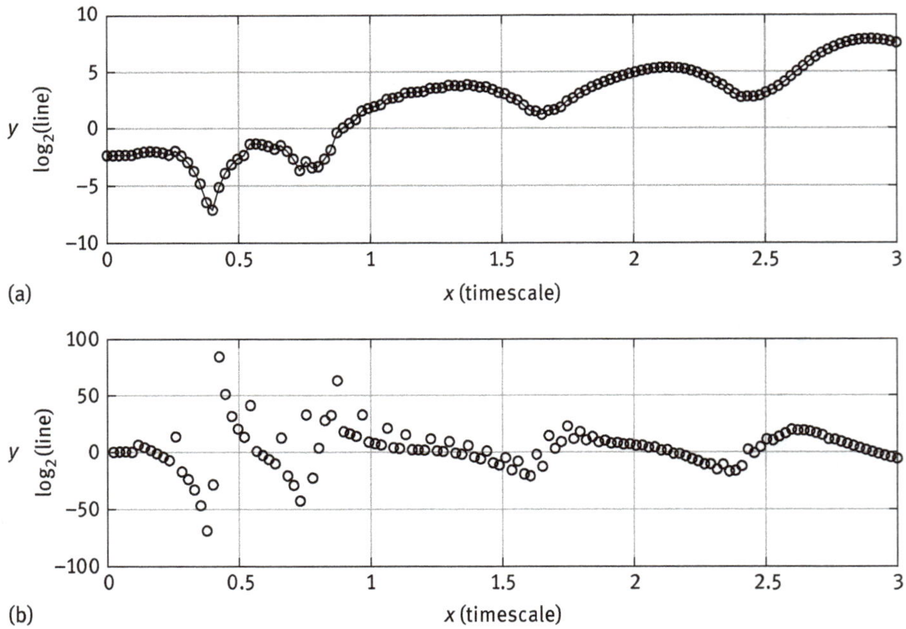

Figure 11.3: Italy Net Domestic Credit (years between 2012 and 2015).

As shown in graphs, the x-axis denotes the time (period from 2012 to 2015 in Italy economy) and the y-axis denotes the Net Domestic Credit (US $). Graphs of Italy Net Domestic Credit (US $) from U.N.I.S. (USA, New Zealand, Italy, Sweden) dataset make up some fractal examples since at any scale they look the same. Do these graphs of the Net Domestic Credit (US $) cover one single year, 1 h or 1 day? If there isn't a label, there is no way that we can provide an answer (Figure 11.4). (Figure 11.3 (a) presents three years' data and Figure 11.3 (b) focused on a small portion of the graph, with Figure 11.3 (a), presenting a period of 6 months.)

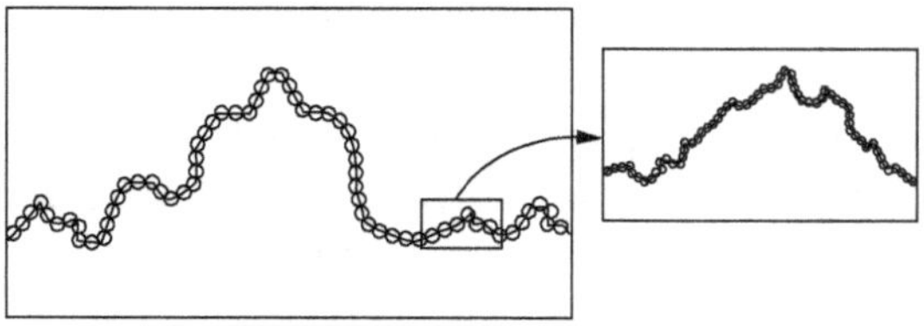

Figure 11.4: Focuses on a tiny part of Figure 11.3(a).

This provides a *stochastic* fractal example, which means that it is generated based on randomness as well as probabilities. Contrary to the deterministic tree-branching structure, it is self-similar statistically [–2–5].

Self-similarity is a central fractal trait. Besides this, it is also of importance to understand that merely self-similarity does not make a fractal. Take, for example, a line which is self-similar. A line seems to be the same at all scales. It can be thought of a pattern that includes many little lines. Despite this, it is not a fractal at all.

George Cantor [6] came up with simple recursive rules for the generation of an infinite fractal set as follows.

Example 11.1 (Cantor set). Erase recursively the middle third of the line segment by the algorithm whose flowchart is provided in Figure 11.5.

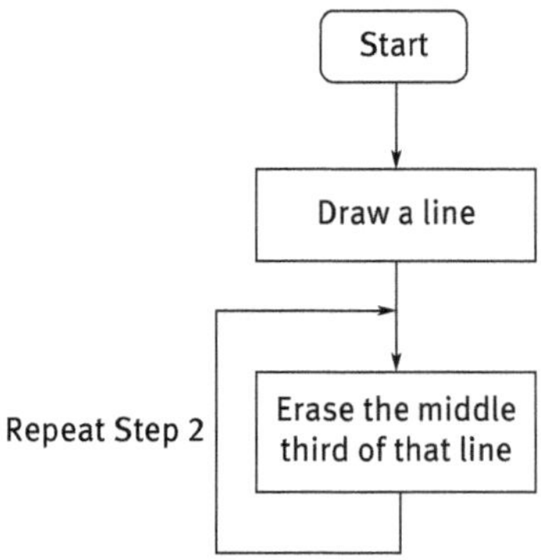

Figure 11.5: Cantor set flowchart for Example 11.1.

Algorithm: Cantor Algorithm

Method:

(1) Create a line
(2) **while** Step 2 for the remaining lines **do** {
(3)   Erase the middle third of that line
(4) }

Figure 11.6: Example 11.1. Algorithm.

Erasing the middle third of that line algorithm has been provided in a step-by-step manner in Figure 11.6.

Take a single line and split it into two. Subsequently, return back to those two lines and implement the same rule by splitting each line into two, and at this point we have four. Next, return back to those four lines and implement the rule. This time, you will have eight. Such a process is called *recursion*, which refers to the repeated application of the same rule for the generation of the fractal.

It is possible for us to apply the steps in the relevant order stage by stage to build the form as depicted in Figure 11.7.

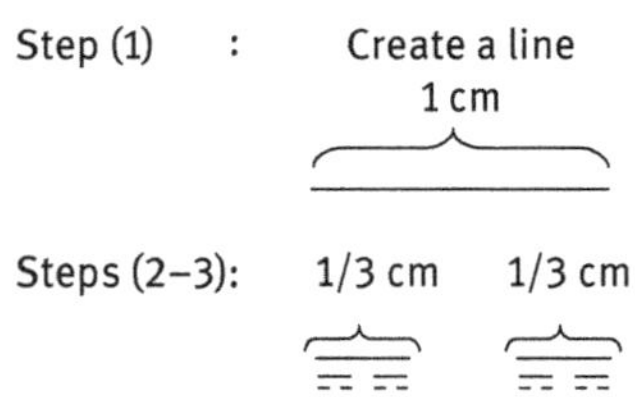

Figure 11.7: Example 11.1. Steps.

Functions called *recursive* are useful for the solution of certain problems.

Example 11.2 Certain mathematical calculations are applied in a recursive manner, the most common example of which is *factorial*. The factorial of any number $\boldsymbol{t}$ is generally written as $n!$. The recursive law is:

$$n! = 1, n = 0$$
$$n! = n(n-1)!, n \geq 1$$

At this point, we can write the flowchart (Figure 11.8).

For instance, the operational flow for 5! is given in Figures 11.9 and 11.10.

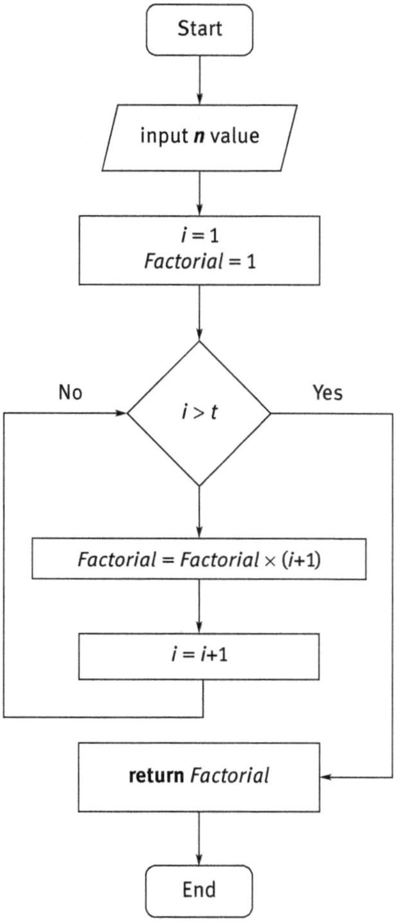

Figure 11.8: *t*-Factorial algorithm flowchart.

Algorithm: Calculate ***t*** Factorial

Input: ***t*** that is calculated factorial

Result: $t!$

Method:

(1) Input ***t* value**
(2) *Factorial* = 1;
(3) **while** $i > t$ **do** {
(4) // using a regular loop to calculate factorial
Factorial = Factorial × (i+1):
(4) }
(5) **return** *Factorial* ;

Figure 11.9: Calculate *t*-factorial algorithm.

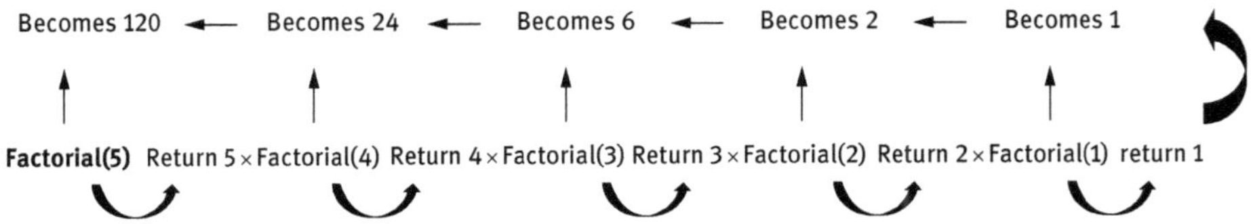

Figure 11.10: The operational flow cycle calculation of 5! using Figure 11.9 algorithm.

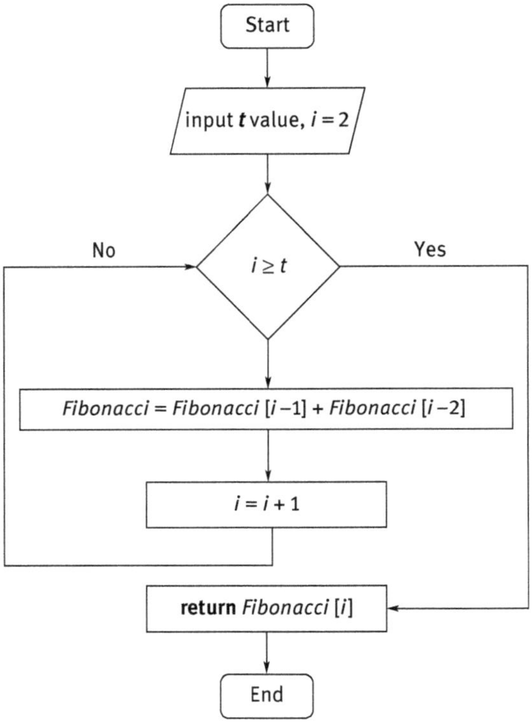

Figure 11.11: Fibonacci series algorithm flowchart for t.

Example 11.3 As a second example let us consider the Fibonacci series. Let's write the flowchart (Figure 11.11) as well as the algorithm (Figure 11.12).

As an example, let us calculate the (orbit) of 2 for $f(x) = 1 + \frac{1}{x}$, as shown in Figure 11.13.

This series is obtained as $\frac{2}{1}, \frac{3}{2}, \frac{5}{3}, \ldots$ Fibonacci series.

Example 11.4 (Devil's staircase). This is a continuous function on the unit interval. It is a no-differentiable piecewise constant which is growing from 0 to 1! (see Figure 11.14) [7].

Consider some number X in the unit interval. Denote it in base 3. Slice off the base 3 expansion following the first 1. Afterwards, all 2's are made in the expansion to 1's. This number at this point has merely 0's or 1's in its expansion, thus, it is to be interpreted as a base 2 number! We can term the new number $f(x)$.

Algorithm: Calculate Fibonacci Series of ***t*** number

Input: ***t*** that is calculated factorial

Result: Fibonacci series of ***t*** number

Method:

```
Input t value
Fibonacci[0] = 0;
  Fibonacci[1] = 1;
while i ≥ t do {
        Fibonacci[i] = Fibonacci[i − 1] + Fibonacci[i − 2];
}
return Fibonacci[i] ;
```

Figure 11.12: Calculate Fibonacci series of t number algorithm.

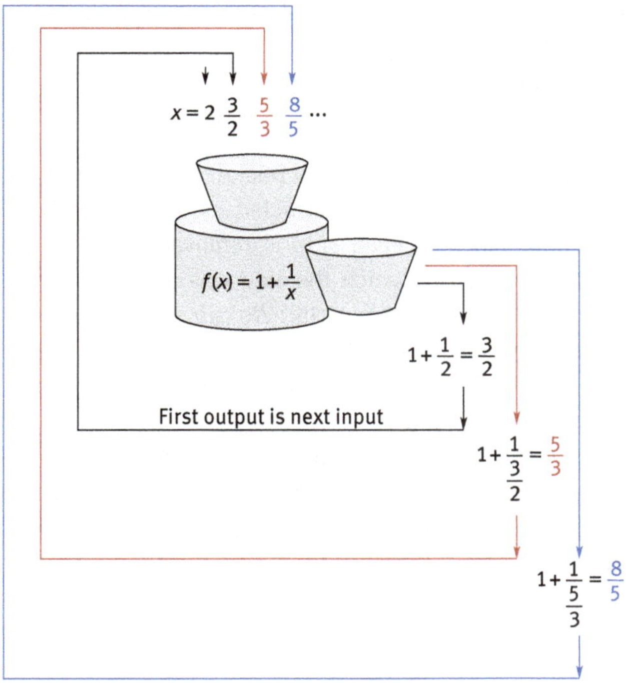

Figure 11.13: Computation of $f(x) = 1 + \frac{1}{x}$.

Should we plot this function, we may obtain the form called the *Devil's staircase* that is concerned with the standard Cantor set [6] as follows: this function is constant with the intervals removed from the Cantor set that is standard. To illustrate:

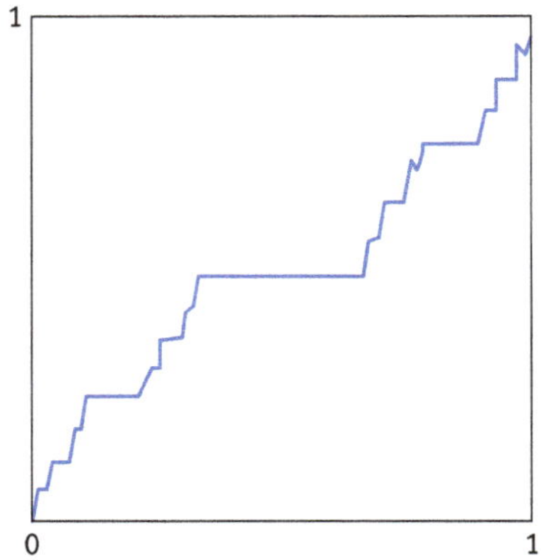

Figure 11.14: Devil's staircase $f(x)$.

$$\textit{if } x \text{ is in } [1/3, 2/3] \textit{ then } f(x) = {}^1\!/_2.$$
$$\textit{if } x \text{ is in } [1/9, 2/9] \textit{then } f(x) = {}^1\!/_4;$$

Should we plot this function, we can realize that this function cannot be differentiated at the Cantor set points. Yet, it has zero derivative everywhere else. However, this function has zero derivative almost everywhere since a Cantor set has measure zero. It "rises" only on Cantor set points.

With the corresponding algorithm:

Example 11.5 (Snow flake). Helge von Koch discovered a fractal, also known as the Koch island, a fractal in the year 1904 [8]. It is constructed by starting with an equilateral triangle. Here, the inner third of each side is removed and another equilateral triangle is built at the location in which the side was removed. Later, the process is to be repeated indefinitely (see Figure 11.15). We can encode the Koch snowflake simply as a Lindenmayer system [9], which as an initial string is "F--F—F"; string rewriting rule is "F" → "F+F--F+F," with an angle of 60°. Zeroth through third iterations of the construction are presented earlier.

N_n is the number of sides, L_n is the length of a single side, l_n is the length of the perimeter and A_n is the snowflake's area after the nth iteration. Additionally, represent the area of the initial $n = 0$ triangle Δ, and the length of an initial $n = 0$ side 1. Then, we have the following:

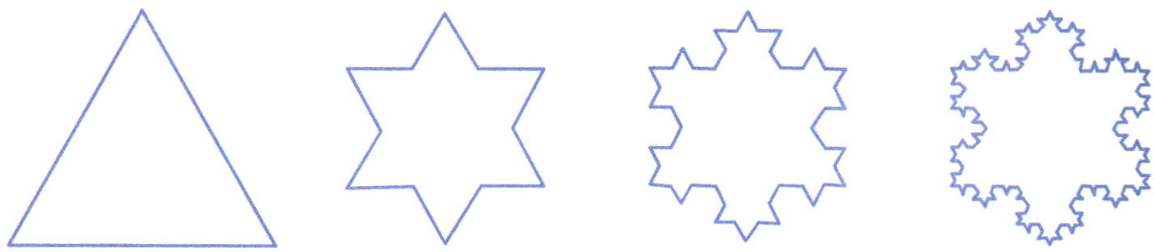

Figure 11.15: Examples of some snowflake types.

$$N_n = 3.8^n$$

$$L_n = \left(\frac{1}{3}\right)^n$$

$$l_n = N_n L_n = 3\left(\frac{8}{3}\right)^n$$

$$A_n = A_{(n-1)} + \frac{1}{4} N_n L_n^2 \Delta = A_{(n-1)} + \frac{1}{3}\left(\frac{8}{9}\right)^{(n-1)} \Delta$$

Solving the recurrence equation with $A_0 = \Delta$ yields

$$A_n = \frac{1}{5}\left[8 - 3\left(\frac{8}{9}\right)^n\right]\Delta,$$

So as $n \to \infty$

$$A_\infty = \frac{8}{5}\Delta.$$

The capacity dimension will then be

$$d_{\text{cap}} = -\lim_{n\to\infty} \frac{\ln N_n}{\ln L_n} = \log_3 8 = \frac{2\ln 2}{\ln 3} = 1.26185...$$

Some good examples of these beautiful tilings shown in Figure 11.16 can be made through iterations toward Koch snowflakes.

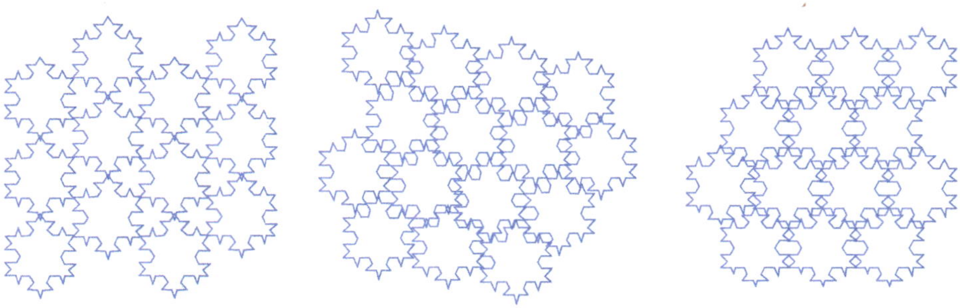

Figure. 11.16: Examples of Koch snowflake type.

Example 11.6 (Peano curves). It is known that the Peano curve passes through the unit square from $(0, 0)$ to $(1, 1)$. Figure 11.17 shows how we can define an orientation and affine transformations mapping each iterate to one of the nine subsquares of the next iterate, at the same time preserving the previous orientations. Figure 11.17 illustrates the procedure for the fist iteration [10].

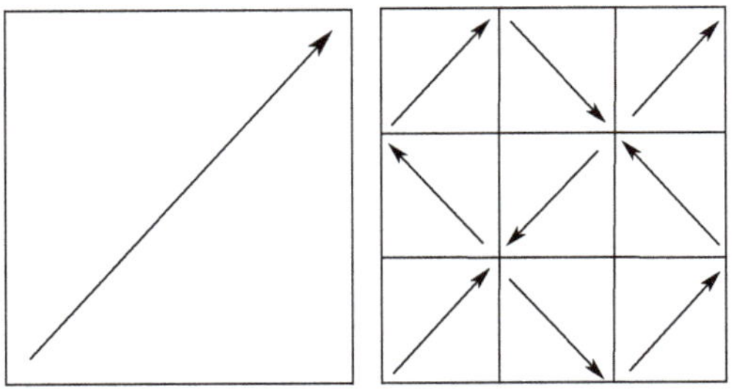

Figure 11.17: The steps of the analytical representation of Peano space-filling curve.

Nine of the transformations can be defined in a similar way as that of Hilbert curve: $p_0 z = \frac{1}{3}z$, $p_1 z = \frac{1}{3}\bar{z} + \frac{1}{3} + \frac{i}{3}$, ... in complex form, $p_j\begin{pmatrix} x_1 \\ x_2 \end{pmatrix} = \frac{1}{3}P_j\begin{pmatrix} x_1 \\ x_2 \end{pmatrix} + \frac{1}{3}p_j$, $j = 0, 1, ..., 8$ in matrix notation. Each p_j maps Ω to the $(j+1)$th subsquare of the first iteration as presented in Figure 11.18.

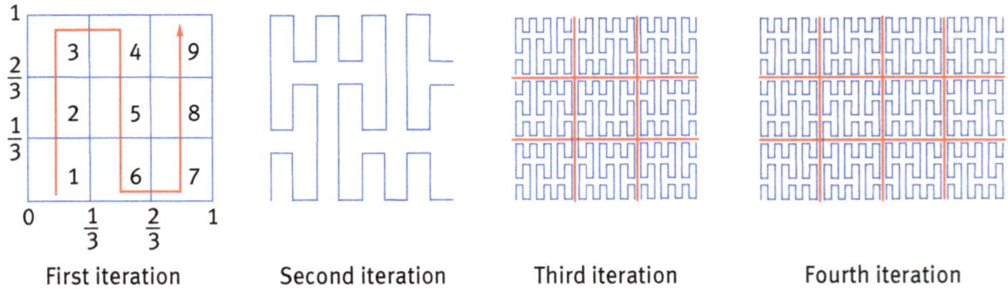

Figure 11.18: Peano space-filling curve and the geometric generation.

Using the ternary representation of t an equality

$$0_3 t_1 t_2 \ldots t_{2n-2} t_{2n} \ldots = 0_9 (3t_1 + t_2)(3t_3 + t_4) \ldots (3t_{2n-1} + t_{2n}) \ldots$$

we can proceed in the same way as with the Hilbert curve. Upon realizing this

$$f_p(t) = \lim_{n \to \infty} p_{3t_1 + t_2} p_{3t_3 + t_4} \cdots p_{3t_{2n-1} + t_{2n}} \Omega \tag{11.1}$$

and apply to finite ternaries, we get an expression for the calculation of image points [9].

Example 11.7 (Sierpiński gasket). In 1915, Sierpiński described the Sierpiński sieve [11]. This is termed as the Sierpiński gasket or Sierpiński triangle. And we can write this curve as a Lindenmayer system [9] with initial string "FXF--FF--FF"; string rewriting rule is "F" → "FF", "X" → "--FXF++FXF++FXF--", with an angle of 60° (Figure 11.19).

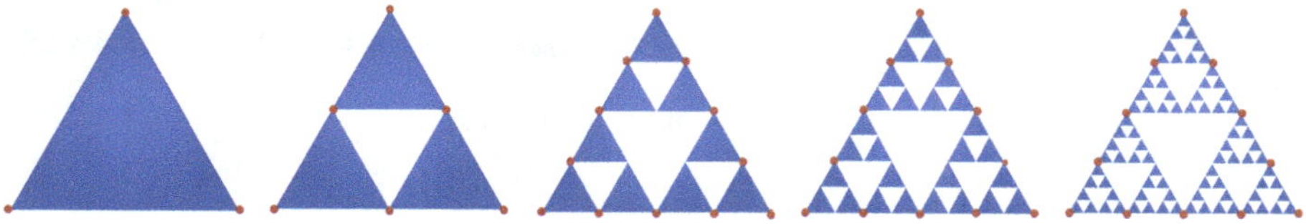

Figure 11.19: Fractal description of Sierpiński sieve.

The nth iteration of the Sierpiński sieve is applied in the Wolfram Language as Sierpiński Mesh [n].

N_n is the black triangles' number following iteration n, L_n is the length of a triangle side and A_n is the fractional area that is black following the n th iteration. Then

$$N_n = 3^{2n}$$

$$L_n = \left(\frac{1}{2}\right)^{2n} = 2^{-2n}$$

$$A_n = N_n L_n^2$$

The capacity dimension is therefore [12]

$d_{cap} = -\lim\limits_{n\to\infty} \frac{\ln N_n}{\ln L_n} = \log_2 3 = \frac{\ln 3}{\ln 2} = 1.5849...$ The Sierpiński sieve is produced by the recursive equation

$$a_n = \prod_{\substack{j \\ e(j,n)=1}} 2^{2^{e(j,n)}} + 1$$

where $e(j,n)$ is the $(j+1)$th least significant bit defined by [13]

$$n = \sum_{j=0}^{t} e(j,n) 2^j$$

and the product is taken over all j such that $e(j,n) = 1$ as shown in Figure 11.20.

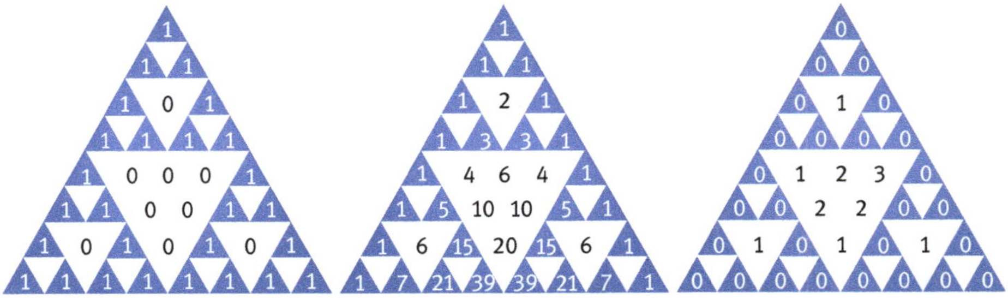

Figure 11.20: Sierpiński sieve is elicited by Pascal triangle (mod 2).

Sierpiński sieve is elicited by the Pascal triangle (mod 2), with the sequence 1; 1, 1; 1, 0, 1; 1, 1, 1, 1; 1, 0, 0, 0, 1; ... [12]. This means coloring all odd numbers blue and even numbers white in Pascal's triangle will generate a Sierpiński sieve.

11.2 Fractal dimension

In this section we discuss about the measure of fractal and of its complexity by the so-called fractal dimension [13].

Now, let's analyze two types of fractal dimensions, which are the self-similarity dimension and the box-counting dimension. Among the many other kinds of dimension, topological dimension, Hausdorff dimension and Euclidean dimension are some types we may name [14, 15].

Prior to the explanation of dimension measuring algorithms, it is important that we enlighten the way a power law works. Essentially, data behave through a power law relationship if they fit this following equation: $y = c \times x^d$, where c is a constant. Plotting the $\log(y)$ versus the $\log(x)$ is one of the ways one can use for determining whether data fit a power law relationship. If the plot happens to be a straight line, then it is a power law relationship with the slope d.

We will see that fractal dimension depends largely on the power law.

Definition 11.1 (Self-similarity dimension). In order to measure the self-similar dimension, a fractal must be self-similar. The power law is valid and it is given as follows:

$$a = \frac{1}{s^D} \tag{11.2}$$

where D is the *self-similar dimension* measure, a is the *number of pieces* and s is the *reduction factor*.

In Figure 11.7, should a line be broken into three pieces, each one will be one-third the length of the original. Thus, $a = 3$, $s = 1/3$ and $D = 1$.

The original triangle is split into half for the Sierpiński gasket, and there exist three pieces. Thus,

$$a = 3, s = 1/2 \text{ and } D = 1.5849.$$

Definition 11.2 (Box-counting dimension). In order to compute the *box-counting dimension*, we have to draw a fractal on a grid. The x-axis of the grid is s being $s = 1/(\text{width of the grid})$. For instance, if the grid is 480 blocks tall by 240 blocks wide, $s = 1/240$. Let $N(s)$ be the number of blocks that intersect with the picture. We can resize the grid by refining the size of the blocks and recur the process. The fractal dimension is obtained by the limit of $N(s)$ with s going to zero. In a log scale, the box-counting dimension equals the slope of a line.

For example:
- For the Koch fractal, $D_F = \frac{\log 16}{\log 9} = 1.26$.
- For the 64-segment quadric fractal it is $D_F = \frac{\log 64}{\log 16} = 1.5$

11.3 Multifractal methods

Multifractals are fractals having multiple scaling rules. The concept of multiscale-multifractality is important for space plasmas since this makes us capable of looking at intermittent turbulence in the solar wind [16, 17]. Kolmogorov [18] and Kraichnan [19] as well as many authors have tried to make progress as to the observed scaling exponents through the use of multifractal phenomenological models of turbulence that describe the distribution of energy flux between surging whirls at several scales [20–22].

Specially, the multifractal spectrum was examined by the use of Voyager [23] (magnetic field fluctuations) data in the outer heliosphere [23–25] and through Helios (plasma) [26] data in the inner heliosphere [27, 28]. Multifractal spectrum was also analyzed directly with respect to the solar wind attractor and it was revealed that it was consistent with the two-scale weighted multifractal measured [21]. The multifractal scaling was tested as well via the use of Ulysses observations [29] and also through the Advanced Composition Explorer and WIND data [29–32].

11.3.1 Two-dimensional fractional Brownian motion

Random walk and Brownian motion are relevant concepts in biological and physical science areas [32, 33]. The Brownian motion theory concentrates on the nonstationary Gaussian process. Or it deems that it is the sum of a stationary pure random process – the white noise in Langevin's approach [27]. Generalization of the classical theory to the fractional Brownian motion (fBm) [34, 35] and the conforming fractional Gaussian noise at present encompasses a wide class of self-similar stochastic processes. Their variances have scale with power law N^{2H} is the number of steps $0 < H < 1$; $H= 0.5$ matches the classical Brownian motion through the independent steps. Mandelbrot [36] proposed the term "fractional" associated with fractional integration and differentiation. The steps in the fBm are correlated strongly and they have long memory. *Long memory process* is defined as the time-dependent process whose empirical autocorrelation shows a decay that is slower than the exponential pattern as arranged by the finitelag processes [37, 38]. Thus, fBm has proven to be an influential mathematical model to study correlated random motion enjoying a wide application in biology and physics [37, 38]. The one-dimensional fBm can be generalized to two dimensions (2D). Not only a stochastic process in 2D, that is, a sequence of 2D random vectors $\{(B_i^x, B_i^y)|i = 0, 1, ...\}$, but also a random field in 2D $\{B_{i,j}|i = 0, 1, ...; j = 0, 1, ...\}$ is considered, a scalar random variable that is defined on a plane. These represent the spin dimension 2 and space dimension 2 in the respective order in critical phenomena theory in the field of statistical physics [39]. The most common fractional Brownian random field is an n-dimensional random vector, the definition of which is made in an m-dimensional space [39].

The fBm function is a Brownian motion $B_H(t)$ such that (see [33]):

$$\operatorname{cov}\{B_H(S), B_H(t)\} = \frac{1}{2}\left\{|S|^{2H} + |t|^{2H} - |S - t|^{2H}\right\} \tag{11.3}$$

$B_H(t)$ is defined by the stochastic integral

$$\begin{aligned} B_H(t) = {} & \frac{1}{\Gamma(H + (1/2))} x \left\{ \int_{-\infty}^{0} \left[(t - S)^{H - \left(\frac{1}{2}\right)} - (-S)(t - S)^{H - \left(\frac{1}{2}\right)} \right] \right\} dW(S) \\ & + \int_{-\infty}^{0} \left[(t - S)^{H - \left(\frac{1}{2}\right)} dW(S) \right] \end{aligned} \tag{11.4}$$

dW represents the Wiener process [40–44]. In addition, the integration is taken in the mean square integral form.

Definition 11.3 (Square integrable in the mean). A random process X is square integrable from a and b if $E\left[\int_a^b X^2(t)dt\right]$ is finite. Notice that if X is bounded on $[a, b]$, in the sense that $|X(t)| \leq M$ with probability 1 for all $a \leq t \leq b$, then X is square integrable from a to b.

Such a kind of a natural extension of *fBm* brings about some sense of a "loss" of properties. The increments of *fBm* in truth are nonstationary, with the process not being self-similar. It is known that singularities and irregularities are the most prominent ones and informative components of a data.

H is the Hurst exponent, $\imath = \alpha/2$ and $\alpha \in (0,1)$ is the fractal dimension [42]. The Hurst exponent H assumes two major roles. The first one is that it makes the determination of the roughness of the sample path besides the long-range dependence properties pertaining to the process. Most of the time, the roughness of the path is based on the location. It may also be necessary to take the multiple regimes into consideration. As for the second role, it can be said that this role can be realized by replacing the Hurst exponent H by a Hölder function (see Section 11.3.2) [43] being:

$$0 < H(t) < 1, \ \forall t \tag{11.5}$$

This kind of a generalization gives the way to multifractional Brownian motion (*mfBm*) term. In such a case, the stochastic integral expression resembles the one pertaining to the *fBm* but the case of H being replaced by the corresponding Hölder function is not applicable. The second integral of eq. (11.4) is with an upper limit of t and the integration is taken in the mean square sense. In addition, for *mfBm*, increments are not stationary anymore and the process is not self-similar anymore either.

In order to generate sample trajectories, it is possible to find both simplified covariance functions and pseudocodes. *2D-fBm* algorithms form time series of spatial coordinates. One can regard this as spatial jumps that are alike off-lattice simulations. However, it is important to remember that these increments are correlated.

11.3.2 Hölder regularity

Assume the Hölder function of exponent β Let (X,d_x) and (Y,d_y) be two metric spaces with corresponding metrics. A function $f: X \to Y$ is a Hölder function of exponent $\beta > 0$; if for each $x, y \in X$, $d_X(x,y) < 1$ there results [44]

$$d_Y(f(x), f(y)) \le C \ d_x(x,y)^{\beta} \tag{11.6}$$

f is a function from an interval $D \subseteq R$ into R. The function f is Hölder continuous on the condition that there are positive constants C and α such that

$$|f(x) - f(y)| \le C|x-y|^{a}, x, y \in D \tag{11.7}$$

The constant α is referred to as the *Hölder exponent*. A continuous function has the property that $f(x) \to f(y)$ as $x \to y$, or equivalently $f(y) - f(x) \to 0$ as $x \to y$. Hölder continuity allows the measurement of the *rate* of convergence. Let us say that if f is Hölder continuous with exponent α, then $f(x) - f(y) \to 0$ as $x \to y$ at least in the same fastness as $|x-y|^{a} \to 0$ [45].

11.3.3 Fractional Brownian motion

Hölder regularity can be used to characterize singular structures in a precise manner [46]. Thus, the regularity of a signal at each point by the pointwise Hölder exponent can be quantified as follows.

Hölder regularity embodies significant singularities [47]. Each point in data is defined pursuant to Hölder exponent (11.6) as has already been mentioned.

Let $f: R \to R, S \in R^{+*}N$ and $x_0 \in R$. $f \in C^S(x_0)$ if and only if $\exists \eta \in R^{+*}$ and polynomial degree P of degree < S and a constant C such that [48]

$$(\forall x \in B(x_0, \eta)|f(x) - P(x - x_0)| \leq C|x - x_0|^S) \tag{11.8}$$

where $B\ (x_0,\ \eta)$ is the local neighborhood around x_0 with a radius η. The point-wise Hölder exponent of f at x_0 is as $\alpha_P(x_0) = \sup_s\{f \in C^s(x_0)\}$.

The Hölder exponent of function $f(t)$ at point t is the sup $(\alpha_P) \in [0, 1]$. For this, a constant C exists with $\forall\ t'$ in a neighborhood of t,

$$|f(t) - f(t')| \ \leq C|t - t'|^{\alpha_P} \tag{11.9}$$

Regarding the *oscillations*, a function $f(t)$ is Hölder with exponent $\alpha_p \in [0, 1]$ at t if $\exists c$ $\forall \tau$ such that $\mathrm{osc}_\tau\ (t) \leq c\tau^{\alpha_P}$, embodying the following:

$$\mathrm{osc}_\tau(t) = \sup_{t', t'' \in |t - \tau, t + \tau|} |f(t') - f(t'')| \tag{11.10}$$

Now, $t = x_0$ and $t' = x_0 + h$ in eq. (11.9) can also be stated as follows:

$$\alpha_p(x_0) = \lim \inf_{h \to 0} \frac{\log|f(x_0\ +\ h)\ -\ f(x_0)|}{\log|h|}$$

Therefore, the problem is to find an α_p that fulfills eqs. (11.11) and (11.12). For practical reasons, in order to simplify the process, we can set $\tau = \beta^r$. At that point, we can write $\mathrm{osc}_\tau \approx c\tau_p^\alpha = \beta^{(\alpha_p r + b)}$, as equivalent to eq. (11.9).

$$\log_\beta(\mathrm{osc}_\tau) \approx (\alpha_p r + b) \tag{11.11}$$

Thus, an estimation of the regularity is generated at each point by calculating the regression slope between the logarithm of the oscillations osc_τ and the logarithm of the dimension of the neighborhood at which oscillations τ are calculated. In the following, we will use the least squares regression for calculating the slope with $\beta = 2$ and $r = 1, 2, 3, 4$. Moreover, it is better not to use all sizes of neighborhoods between the two values $\tau_{\min}$ and $\tau_{\max}$. Here, the point x_0 is computed merely on intervals of the form $[x_0 - \tau_r\text{: } x_0 + \tau_r]$. For 2D MS dataset, economy (U.N.I.S.) dataset and Wechsler Adult Intelligence Scaled – Revised (WAIS-R) dataset, x_0 defines a point in 2D space and τ_r a radius around x_0, such that $d(t', t) \leq \tau_r$: and $d(t'', t) \leq \tau_r$, where $d(a, b)$ is the Euclidean distance between a and b [49].

In this section, applications are within this range $0 < H < 1$ and we use 2D fBm Hölder different functions (polynomial [49–51] and exponential [49–51]) that take as input the points (x, y) of the dataset and provide as output the desired regularity; these functions are used to show the dataset singularities:

(a) A polynomial of degree n is a function of the form

$$H_1(x,y) = a_n x^n y^n + a_{n-1} x^{n-1} y^{n-1} + \cdots + a_2 x^2 y^2 + a_1 xy + a_0 \tag{11.12}$$

where $a_{0,\dots,n}$ are constant values.

(b) The general definition of the exponential function is [52]

$$H_2(x,y) = \frac{L}{1 + e^{-k(x - x_0 y)}} \tag{11.13}$$

where e is the natural logarithm base, x_0 is the x-value of the sigmoid midpoint, L is the curve maximum value and k the curve steepness.

2D fBm Hölder function coefficient value interval: Notice that, in the case $H = \frac{1}{2}$, B is a planar Brownian motion. We will always assume that $H > \frac{1}{2}$. In this case, it is known that B is a transient process (see [53]). In this respect, study of the 2D fBm with Hurst parameter $H > \frac{1}{2}$ could seem much simpler than the study of the planar Brownian motion [54–56].

11.3.3.1 Diffusion-limited aggregation

Witten and Sander [50] introduced the diffusion-limited aggregation (DLA) model, also called Brownian trees, in order to simulate cluster formation processes that relied on a sequence of computer-generated random walks on a lattice that simulated diffusion of a particle. Tacitly, each random walk's step size was fixed as one lattice unit. The DLA simulation's nonlattice version was conducted by Meakin [57], who limited the step size to one particle diameter. However, diffusion in real cluster formation processes might go beyond this diameter, when it is considered that the step size depends on physical parameters of the system such as concentration, temperature, particle size and pressure. For this reason, it has become significant to define the effects of the step size on the manifold processes regarding cluster formation.

The flowchart for the DLA algorithm is given in Figure 11.21.

Based on the flowchart for the DLA algorithm as shown in Figure 11.21, you can find the steps for the DLA algorithm. The steps applied in DLA algorithm are as follows:

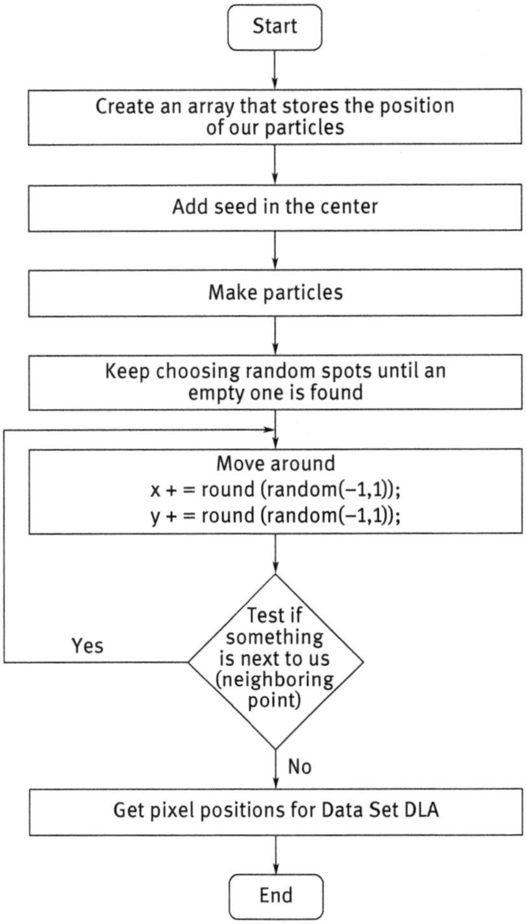

Figure 11.21: The flowchart for the DLA algorithm.

The general structure of the DLA algorithm is shown in Figure 11.22. Let us try to figure out how the DLA algorithm is applied to the dataset. Figure 11.23 shows the steps of this.

Step (1) DLA algorithm for dataset particle count is chosen.

Steps (2–18) Set the random number seeds based on the current time and points.

Steps (19–21) If you do not strike an existing point, save intermediate images for your dataset.

DLA model has a highly big step size. In this book, we concentrate on the step-size effects on the structure of clusters, formed by DLA model concerning the following

Algorithm: Diffusion Limited Aggregation Algorithm

Input: An array that stores the count of Data Set particles

Output: Number of points to introduce

Method:

```
// make particles count
//add seed in the center
CenterofX = width / 2;
CenterofY = height / 2;
while (i < particlescount) do
{
// move around particle
// draw particle x and y
if (!stuck) {
  x += round(random(−1, 1));
  y += round(random(−1, 1));
   if (x < 0 || y < 0 || x ≥ width || y ≥ height) {
  // keep choosing random spots till an empty one is found
  do {
    x = floor(random(width));
    y = floor(random(height));
  } while (field[y × width + x]);
}
if (particles[i].stuck){
  pixels[particles[i].y × width + particles[i].x] = color(0);
  // returns true if no neighboring pixels and get positions for points
```

Figure 11.22: General DLA algorithm.

Figure 11.23: DLA particle clusters are formed randomly by moving them in the direction of a target area till they strike with the existing structures, and then become part of the clusters.

datasets: MS dataset, economy (U.N.I.S.) dataset and WAIS-R dataset. We have also considered the two-length-scale concept as was suggested by Bensimon et al. [58]. A diffusing particle is deemed as adhering to an aggregate upon the center of the diffusing particle being at a distance from any member particle of the aggregate . The particle diameter is normally taken as the sticking length on the basis of hard-sphere concept. Specific for diffusion related processes, the other length scale is the mean free path l_0 belonging to the diffusing particle termed as the step size of random walks. We look at the case of $l_0 \gg a$ along with the extreme case of $10 \gg a$ (studied by Bensimon et al. [58]) due to the fact that the constraint of $l_0 \le a$ might not be valid for all physical systems.

DLA is classified into family of fractals that are termed as stochastic fractals due to their formation by random processes.

DLA method in data analysis: Let us apply the DLA method on the datasets comprising numeric data. They are economy (U.N.I.S.) dataset (Table 2.8), MS dataset (see Table 2.12) and WAIS-R dataset (see Table 2.19). The nonlattice DLA computer simulations were performed with various step sizes ranging from 128 to 256 points for 10,000 particles.

The cluster in Figures 11.25(a) and (b), 11.26(a) and (b) and 11.27(a) and (b) has a highly open structure so you can imagine that a motion with a small step size can move a particle into the central region of the cluster easily. On the other hand, the direction of each movement is random so it is very improbable for a particle to retain its incident direction through a number of steps. It will travel around in a certain domain producing a considerably extensive incident beam. Thus, a particle after short random walks in general traces the outer branches of the growing cluster rather than entering into the cluster core. Through a larger step size, the particle may travel a lengthier distance in a single direction.

Here we can have a look at our implementation of DLA for MS dataset, economy (U.N.I.S.) dataset and WAIS-R dataset for FracLab [51]:

(a) For each step, the particles vary their position by the use of these two motions. A random motion is selected from the economy (U.N.I.S.) dataset, MS dataset and WAIS-R dataset of 2D numeric matrices that were randomly produced. The fixed set of transforms makes it faster in a way, allowing for more evident structures to form compared to what may be possible if it was merely random.
(b) An attractive motion is the particle attracted toward the center of the screen, or else the rendering time would be excessively long.

During the handling of the set of transforms, when one adds (a, b), it is also important to insert (-a, -b). In this way, the attraction force toward the center is allowed to be the primary influence on the particle. This is also done for another reason, which is due to the fact that without this provision, there are substantial number of cases in which the particle never reaches close enough toward the standing cluster.

Algorithm: Diffusion Limited Aggregation Algorithm for Economy Data Set
Input: An array that stores the count of Economy Data Set (228 × 18) particles for 10.000 particles
Output: Number of 128 and 256 points to introduce
Method:

```
(1)  // make particles count
(2)  //add seed in the center
(3)  Center of X = width / 2;
(4)  Center of Y = height / 2;
(5)  while (i < 10000) do
(6)  {
(7)  // move around particle
(8)  // draw particle x and y
(9)  if (!stuck) {
(10)     x += round ( random(−1, 1));
(11)     y += round ( random(−1, 1));
(12)      if (x < 0 || y < 0 || x ≥ width || y ≥ height) {
(13)     // keep choosing random spots till an empty one is found
(14)     do {
(15)        x = floor (random(width));
(16)        y = floor (random(height));
(17)     } while (field [y × width + x] );
(18) }
(19) if (particles[i].stuck){
(20)    pixels [ particles [i].y × width + particles [i].x ] = color (0);
(21)    // returns true if no neighboring pixels and get positions for points
```

Figure 11.24: General DLA algorithm for the economy dataset.

Analogously for the economic dataset we have the following algorithm (Figure 11.24).

The DLA algorithm steps are as follows:

Step (1) DLA algorithm for economy dataset particle count is selected 10,000 for economy dataset.

Steps (2–18) Set the random number seeds based on the current time, and 128 points and 256 points are introduced for application of economy dataset DLA. We repeat till we hit another pixel. Our objective is to move in a random direction. x is represented as DLA 128 and 256 point image's weight and y is represented as DLA 128 and 256 point image's height.

Steps (19–21) If we do not structure an existing point, save intermediate images for economy dataset. Returns true if no neighboring pixels and get positions for 128 points and 256 points.

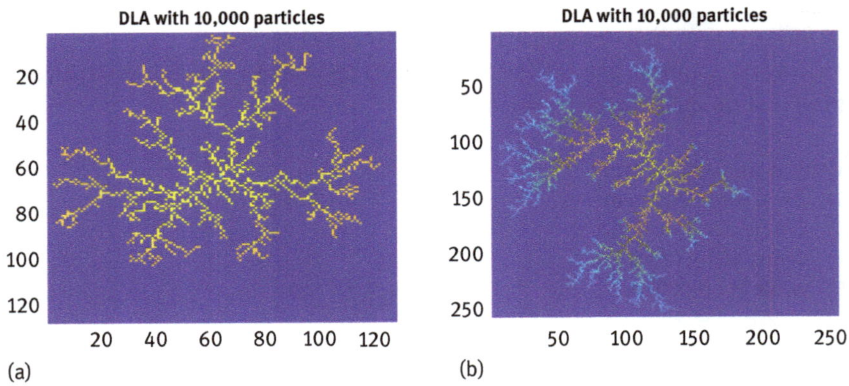

Figure 11.25: Economy (U.N.I.S.) dataset random cluster of 10,000 particles generated using the DLA model with the step size of (a) 128 points and (b) 256 points. (a) Economy (U.N.I.S.) dataset 128 points DLA and (b) economy (U.N.I.S.) dataset 256 points DLA.

Figure 11.25 (a) and –(b) shows the structure of 10,000-particle clusters produced by the DLA model with step size of 128 points and 256 points for the economy (U.N.I.S.) dataset.

It is seen through the figures that cluster structure becomes more compact when the step size is increased. Figure 11.26(a) shows the radius of cluster as a function of the launched 10,000 particles by the DLA model with step size of 128 points concerning the economy (U.N.I.S.) dataset. It is seen through the figures that cluster structure becomes more compact when the step size is increased. Figure 11.26(b) shows the radius of cluster as a function of the launched 10,000 particles by the DLA model with step size of 256 points concerning the economy (U.N.I.S.) dataset.

Economy (U.N.I.S.) dataset numerical simulations of the DLA process were initially performed in 2D, having been reported in plots of a typical aggregate (Figure11.26(a) and (b)). Nonlattice DLA computer simulations were conducted with various step sizes that range from 128 to 256 points for 10,000 particles. Figure 11.27(a) and (b) presents the structure of 10,000-particle clusters produced by the DLA model with step size of 128 points and 256 points for the economy (U.N.I.S.) dataset. You can see that the cluster structure becomes more compact when the step size is increased. Figure 11.27(a) and (b) shows the radius of cluster as a function of the launched 10,000 particles by the DLA model with step size of 128 points and 256 points for economy (U.N.I.S.) dataset. At all steps of the calculation, it is required to launch particles randomly, allowing them to walk in a random manner, but stop them when they reach a perimeter site, or in another case, it is essential to start them over when they walk away. It is adequate to start walkers on a small circle defining the existing cluster for the launching. We select another circle twice the radius of the cluster

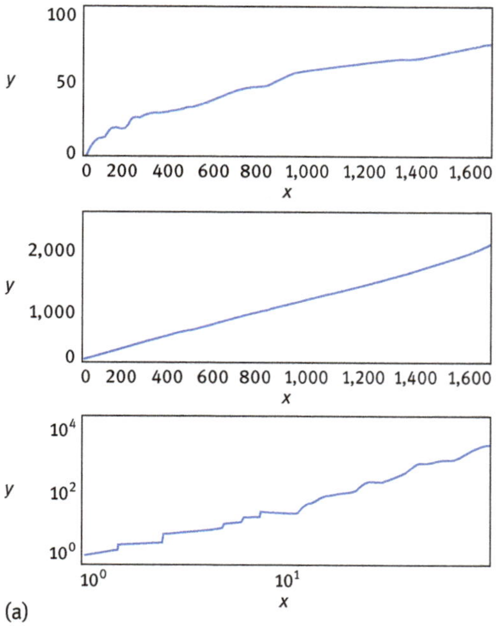

Radius of the cluster as a function (on *x*-axis)	Launched particles (on *y*-axis)
0	0
800	50
1,600	60

Particles in the cluster as a function (on *x*-axis)	Launched particles (on *y*-axis)
0	0
1,000	1,000
1,600	1,600

Logarithm of the radius (on *x*-axis)	Particles are as a function of the logarithm of the radius (on *y*-axis)
10^0	0
10^2	10^1

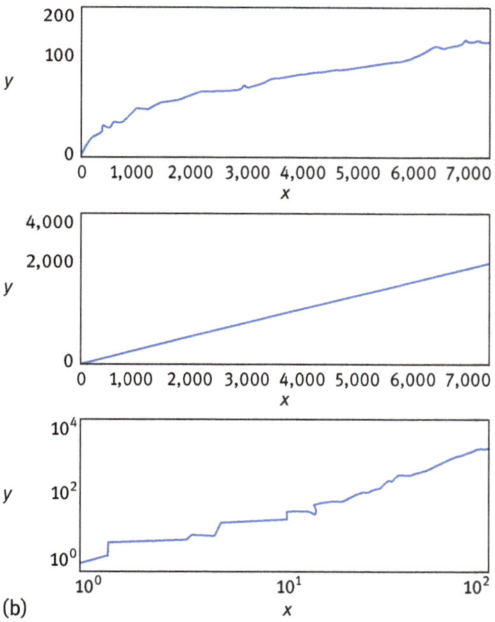

Radius of the cluster as a function (on *x*-axis)	Launched particles (on *y*-axis)
0	0
4,000	100
7,000	150

Particles in the cluster as a function (on *x*-axis)	Launched particles (on *y*-axis)
0	0
4,000	2,000
7,000	3,500

Logarithm of the radius (on *x*-axis)	Particles are as a function of the logarithm of the radius (on *y*-axis)
10^0	0
10^1	10^2

Figure 11.26: Radius of cluster as a function of the launched 10,000 particles by the DLA model with step size of (a) 128 points and (b) 256 points for economy (U.N.I.S.) dataset.

for the purpose of grabbing the errant particles. Figure 11.27(a) and (b) shows that 128 points DLA logarithm of the radius cluster as a function is increased according to 256 points DLA for economy (U.N.I.S.) dataset.

The DLA algorithm is given in Figure 11.27.

Algorithm: Diffusion Limited Aggregation Algorithm for MS Data Set
Input: An array that stores the count of MS Data Set (304 × 112) particles as 10.000 particles
Output: Number of 128 and 256 points to introduce
Method:

```
(1)  // make particles count
(2)  //add seed in the center
(3)  Center of X = width / 2;
(4)  Center of Y = height / 2;
(5)  while (i < 10000) do
(6)  {
(7)  // move around particle
(8)  // draw particle x and y
(9)  if (!stuck) {
(10)     x += round ( random(−1, 1));
(11)     y += round ( random(−1, 1));
(12)      if (x < 0 || y < 0 || x ≥ width || y ≥ height) {
(13)     // keep choosing random spots till an empty one is found
(14)     do {
(15)        x = floor (random(width));
(16)        y = floor (random(height));
(17)     } while (field [y × width + x] );
(18) }
(19) if (particles[i].stuck){
(20)    pixels [ particles [i].y × width + particles [i].x ] = color (0);
(21)    // returns true if no neighboring pixels and get positions for points
```

Figure 11.27: General DLA algorithm for the MS dataset.

The MS dataset provided in Table 2.12 in line with the DLA algorithm steps (Figure 11.24) is based on the following steps:

Step (1) DLA algorithm for MS dataset particle count is selected 10,000 for the MS dataset.

Steps (2–18) Set the random number seeds based upon the current time, and 128 points and 256 points are introduced for application of MS dataset DLA. We repeat till we hit another pixel. Our objective is to move in a random direction. x is represented as DLA 128 and 256 point image's weight and y is represented as DLA 128 and 256 point image's height.

Steps (19–21) If we do not structure an existing point, save intermediate images for the MS dataset. Returns true if no neighboring pixels and obtain positions for 128 points and 256 points.

Figure 11.28(a) and (b) shows the structure of 10,000-particle clusters produced by the DLA model with step size of 128 points and 256 points for the MS dataset.

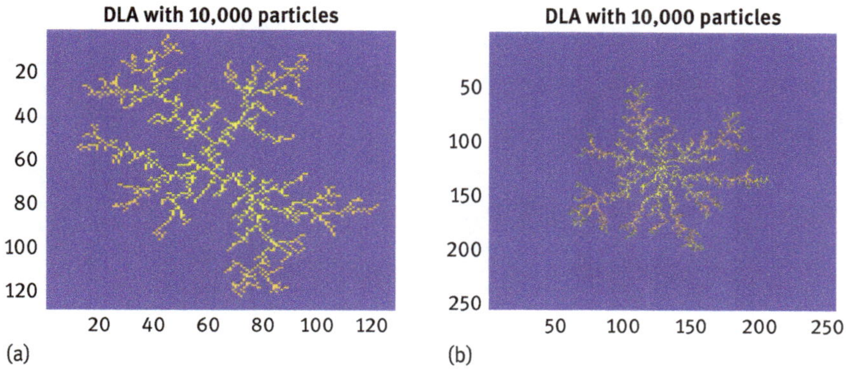

Figure 11.28: MS dataset random cluster of 10,000 particles produced using the DLA model with the step size of (a) 128 points and (b) 256 points. (a) MS dataset 128 points DLA and (b) MS dataset 256 points DLA

It is seen through the figures that cluster structure becomes more compact when the step size is increasing. Figure 11.29(a) shows the radius of cluster as a function of the launched 10,000 particles by the DLA model with step size of 128 points concerning the MS dataset. It is seen that cluster structure becomes more compact when the step size is increasing. Figure 11.29 (b) shows the radius of cluster as a function of the launched 10,000 particles by the DLA model with step size of 256 points concerning the MS dataset.

MS dataset numerical simulations of the DLA process were initially performed in 2D, having been reported in plot of a typical aggregate (Figure 11.25(a) and (b)). Nonlattice DLA computer simulations were conducted with various step sizes that range from 128 to 256 points for 10,000 particles. Figure 11.28(a) and (b) presents the structure of 10,000-particle clusters produced by the DLA model with step size of 128 points and 256 points for the MS dataset. You can see that the cluster structure becomes more compact when the step size is increased. Figure 11.28(a) and (b) shows the radius of cluster as a function of the launched 10,000 particles by the DLA model with step size of 128 points and 256 points MS dataset. At all steps of the calculation, it is required to launch particles randomly, allowing them to walk in a random manner, but stop them when they reach a perimeter site, or in another case, it is essential to start them over when they walk away. It is adequate to start walkers on a small circle defining the existing cluster for the launching.

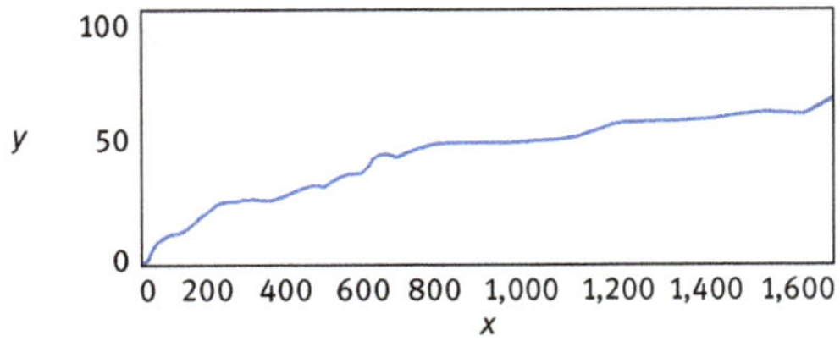

Radius of the cluster as a function (on *x*-axis)	Launched particles (on *y*-axis)
0	0
50	600
100	1,600

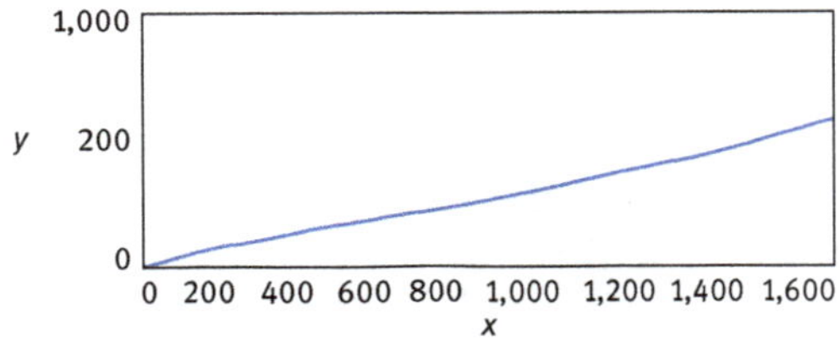

Particles in the cluster as a function (on *x*-axis)	Launched particles (on *y*-axis)
0	0
1,000	800
1,100	1,600

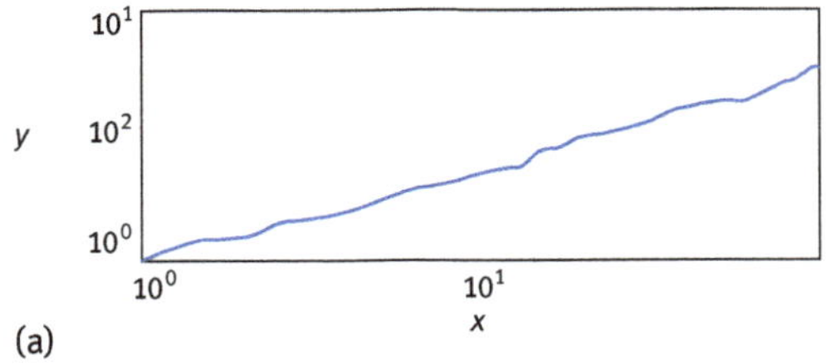

Logarithm of the radius (on *x*-axis)	Particles are as a function of the logarithm of the radius (on *y*-axis)
10^0	0
10^2	10^1

(a)

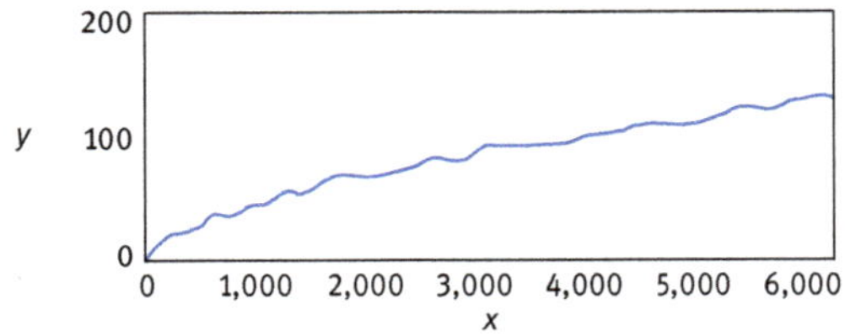

Radius of the cluster as a function (on *x*-axis)	Launched particles (on *y*-axis)
0	0
100	3,000
150	6,000

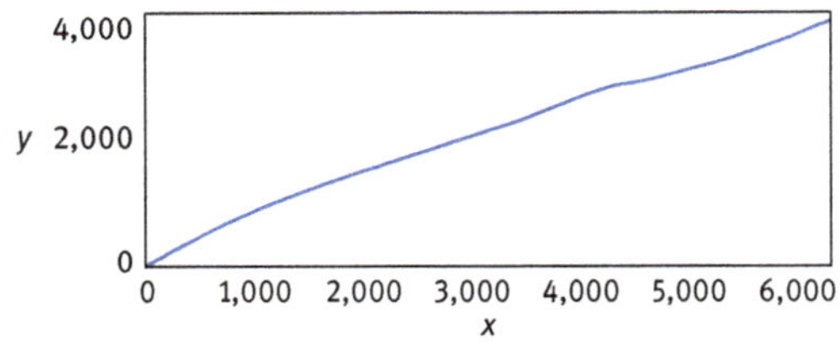

Particles in the cluster as a function (on *x*-axis)	Launched particles (on *y*-axis)
0	0
1,000	800
2,000	6,000

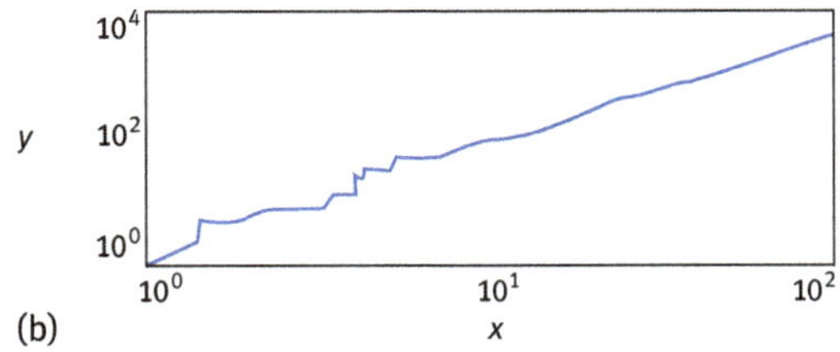

Logarithm of the radius (on *x*-axis)	Particles are as a function of the logarithm of the radius (on *y*-axis)
10^0	0
10^2	10^1

(b)

Figure 11.29: Radius of cluster as a function of the launched 10,000 particles by the DLA model with step size of (a) 128 points and (b) 256 points for the MS dataset.

We select another circle twice the radius of the cluster for the purpose of grabbing the errant particles. Figure 11.28(a) and (b) shows that 128 points DLA logarithm of the radius cluster as a function is increased according to 256 points DLA for MS dataset (Figure 11.30).

Algorithm: Diffusion Limited Aggregation Algorithm for WAIS-R Data Set
Input: An array that stores the count of WAIS-R Data Set (400 × 21) particles for 10.000 particles
Output: Number of 128 and 256 points to introduce
Method:

```
(1)  // make particles count
(2)  //add seed in the center
(3)  Center of X = width / 2;
(4)  Center of Y = height / 2;
(5)  while (i < 10000) do
(6)  {
(7)  // move around particle
(8)  // draw particle x and y
(9)  if (!stuck) {
(10)     x += round ( random(−1, 1));
(11)     y += round ( random(−1, 1));
(12)      if (x < 0 || y < 0 || x ≥ width || y ≥ height) {
(13)     // keep choosing random spots till an empty one is found
(14)     do {
(15)         x = floor (random(width));
(16)         y = floor (random(height));
(17)     } while (field [y × width + x] );
(18) }
(19) if (particles[i]. stuck){
(20)    pixels [ particles [i].y × width + particles [i].x ] = color (0);
(21)    // returns true if no neighboring pixels and get positions for points
```

Figure 11.30: General DLA algorithm for the WAIS-R dataset.

The steps of DLA algorithm are as follows:

Step (1) DLA algorithm for WAIS-R dataset particle count is selected 10,000 for WAIS-R dataset.

Steps (2–18) Set the random number seeds based on the current time, and 128 points and 256 points are introduced for application of WAIS-R dataset DLA. We repeat till we hit another pixel. Our aim is to move in a random direction. x is represented as DLA 128 and 256 point image's weight and y is represented as DLA 128 and 256 point image's height.

Steps (19–21) If we do not structure an existing point, save intermediate images for WAIS-R dataset. Returns true if no neighboring pixels and obtain positions for 128 points and 256 points.

Figure 11.31(a) and (b) shows the structure of 10,000-particle clusters produced by the DLA model with step size of 128 points and 256 points for the WAIS-R dataset.

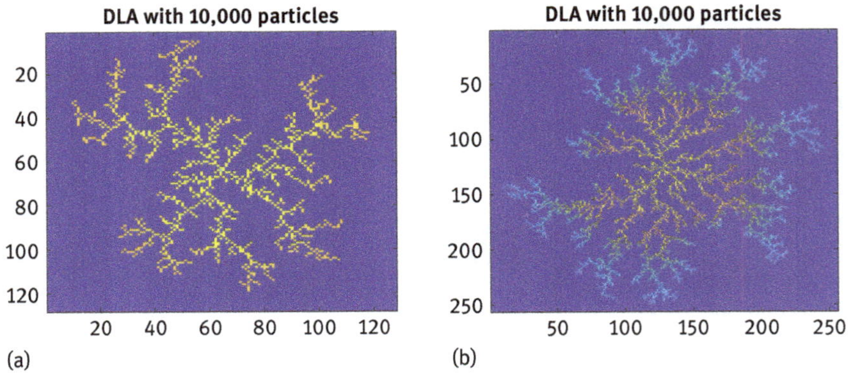

Figure 11.31: WAIS-R dataset random cluster of 10,000 particles produced using the DLA model with the step size of (a) 128 points and (b) 256 points. (a) WAIS-R dataset 128 points DLA and (b) WAIS-R dataset 256 points DLA.

It is seen through the figures that cluster structure becomes more compact when the step size is increased. Figure 11.31(a) shows the radius of cluster as a function of the launched 10,000 particles by the DLA model with step size of 128 points concerning the WAIS-R dataset. It is seen through the figures that cluster structure becomes more compact when the step size is increased. Figure 11.28(b) shows the radius of cluster as a function of the launched 10,000 particles by the DLA model with step size of 256 points concerning the WAIS-R dataset.

WAIS-R dataset numerical simulations of the DLA process were initially performed in 2D, having been reported in plot of a typical aggregate (Figure 11.32(a) and (b)). Nonlattice DLA computer simulations were conducted with various step sizes that range from 128 to 256 points for 10,000 particles. Figure 11.32(a) and (b) presents the structure of 10,000-particle clusters produced by the DLA model with step size of 128 points and 256 points for the WAIS-R dataset. You can see that the cluster structure becomes more compact when the step size is increased. Figure 11.32(a) and (b) shows the radius of cluster as a function of the launched 10,000 particles by the DLA model with step size of 128 points and 256 points for the WAIS-R dataset. At all steps of the calculation, it is required to launch particles randomly, allowing them to walk in a random manner, but stop them when they reach a perimeter site, or in another case, it is essential to start them over when they walk away. It is adequate to start walkers on a

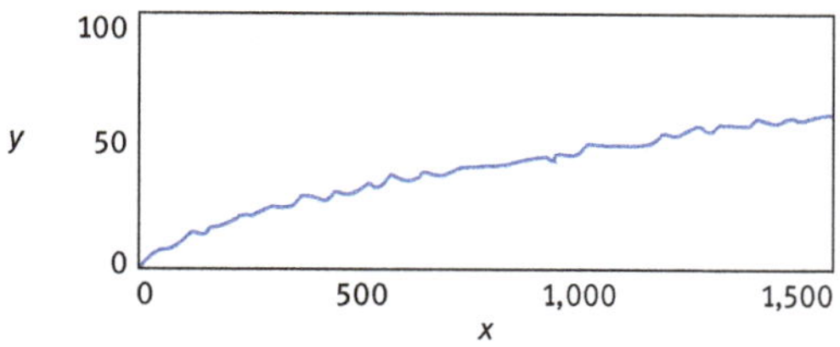

Radius of the cluster as a function (on *x*-axis)	Launched particles (on *y*-axis)
0	0
500	50
1,500	75

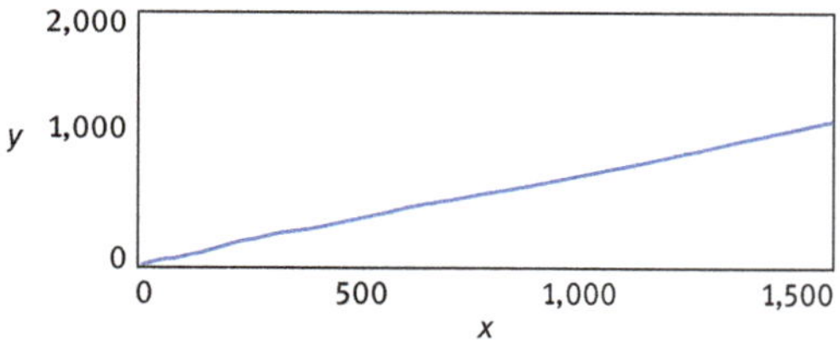

Particles in the cluster as a function (on *x*-axis)	Launched particles (on *y*-axis)
0	0
1,000	8,00
1,500	1,500

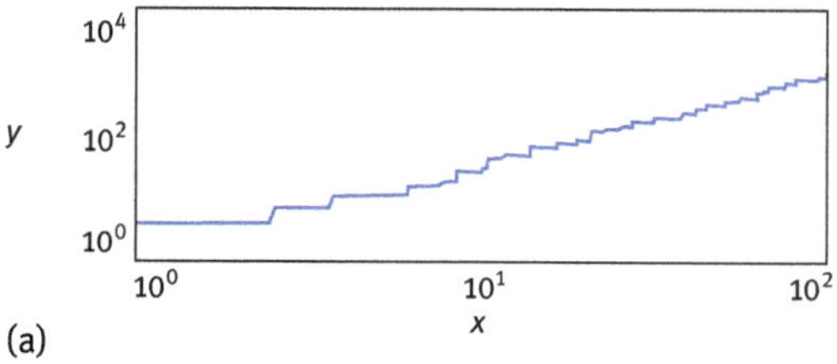

Logarithm of the radius (on *x*-axis)	Particles are as a function of the logarithm of the radius (on *y*-axis)
10^1	10^2

(a)

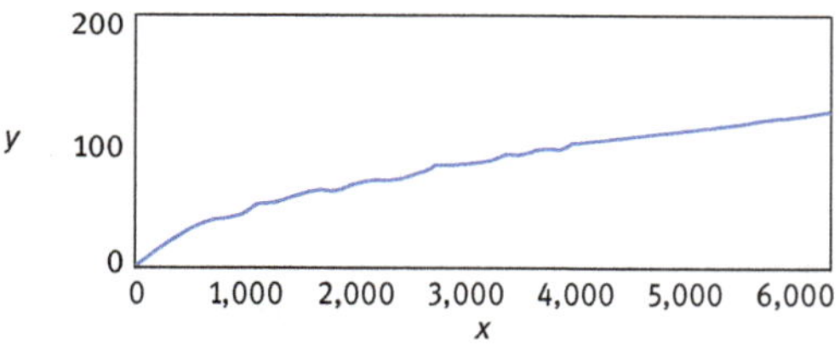

Radius of the cluster as a function (on *x*-axis)	Launched particles (on *y*-axis)
0	0
1,000	50
6,000	100

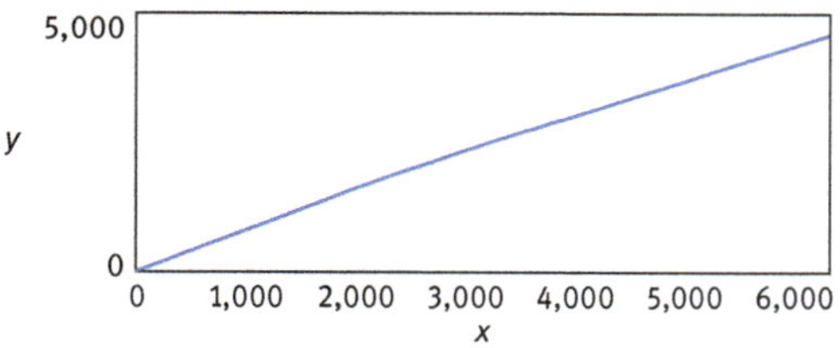

Particles in the cluster as a function (on *x*-axis)	Launched particles (on *y*-axis)
0	0
6,000	4,000

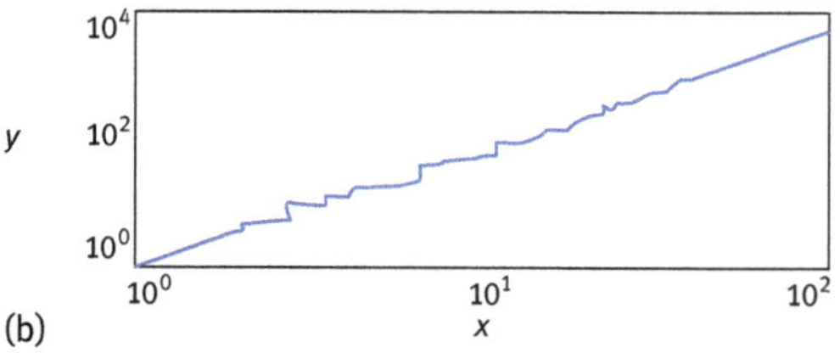

Logarithm of the radius (on *x*-axis)	Particles are as a function of the logarithm of the radius (on *y*-axis)
10^1	10^2
10^2	10^3

(b)

Figure 11.32: Radius of cluster as a function of the launched 10,000 particles by the DLA model with step size of 128 points (a) and 256 points (b) WAIS-R dataset.

small circle defining the existing cluster for the launching. We select another circle twice the radius of the cluster for the purpose of grabbing the errant particles. Figure 11.32(a) and (b) shows that 128 points DLA logarithm of the radius cluster as a function is increased according to 256 points DLA for WAIS-R dataset.

11.4 Multifractal analysis with LVQ algorithm

In this section, we consider the classification of a dataset through the multifractal Brownian motion synthesis 2D multifractal analysis as follows:

(a) Multifractal Brownian motion synthesis 2D is applied to the data.
(b) Brownian motion Hölder regularity (polynomial and exponential) for the analysis is applied to the data for the purpose of identifying the data.
(c) Singularities are classified through the Kohonen learning vector quantization (LVQ) algorithm. The best matrix size (which yields the best classification result) is selected for the new dataset while applying the Hölder functions (polynomial and exponential) on the dataset.

Two multifractal Brownian motion synthesis 2D functions are provided below:

General polynomial Hölder function:

$$H_1(x,y) = a_n x^n y^n + a_{n-1} x^{n-1} y^{n-1} + \cdots + a_2 x^2 y^2 + a_1 xy + a_0 \tag{11.12}$$

The coefficient values in eq. (11.12) (a_n, a_{n-1},...,a_0) are derived based on the results of the LVQ algorithm according to *2D fBm Hölder function coefficient value*. They are the coefficient values that yield the most accurate result. With these values, the definition is made as follows:

$$(a_1 = 0.8, a_0 = 0.5); H_1(x,y) = (0.5 + 0.8 \times x \times y) \tag{11.13}$$

General exponential Hölder function:

$$H_2(x,y) = \frac{L}{1 + e^{-k(x - x_0 y)}} \tag{11.14}$$

The coefficient values in eq. (11.14) (x_0, L) are derived based on the results of the LVQ algorithm according to *2D fBm Hölder function coefficient value*. They are the coefficient values that yield the most accurate result. With these values, the definition is made as follows:

$$(L = 0.6, k = 100, x_0 = 0.5\ ; H_2(x,y) = 0.6 + \frac{0.6}{1 + \exp(-100 \times (x - 0.5y))} \tag{11.15}$$

Two-dimensional fBm Hölder function (see eqs. (11.12) and (11.13)) coefficient values are applied with *2D fBm Hölder function coefficient value interval* and chosen accordingly. In eqs. (11.13) and (11.15) (the selected coefficient values), LVQ algorithm was applied to the new singled out datasets (p_New Dataset and e_New Dataset). According to the 2D fBm Hölder function coefficient value interval, coefficient values are chosen among the coefficients that yield the most accurate results in the LVQ algorithm application. Equations (11.13) and (11.15) show that the coefficient values change based on the accuracy rates as derived from the algorithm applied to the data.

How can we single out the singularities when the multifractal Brownian motion synthesis 2D functions (polynomial and exponential) are applied onto the dataset (D)?

As explained in Chapter 2, for $i = 0, 1, \ldots, K$ in $x : i_0, i_1, \ldots, i_K$; $j = 0, 1, \ldots, L$ in $y : j_0, j_1, \ldots, j_M$, $D(K \times L)$ represents a dataset. $x : i.$ is the entry on the line and $y: j.$ is the matrix that represents the attribute in the column (see Figure 11.33). A matrix is a 2D array of numbers. When the array has K rows and L columns, it is said that the matrix size is $K \times L$.

Figure 11.33 shows the D dataset (4×5) as depicted with row (x) and column (y) numbers. As the dataset we consider the MS dataset, economy (U.N.I.S.) dataset and WAIS-R dataset.

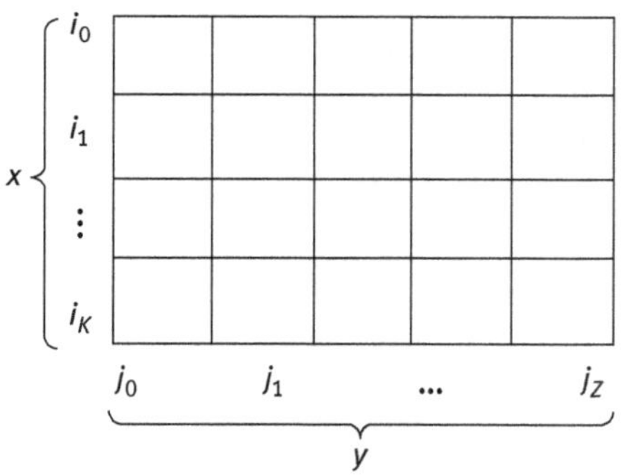

Figure 11.33: Presented with x row and y column as an example regarding D dataset.

The process to single out the singularities that will be obtained from the application of the multifractal Hölder regularity functions on these three different data can be summarized as follows:

The steps in Figure 11.34 are described in detail as follows:

Step (i) Application to data multifractal Brownian motion synthesis 2D. Multifractal Brownian motion synthesis 2D is described as follows: primarily you have to calculate the fractal dimension $D(D = [x_1, x_2, \ldots, x_m])$ of the original dataset by using eqs. (11.12) and (11.13). As the next step, you have the new dataset (p_New Datasets and e_New Datasets) formed by selecting attributes that bring about the minimum change of current

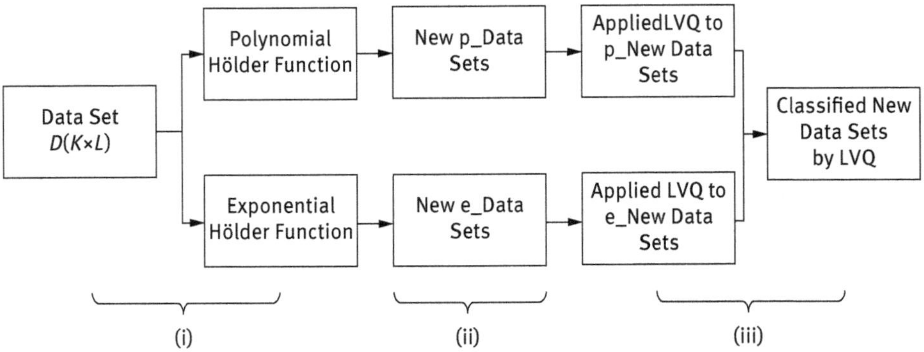

Figure 11.34: Classification of the dataset with the application of multifractal Hölder regularity through LVQ algorithm.

dataset's fractal dimension until the number of remaining features is the upper bound of fractal dimension D. D^{new} ($D^{\text{new}} = [x_1^{\text{new}}, x_2^{\text{new}}, ..., x_n^{\text{new}}]$).

Step (ii) Application on the new data of the multifractal Brownian motion synthesis 2D algorithm (see [50] for more information).

It is possible for us to perform the classification of the new dataset. For this, let us have a closer look at the steps provided in Figure 11.35.

Algorithm: Multifractals Analysis with LVQ Algorithm

Input: New training set that include n number of samples obtained ($D^{new} = [x_1^{new}, x_2^{new}, \dots, x_n^{new}])(x_i \in R^d)$

Output: label data that corresponds to m number of samples ($Y = [y_1, y_2, \dots, y_m]$)

Method:

(1) //Initialize learning rate γ
(2) **while** (stopping condition is false){
(3) **foreach** input vector $x_1^{new}, x_2^{new}, \dots, x_n^{new}$
(4) **for** each weight w_{ij} in network {
(5) Euclidian distance between the input vector and (weight vector for) the *jth* output unit.} //update w_j
(6) $if(Y = y_j)$
(7) $w_j(new) = w_j(old) + \gamma[xnew - w_j(old)]$
(8) **else**
(9) $w_j(new) = w_j(old) - \gamma[x^{new} - w_j(old)]$Update learning rate
(10) //Predict each test sample

Figure 11.35: General LVQ algorithm for the newly produced data.

Step (iii) The new datasets produced by the singled out singularities (p_New Dataset and e_New Dataset) are as a final point classified by using the Kohonen LVQ algorithm Figure 11.35.

Figure 11.35 explains the steps of LVQ algorithm as follows:

Step (1) Initialize reference vectors (several strategies are discussed shortly). Initialize learning rate, y is default (0).

Step (2) While stopping condition is false, follow Steps (2–6).

Step (3) *foreach* training input vector x^{new}, follow Steps (3–4).

Step (4) Find j so that $\left\|x^{new} - w_j\right\|$ is a minimum.

Step (5) Update w_{ij}.

Steps (6–9) *if* $(Y = y_j)$

$$w_j(\text{new}) = w_j(\text{old}) + \gamma[x^{\text{new}} - w_j(\text{old})]$$

else

$$w_j(\text{new}) = w_j(\text{old}) - \gamma[x^{\text{new}} - w_j(\text{old})]$$

Step (10) The condition may assume a fixed number of iterations (i.e., executions of Step (2)) or learning rate reaching a satisfactorily small value.

11.4.1 Polynomial Hölder function with LVQ algorithm for the analysis of various data

LVQ algorithm has been applied on economy (U.N.I.S.) dataset, multiple sclerosis (MS) dataset and WAIS-R dataset in Sections 11.4.1.1, 11.4.1.2 and 11.4.1.3, respectively.

11.4.1.1 Polynomial Hölder function with LVQ algorithm for the analysis of economy (U.N.I.S.)

For the second set of data, in the following sections, we shall use some data related to USA (abbreviated by the letter U), New Zealand (represented by the letter N), Italian (shown with the letter I) and Sweden (signified by the letter S) economies for U.N.I.S. countries. The attributes of these countries' economies are data that are concerned with *years, unemployment, GDP per capita (current international $), youth male (% of male labor force ages 15–24) (national estimate), ...,GDP growth (annual %)*. Data consist of a total of 18 attributes. Data that belong to U.N.I.S. economies from 1960 to 2015 are defined based on the attributes provided in Table 2.8 economy dataset (http://data.worldbank.org) [12], to be used in the following sections.

Our aim is to classify the economy dataset through the multifractal Brownian motion synthesis 2D multifractal analysis of the data. Our method is based on the following steps provided in Figure 11.36:

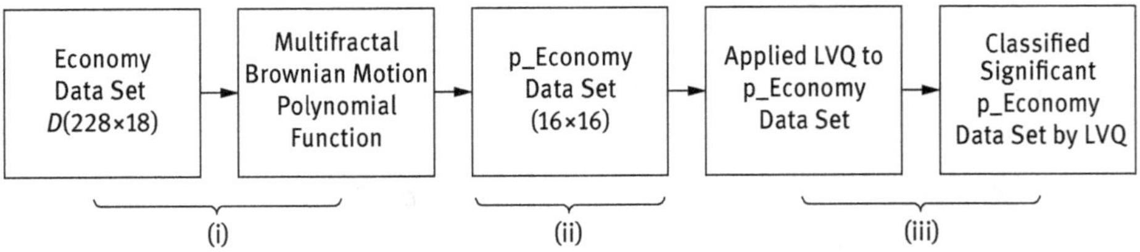

Figure 11.36: Classification of the economy dataset with the application of multifractal Hölder polynomial regularity function through LVQ algorithm.

i. Multifractal Brownian motion synthesis 2D is applied to the economy (U.N.I.S.) dataset (228 × 18).
ii. Brownian motion Hölder regularity (polynomial) for analysis is applied to the data for the purpose of identifying on the data and some new datasets comprise the singled out singularities regarding p_Economy dataset (16 × 16) (see eq. (11.13)):

$$H_1(x,y) = (0.5 + 0.8 \times x \times y).$$

iii. The new datasets (p_Economy dataset) comprised of the singled out singularities, with the matrix size (16 × 16) yielding the best classification result, are finally classified by using the Kohonen LVQ algorithm.

The steps of Figure 11.36 are explained as follows:

Step (i) Application to economy dataset multifractal Brownian motion synthesis 2D. Multifractal Brownian motion synthesis 2D can be described as follows: primarily you have to calculate the fractal dimension $D(D = [x_1, x_2, \ldots, x_{228}])$ of the original economy dataset. Fractal dimension is computed with eq. (11.13). As the next step, you have the new dataset (p_Economy Dataset) formed by selecting attributes that bring about the minimum change of current dataset's fractal dimension until the number of remaining features is the upper bound of fractal dimension D. The newly produced multifractal Brownian motion synthesis 2D dataset is $D^{\text{new}}(D^{\text{new}} = [x_1^{\text{new}}, x_2^{\text{new}}, \ldots, x_{16}^{\text{new}}])$ (see [50] for more information).

Step (ii) Application to data multifractal Brownian motion synthesis 2D (see [50] for more information).

It is possible for us to perform the classification of the p_ Economy dataset. For this, we can have a closer glance at the steps presented in Figure 11.37:

Step (iii) The singled out singularity p_Economy datasets are finally classified by using the Kohonen LVQ algorithm in Figure 11.37.

Figure 11.37 explains the steps of LVQ algorithm as follows:

Step (1) Initialize reference vectors (several strategies are discussed shortly). Initialize learning rate, y is default (0).

Algorithm: Multifractals Analysis with LVQ Algorithm

Input: New training set that include n number of samples obtained from **p_ Economy Data Set**$(D^{new} = [x_1^{new}, x_2^{new}, \ldots, x_{16}^{new}])(x_i \in R^d)$

Output: label data that corresponds to m number of samples C = [USA, New Zealand, Italy, Sweden]

Method:

(1) //Initialize learning rate γ
(2) **while** (stopping condition is false){
(3) Q **foreach** input vector $x_1, x_2, \ldots, x_{128}$
(4) **for** each weight w_{ij} in network {
(5) Euclidian distance between the input vector and (weight vector for) the *jth* output unit.}
//update w_j
(6) $if(Y = y_j)$
(7) $w_j(new) = w_j(old) + \gamma[x^{new} - w_j(old)]$
(8) **else**
(9) $w_j(new) = w_j(old) - \gamma[x - w_j(old)]$Update learning rate
(10) //Predict each test sample

Figure 11.37: General LVQ algorithm for p_ Economy dataset.

Step (2) While stopping condition is false, follow Steps (2–6).

Step (3) *foreach* training input vector x^{new} follow Steps (3–4).

Step (4) Find j so that $\|x^{\text{new}} - w_j\|$ is a minimum.

Step (5) Update w_{ij}.

Steps (6–8) $if\ (Y = y_j)$

$$w_j(\text{new}) = w_j(\text{old}) + \gamma[x^{\text{new}} - w_j(\text{old})]$$

else

$$w_j(\text{new}) = w_j(\text{old}) - \gamma[x^{\text{new}} - w_j(\text{old})]$$

Step (9) The condition is likely to postulate a fixed number of iterations (i.e., executions of Step (2)) or learning rate reaches an acceptably small value.

Therefore, the data in the new p_Economy dataset with 33.33% portion is allocated for the test procedure, being classified as USA, New Zealand, Italy, Sweden yielding an accuracy rate of *84.02%* based on the LVQ algorithm.

11.4.1.2 Polynomial Hölder function with LVQ algorithm for the analysis of multiple sclerosis

As presented in Table 2.12, multiple sclerosis dataset has data from the following groups: 76 samples belonging to relapsing remitting multiple sclerosis (RRMS),

76 samples to secondary progressive multiple sclerosis (SPMS), 76 samples to primary progressive multiple sclerosis (PPMS) and 76 samples to healthy subjects of control group. The attributes of the control group are data regarding *brain stem (MRI [magnetic resonance imaging] 1), corpus callosum periventricular (MRI 2), upper cervical (MRI 3) lesion diameter size (mm) in the MRI image* and Expanded Disability Status Scale (*EDSS*) score. Data are made up of a total of 112 attributes. By using these attributes of 304 individuals, we can know whether the data belong to the MS subgroup or healthy group is known. How can we make the classification as to which MS patient belongs to which subgroup of MS including healthy individuals and those diagnosed with MS (based on the lesion diameters (*MRI 1, MRI 2, MRI 3*), number of lesion size for (*MRI 1, MRI 2, MRI 3*) as obtained from MRI images and EDSS scores)? The dimension of D matrix is 304×112, which means it contains an MS dataset of 304 individuals with their 112 attributes (see Table 2.12) for the MS dataset:

Our purpose is to ensure the classification of the MS dataset by the multifractal Brownian motion synthesis 2D multifractal analysis of the data. Our method follows the steps mentioned in Figure 11.38:

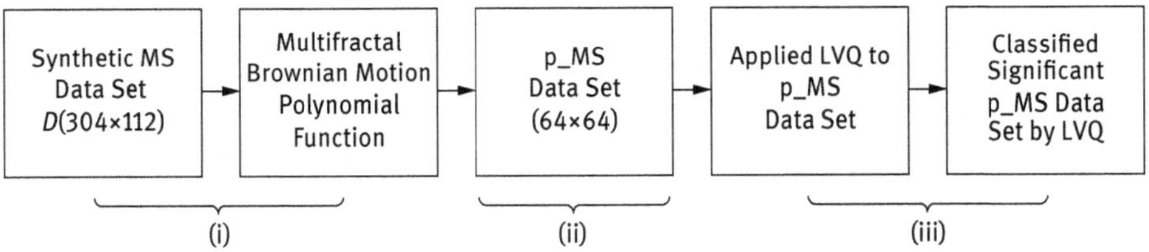

Figure 11.38: Classification of the MS dataset with the application of multifractal Hölder polynomial regularity function through LVQ algorithm.

(i) Multifractal Brownian motion synthesis 2D is applied to the MS dataset (304×112).

(ii) Brownian motion Hölder regularity (polynomial) for analysis is applied to the data for the purpose of identifying on the data and some new datasets comprised of the singled out singularities regarding p_MS dataset (64×64) (see eq. (11.13)):

$$H_1(x, y) = (0.5 + 0.8 \times x \times y).$$

(iii) The new datasets (p_MS dataset) comprised of singled out singularities, with the matrix size (64×64) yielding the best classification result, are finally classified using the Kohonen LVQ algorithm.

The steps in Figure 11.38 are explained as follows:

Step (i) Application to MS dataset multifractal Brownian motion synthesis 2D. Multifractal Brownian motion synthesis 2D can be described as follows: primarily you have to calculate the fractal dimension $D(D = [x_1, x_2, \ldots, x_{304}])$ of the original MS dataset. Fractal dimension is calculated with eq. (11.13). As the next step, you have the new dataset (p_MS dataset) formed by selecting attributes that bring about the minimum change of current dataset's fractal dimension until the number of remaining features is the upper bound of fractal dimension D. The newly produced multifractal Brownian motion synthesis 2D dataset is $D^{\text{new}}(D^{\text{new}} = [x_1^{\text{new}}, x_2^{\text{new}}, \ldots, x_{64}^{\text{new}}])$.

Step (ii) Application to data multifractal Brownian motion synthesis 2D (see [50] for more information).

It is possible for us to perform the classification of p_MS dataset. For this, let us have a closer look at the steps provided in Figure 11.39:

Algorithm: Multifractals Analysis with LVQ Algorithm

Input: New training set that include n number of samples obtained from **p_MS Dataset** $(D^{new} = [x_1^{new}, x_2^{new}, \ldots, x_{64}^{new}])(x_i \in R^d)$

Output: label data that corresponds to m number of samples $C =$ [PPMS, RRMS, SPMS, Healthy]

Method:

(1) //Initialize learning rate γ
(2) **while** (stopping condition is false){
(3) **foreach** input vector $x_1, x_2, \ldots, x_{64}$
(4)   **for** each weight w_{ij} in network {
(5) // Euclidian distance between the input vector and (weight vector for) the *jth* output unit.}
//update w_j
(6) $if(Y = y_j)$
(7) $w_j(new) = w_j(old) + \gamma[xnew - w_j(old)]$
(8) **else**
(9) $w_j(new) = w_j(old) - \gamma[x - w_j(old)]$ Update learning rate
(10) //Predict each test sample

Figure 11.39: General LVQ algorithm for p_MS dataset.

Step (iii) The singled out singularity p_MS datasets are finally classified using the Kohonen LVQ algorithm in Figure 11.39.

Figure 11.39 explains the steps of LVQ algorithm as follows:

Step (1) Initialize reference vectors (several strategies are discussed shortly). Initialize learning rate, y is default (0).

Step (2) While stopping condition is false, follow Steps (2–6).

Step (3) *foreach* training input vector x^{new} follow (Steps 3–4).

Step (4) Find j so that $\|x^{new} - w_j\|$ is a minimum.

Step (5) Update w_{ij}.

Steps (6–8) $if(Y = y_j)$

$$w_j(\text{new}) = w_j(\text{old}) + \gamma[x^{new} - w_j(\text{old})]$$

else

$$w_j(\text{new}) = w_j(\text{old}) - \gamma[x^{new} - w_j(\text{old})]$$

Step (9) The condition is likely to postulate a fixed number of iterations (i.e., executions of Step (2)) or learning rate reaches an acceptably small value.

Consequently, the data in the new p_MS dataset (we singled out the singularities) with 33.33% portion are allocated for the test procedure, being classified as *RRMS, SPMS, PPMS* and *healthy* with an accuracy rate of *80%* based on the LVQ algorithm.

11.4.1.3 Polynomial Hölder function with LVQ algorithm for the analysis of mental functions

As presented in Table 2.19, the WAIS-R dataset has data 200 belonging to patient and 200 samples to healthy control group. The attributes of the control group are data regarding school education, gender, ..., D.M. Data are made up of a total of 21 attributes. By using these attributes of 400 individuals, it is known that whether the data belong to patient or healthy group. How can we make the classification as to which individual belongs to which patient or healthy individuals and those diagnosed with WAIS-R test (based on the school education, gender, the DM, vocabulary, QIV, VIV, ..., D.M see Chapter 2, Table 2.18)? D matrix has a dimension of 400×21. This means D matrix includes the WAIS-R dataset of 400 individuals along with their 21 attributes (see Table 2.19) for the WAIS-R dataset. For the classification of D matrix through LVQ the first step training procedure is to be employed.

Our purpose is to ensure the classification of the WAIS-R dataset by the multifractal Brownian motion synthesis 2D multifractal analysis of the data. Our method follows the steps given in Figure 11.40:

(i) Multifractal Brownian motion synthesis 2D is applied to the WAIS-R dataset (400×21).

(ii) Brownian motion Hölder regularity (polynomial) for analysis is applied to the data for the purpose of identifying on the data and some new datasets comprised of the singled out singularities regarding p_WAIS-R dataset.

(iii) The new datasets (p_WAIS-R dataset) comprised of the singled out singularities, with the matrix size (16 × 16) yielding the best classification result, are finally classified using the Kohonen LVQ algorithm.

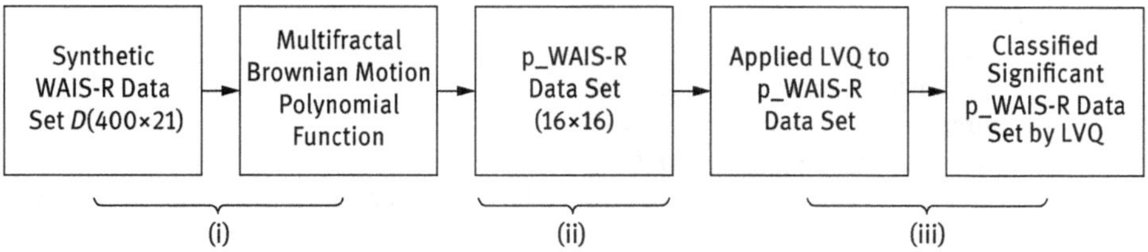

Figure 11.40: Classification of the WAIS-R dataset with the application of multifractal polynomial Hölder regularity function through LVQ algorithm.

The steps of Figure 11.40 are explained as follows:

Step (i) Application to WAIS-R dataset multifractal Brownian motion synthesis 2D. Multifractal Brownian motion synthesis 2D can be described as follows: primarily you have to compute the fractal dimension $D(D = [x_1, x_2, \ldots, x_{400}])$ of the original WAIS-R dataset. Fractal dimension is computed with eq. (11.13). As the next step, you have the new dataset (p_WAIS-R dataset) formed by selecting attributes that bring about the minimum change of current dataset's fractal dimension until the number of remaining features is the upper bound of fractal dimension D. The .newly produced multifractal Brownian motion synthesis 2D dataset is $D^{\text{new}}(D^{\text{new}} = [x_1^{\text{new}}, x_2^{\text{new}}, \ldots, x_{16}^{\text{new}}])$.

Step (ii) Application to data multifractal Brownian motion synthesis 2D (see [50] for more information).

We can perform the classification of p_ WAIS-R dataset. Hence, study the steps provided in Figure 11.41:

Step (iii) The singled out singularities p_WAIS-R dataset are finally classified through the use of the Kohonen LVQ algorithm in Figure 11.42.

Figure 11.41 explains the steps of LVQ algorithm as follows:

Step (1) Initialize reference vectors (several strategies are discussed shortly). Initialize learning rate, y is default (0).

Step (2) While stopping condition is false, follow Steps (2–6).

Step (3) *foreach* training input vector x^{new} follow Steps (3–4).

Step (4) Find j so that $\|x^{\text{new}} - w_j\|$ is a minimum.

Step (5) Update w_{ij}.

Algorithm: Multifractals Analysis with LVQ Algorithm

Input: New training set that include n number of samples obtained from **p_ WAIS-R Dataset**$(D^{new} = [x_1^{new}, x_2^{new}, \ldots, x_{16}^{new}])(x_i \in R^d)$

Output: label data that corresponds to m number of samples C = [Healthy, Patient]

Method:

(1) //Initialize learning rate γ
(2) **while** (stopping condition is false){
(3) **foreach** input vector $x_1, x_2, \ldots, x_{400}$
(4) **for** each weight w_{ij} in network {
(5) Euclidian distance between the input vector and (weight vector for) the *jth* output unit.}
//update w_j
(6) $if(Y = y_j)$
(7) $w_j(new) = w_j(old) + \gamma[xnew - w_j(old)]$
(8) **else**
(9) $w_j(new) = w_j(old) - \gamma[x - w_j(old)]$Update learning rate
(10) //Predict each test sample

Figure 11.41: General LVQ algorithm for p_ WAIS-R dataset.

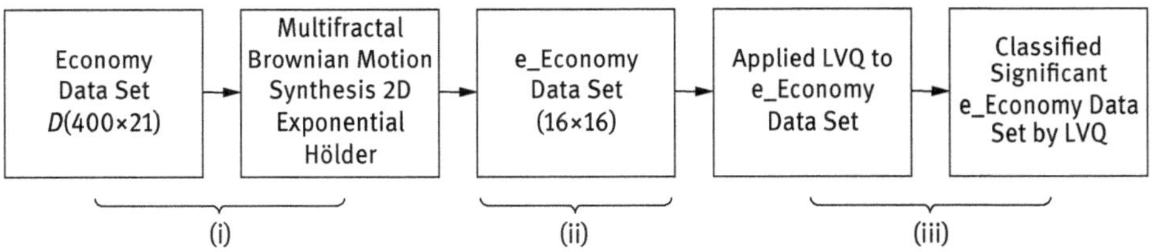

Figure 11.42: Classification of the economy dataset with the application of multifractal Brownian motion synthesis 2D exponential Hölder function through LVQ algorithm.

Steps (6–8) $\text{if}(Y = y_j)$

$$w_j(\text{new}) = w_j(\text{old}) + \gamma[x^{\text{new}} - w_j(\text{old})]$$

else

$$w_j(\text{new}) = w_j(\text{old}) - \gamma[x^{\text{new}} - w_j(\text{old})]$$

Steps (9–10) The condition might postulate a fixed number of iterations (i.e., executions of Step (2)) or learning rate reaching an acceptably small value.

Therefore, the data in the new p_WAIS-R dataset with 33.33% portion are allocated for the test procedure, being classified as Y = [Patient, Healthy] with an accuracy rate of *81.15%* based on the LVQ algorithm.

11.4.2 Exponential Hölder function with LVQ algorithm for the analysis of various data

LVQ algorithm has been applied on economy (U.N.I.S.) dataset, multiple sclerosis (MS) dataset and WAIS-R dataset in Sections 11.4.2.1, 11.4.2.2 and 11.4.2.3, respectively.

11.4.2.1 Exponential Hölder function with LVQ algorithm for the analysis of economy (U.N.I.S.)

As the second set of data, we will use in the following sections, some data related to U.N.I.S. countries' economies. The attributes of these countries' economies are data regarding *years, unemployment, GDP per capita (current international $), youth male (% of male labor force ages 15–24) (national estimate), ...,GDP growth (annual %).* Data are made up of a total of 18 attributes. Data belonging to U.N.I.S. economies from 1960 to 2015 are defined based on the attributes given in Table 2.8 economy dataset (http://data.worldbank.org) [12], which will be used in the following sections.

Our aim is to classify the economy dataset through the Hölder exponent multifractal analysis of the data. Our method is based on the following steps in Figure 11.42:

(i) Multifractal Brownian motion synthesis 2D is applied to the economy dataset (228×18). Brownian motion Hölder regularity (exponential) for analysis is applied to the data for the purpose of identifying on the data and some new datasets comprised of the singled out singularities regarding the e_ Economy dataset (16×16) as eq. (11.15):

$$H_2(x, y) = 0.6 + \frac{0.6}{1 + \exp(-100 \times (x - 0.5y))}$$

(ii) The matrix size with the best classification result is (16×16) for the new dataset (e_Economy dataset), which is made up of the singled out singularities.

The steps of Figure 11.42 are explained as follows:

Step (i) Application to economy dataset multifractal Brownian motion synthesis 2D. Multifractal Brownian motion synthesis 2D can be described as follows: primarily you have to calculate the fractal dimension $D(D = [x_1, x_2, ..., x_{128}])$ of the original economy dataset. Fractal dimension is computed with eq. (11.15). As the next step, you have the new dataset (e_Economy Dataset) formed by selecting attributes that bring about the minimum change of current dataset's fractal dimension until the number of remaining features is the upper bound of fractal dimension D. The newly produced multifractal Brownian motion synthesis 2D dataset is $D^{\text{new}}(D^{\text{new}} = [x_1^{\text{new}}, x_2^{\text{new}}, ..., x_{128}^{\text{new}}])$. Let us have a closer look at the steps provided in Figure 11.43:

Step (ii) Application to data multifractal Brownian motion synthesis 2D (see [50] for more information).

It is possible for us to perform the classification of the e_Economy dataset. For this, let us have a closer look at the steps provided in Figure 11.43:

Algorithm: Multifractals Analysis with LVQ Algorithm

Input: New training set that include n number of samples obtained from **e_Economy Data Set**$(D^{new} = [x_1^{new}, x_2^{new}, \ldots, x_{16}^{new}])(x_i \in R^d)$

Output: label data that corresponds to m number of samples C = [USA, New Zealand, Italy, Sweden]

Method:

(1) //Initialize learning rate γ
(2) **while** (stopping condition is false){
(3) **foreach** input vector $x_1, x_2, \ldots, x_{128}$
(4) **for** each weight w_{ij} in network {
(5) Euclidian distance between the input vector and (weight vector for) the *jth* output unit.} //update w_j
(6) $if(Y = y_j)$
(7) $w_j(new) = w_j(old) + \gamma[x^{new} - w_j(old)]$
(8) **else**
(9) $w_j(new) = w_j(old) - \gamma[x - w_j(old)]$Update learning rate
(10) //Predict each test sample

Figure 11.43: General LVQ algorithm for e_Economy dataset.

Step (iii) The singled out singularities e_Economy dataset are finally classified by using the Kohonen LVQ algorithm in Figure 11.43.

Figure 11.44 explains the steps of LVQ algorithm as follows:

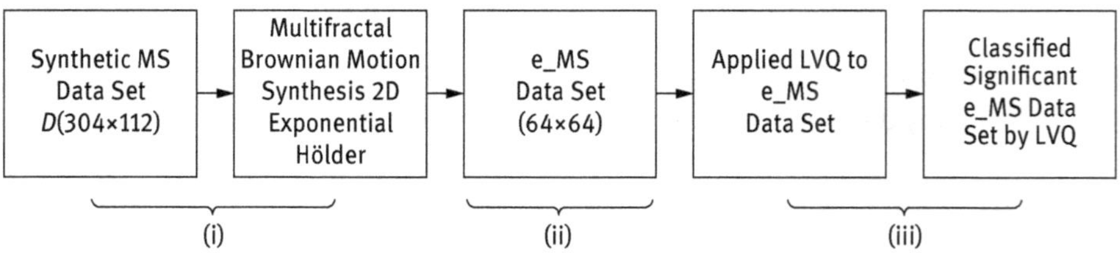

Figure 11.44: Classification of the MS dataset with the application of multifractal Brownian motion synthesis 2D exponential Hölder regularity function through LVQ algorithm.

Step (1) Initialize reference vectors (several strategies are discussed shortly). Initialize learning rate, y is default (0).

Step (2) While stopping condition is false, follow Steps (2–6).

Step (3) *foreach* training input vector $\boldsymbol{x}^{new}$ follow (Steps 3–4).

Step (4) Find j so that $\left\|x^{\text{new}} - w_j\right\|$ is a minimum.

Step (5) Update w_{ij}.

Steps (6–8) if $(Y = y_j)$

$$w_j(\text{new}) = w_j(\text{old}) + \gamma[x^{\text{new}} - w_j(\text{old})]$$

else

$$w_j(\text{new}) = w_j(\text{old}) - \gamma[x^{\text{new}} - w_j(\text{old})]$$

Step (9) The condition might postulate a fixed number of iterations (i.e., execution of Step (2)) or learning rate reaches an acceptably small value.

As a result, the data in the new e_Economy dataset with 33.33% portion are allocated for the test procedure, being classified as $Y = [\text{USA, New Zealand, Italy, Sweden}]$ with an accuracy rate of *80.30%* based on the LVQ algorithm.

11.4.2.2 Exponential Hölder function with LVQ algorithm for the analysis of multiple sclerosis

As presented in Table 2.12, multiple sclerosis dataset has data from the following groups: 76 samples belonging to RRMS, 76 samples to SPMS, 76 samples to PPMS, 76 samples to healthy subjects of control group. The attributes of the control group are data regarding *brain stem (MRI 1), corpus callosum periventricular (MRI 2), upper cervical (MRI 3) lesion diameter size (mm) in the MRI image and EDSS* score. Data are made up of a total of 112 attributes. By using these attributes of 304 individuals, we can know whether the data belong to the MS subgroup or healthy group. How can we make the classification as to which MS patient belongs to which subgroup of MS including healthy individuals and those diagnosed with MS (based on the lesion diameters (*MRI 1, MRI 2, MRI 3*), number of lesion size for (*MRI 1, MRI 2, MRI 3*) as obtained from MRI images and EDSS scores)? D matrix has a dimension of 304×112. This means D matrix includes the MS dataset of 304 individuals along with their 112 attributes (see Table 2.12 for the MS dataset):

Our purpose is to ensure the classification of the MS dataset by the multifractal Brownian motion synthesis 2D multifractal analysis of the data. Our method follows the steps shown in Figure 11.44:

(i) Multifractal Brownian motion synthesis 2D is applied to the MS dataset (400×21). Brownian motion Hölder regularity (exponential) for analysis is applied to the data for the purpose of identifying on the data and some new datasets comprised of the singled out singularities regarding the e_MS dataset (64×64) (see eq. (11.15)):

$$H_2(x,y) = 0.6 + \frac{0.6}{1 + \exp(-100 \times (x - 0.5y))}$$

The singled out singularities e_MS dataset are finally classified by using the Kohonen LVQ algorithm. The new datasets (e_MS dataset) comprised of the singled out singularities, with the matrix size (64×64), are finally classified by using the Kohonen LVQ algorithm.

The steps of Figure 11.44 are explained as follows:

Step (i) Application to MS dataset multifractal Brownian motion synthesis 2D. Multifractal Brownian motion synthesis 2D can be described as follows: primarily you have to calculate the fractal dimension $D(D = [x_1, x_2, ..., x_{304}])$ of the original MS dataset. Fractal dimension is computed with eq. (11.15). As the next step, you have the new dataset (e_MS Dataset) formed by selecting attributes that bring about the minimum change of current dataset's fractal dimension until the number of remaining features is the upper bound of fractal dimension D. The newly produced multifractal Brownian motion synthesis 2D dataset is $D^{\text{new}}(D^{\text{new}} = [x_1^{\text{new}}, x_2^{\text{new}}, ..., x_{64}^{\text{new}}])$.

Step (ii) Application to data multifractal Brownian MOTION SYNTHESIS 2D (see [50] for more information).

We can carry out the classification of the e_MS dataset. For this, let us have a closer glance at the steps provided in Figure 11.45:

Algorithm: Multifractals Analysis with LVQ Algorithm

Input: New training set that include n number of samples obtained from **e_MS Data Set**$(D^{new} = [x_1^{new}, x_2^{new}, \ldots, x_{64}^{new}])(x_i \in R^d)$

Output: label data that corresponds to m number of samples C = [PPMS, RRMS, SPMS, Healthy]

Method:

(1) //Initialize learning rate γ
(2) **while** (stopping condition is false){
(3) **foreach** input vector $x_1, x_2, \ldots, x_{128}$
(4) **for** each weight w_{ij} in network {
(5) Euclidian distance between the input vector and (weight vector for) the *jth* output unit.}
//update w_j
(5) **if** $(Y = y_j)$
(6) $w_j(new) = w_j(old) + \gamma[x^{new} - w_j(old)]$
(7) **else**
(8) $w_j(new) = w_j(old) - \gamma[x - w_j(old)]$ Update learning rate
(9) //Predict each test sample

Figure 11.45: General LVQ algorithm for e_MS dataset.

Step (iii) The new dataset made up of the singled out singularities, e_MS dataset, is finally classified using the Kohonen LVQ algorithm in Figure 11.45.

Figure 11.45 explains the steps of LVQ algorithm as follows:

Step (1) Initialize reference vectors (several strategies are discussed shortly). Initialize learning rate, y is default (0).

Step (2) While stopping condition is false, follow Steps (2–6).

Step (3) *foreach* training input vector x^{new} follow Steps (3–4).

Step (4) Find j so that $\|x^{\text{new}} - w_j\|$ is a minimum.

Step (5) Update w_{ij}.

Steps (6–8) *if* $(Y = y_j)$

$$w_j(\text{new}) = w_j(\text{old}) + \gamma[x^{\text{new}} - w_j(\text{old})]$$

else

$$w_j(\text{new}) = w_j(\text{old}) - \gamma[x^{\text{new}} - w_j(\text{old})]$$

Step (9) The condition is likely to postulate a fixed number of iterations (i.e., executions of Step (2)) or learning rate reaches an acceptably small value.

Thus, the data in the new MS dataset with 33.33% portion are allocated for the test procedure, being classified as $Y = [\text{PPMS, SPMS, RRMS, healthy}]$ with an accuracy rate of *83.5%* grounded on the LVQ algorithm.

11.4.2.3 Exponential Hölder function with LVQ algorithm for the analysis of mental functions

As presented in Table 2.19, the WAIS-R dataset has data with 200 belonging to patients and 200 samples to healthy control group. The attributes of the control group are data regarding school education, gender, ..., D.M. Data are made up of a total of 21 attributes. By using these attributes of 400 individuals, we know whether the data belong to patient or healthy group. How can we make the classification as to which individual belongs to which patient or healthy individuals and those diagnosed with WAIS-R test (based on the school education, gender, the DM, vocabulary, QIV, VIV, ..., D.M)? D matrix has a dimension of 400×21. This means D matrix includes the WAIS-R dataset of 400 individuals along with their 21 attributes (see Table 2.19 for the WAIS-R dataset). For the classification of D matrix through LVQ the first step training procedure is to be employed.

Our purpose is to ensure the classification of the WAIS-R dataset by the multifractal Brownian motion synthesis 2D multifractal analysis of the data. Our method follows the steps shown in Figure 11.46:

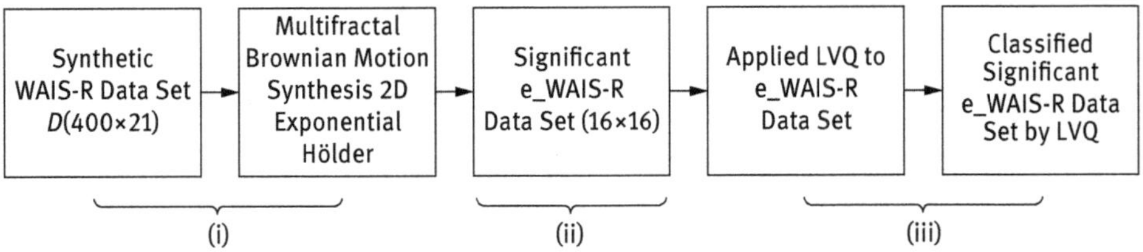

Figure 11.46: Classification of the WAIS-R dataset with the application of multifractal Brownian motion synthesis 2D exponential Hölder function through LVQ algorithm.

(i) Multifractal Brownian motion synthesis 2D is applied to the WAIS-R dataset (400×21).

(ii) Brownian motion Hölder regularity (exponential) for analysis is applied to the data for the purpose of identifying on the data and new dataset made up of the singled out singularities regarding e_WAIS-R dataset:

$$H_2(x, y) = 0.6 + \frac{0.6}{1 + \exp(-100 \times (x - 0.5y))}$$

(iii) The new dataset made up of singled out singularities e_WAIS-R dataset has the matrix size (16×16) yielding the best classification result and they are classified using the Kohonen LVQ algorithm.

The steps in Figure 11.46 are explained as follows:

Step (i) Application to WAIS-R dataset multifractal Brownian motion synthesis 2D. Multifractal Brownian motion synthesis 2D can be described as follows: primarily you have to calculate the fractal dimension $D(D = [x_1, x_2, \ldots, x_{400}])$ of the original WAIS-R dataset. Fractal dimension is computed with eq. (11.15). As the next step, you have the new dataset (e_WAIS-R dataset) formed by selecting attributes that bring about the minimum change of current dataset's fractal dimension until the number of remaining features is the upper bound of fractal dimension D. The newly produced multifractal Brownian motion synthesis 2D dataset is $D^{\text{new}}(D^{\text{new}} = [x_1^{\text{new}}, x_2^{\text{new}}, \ldots, x_{16}^{\text{new}}])$.

Step (ii) Application to data multifractal Brownian motion synthesis 2D (see [50] for more information).

We can do the classification of the e_WAIS-R dataset. For this, let us have a closer look at the steps provided in Figure 11.47:

Algorithm: Multifractals Analysis with LVQ Algorithm

Input: New training set that include n number of samples obtained from **e_WAIS-R Data Set**$(D^{new} = [x_1^{new}, x_2^{new}, \ldots, x_{16}^{new}])(x_i \in R^d)$

Output: label data that corresponds to m number of samples C = [Healthy, Patient]

Method:

(1) //Initialize learning rate γ
(2) **while** (stopping condition is false){
(3) **foreach** input vector $x_1, x_2, \ldots, x_{400}$
(4)   **for** each weight w_{ij} in network {
(5) Euclidian distance between the input vector and (weight vector for) the *jth* output unit.}
//update w_j
(5) **if** $(Y = y_j)$
(6) $w_j(new) = w_j(old) + \gamma[x^{new} - w_j(old)]$
(7) **else**
(8) $w_j(new) = w_j(old) - \gamma[x - w_j(old)]$Update learning rate
(9) //Predict each test sample

Figure 11.47: General LVQ algorithm for e_WAIS-R dataset.

Step (iii) The singled out singularities e_WAIS-R dataset are finally classified by using the Kohonen LVQ algorithm in Figure 11.47.

Figure 11.47 explains the steps of LVQ algorithm as follows:

Step (1) Initialize reference vectors (several strategies are discussed shortly). Initialize learning rate, γ is default (0).

Step (2) While stopping condition is false, follow Steps (2–6).

Step (3) *foreach* training input vector x^{new} follow Steps (3–4).

Step (4) Find j so that $\|x^{\text{new}} - w_j\|$ is a minimum.

Step (5) Update w_{ij}.

Steps (6–8) *if* $(Y = y_j)$

$$w_j(\text{new}) = w_j(\text{old}) + \gamma[x^{\text{new}} - w_j(\text{old})]$$

else

$$w_j(\text{new}) = w_j(\text{old}) - \gamma[x^{\text{new}} - w_j(\text{old})]$$

Step (9) The condition is likely to postulate a fixed number of iterations (i.e., execution of Step (2)) or learning rate reaches an acceptably small value.

Thus, the data in the new e_WAIS-R dataset with 33.33% portion are allocated for the test procedure, being classified as Y = [Patient, Healthy] with an accuracy rate of *82.5%* based on the LVQ algorithm.

New datasets have been obtained by the application of multifractal method Hölder regularity functions (polynomial and exponential function) on MS dataset, economy (U.N.I.S.) dataset and WAIS-R datasets.

- New datasets, comprised of the singled out singularities, p_MS Dataset, p_Economy Dataset and p_WAIS-R Dataset, have been obtained from the application of polynomial Hölder functions.
- New datasets, comprised of the singled out singularities, e_MS Dataset, e_Economy Dataset and e_WAIS-R Dataset, have been obtained by applying exponential Hölder functions.

  LVQ algorithm has been applied in order to get the classification accuracy rate results for our new datasets as obtained from the singled out singularities (see Table 11.1).

Table 11.1: The LVQ classification accuracy rates of multifractal Brownian motion Hölder functions.

Datasets	Hölder functions	
	Polynomial (%)	**Exponential (%)**
Economy (U.N.I.S.)	84.02	80.30
MS	80	83.5
WAIS-R	81.15	82.5

As shown in Table 11.1, the classification of datasets such as the p_Economy dataset that can be separable linearly through multifractal Brownian motion synthesis 2D Hölder functions can be said to be more accurate compared to the classification of other datasets (p_MS Dataset, e_MS Dataset; p_Economy Dataset, e_Economy Dataset; and e_WAIS-R Dataset).

References

[1] Mandelbrot BB. Stochastic models for the Earth's relief, the shape and the fractal dimension of the coastlines, and the number-area rule for islands. *Proceedings of the National Academy of Sciences*, 1975, 72(10), 3825–3828.

[2] Barnsley MF, Hurd LP. Fractal image compression. New York, NY, USA: AK Peters, Ltd., 1993.

[3] M.F. Barnsley. Fractals Everywhere. Wellesley, MA: Academic Press, 1998.

[4] Darcel C, Bour O, Davy P, De Dreuzy JR. Connectivity properties of two-dimensional fracture networks with stochastic fractal correlation. *Water Resources Research, American Geophysical Union (United States)*, 2003, 39(10).

[5] Uppaluri R, Hoffman EA, Sonka M, Hartley PG, Hunninghake GW, McLennan G. Computer recognition of regional lung disease patterns. *American Journal of Respiratory and Critical Care Medicine*, US, 1999, 160 (2), 648–654.

[6] Schroeder M. Fractals, chaos, power laws: Minutes from an infinite paradise. United States: Courier Corporation, 2009.

[7] Created, developed, and nurtured by Eric Weisstein at Wolfram Research, http://mathworld.wolfram.com/DevilsStaircase.html, 2017.

[8] Sugihara G, May RM. Applications of fractals in ecology. *Trends in Ecology & Evolution, Cell Press, Elsevier*, 1990, 5(3), 79–86.

[9] Ott E. Chaos in dynamical systems. United Kingdom: Cambridge University Press, 1993.

[10] Wagon S. Mathematica in action. New York, USA: W. H. Freeman, 1991.

[11] Sierpinski W. Elementary theory of numbers. Poland: Polish Scientific Publishers, 1988.

[12] Sierpiński W. General topology. Toronto: University of Toronto Press, 1952.

[13] Niemeyer L, Pietronero L, Wiesmann HJ. Fractal dimension of dielectric breakdown. *Physical Review Letters*, 1984, 52(12), 1033.

[14] Morse DR, Lawton JH, Dodson MM, Williamson MH. Fractal dimension of vegetation and the distribution of arthropod body lengths. *Nature* 1985, 314 (6013), 731–733.

[15] Sarkar N, Chaudhuri BB. An efficient differential box-counting approach to compute fractal dimension of image. *IEEE Transactions on Systems, Man, and Cybernetics*, 1994, 24(1), 115–120.

[16] Chang T. Self-organized criticality, multi-fractal spectra, sporadic localized reconnections and intermittent turbulence in the magnetotail. *Physics of Plasmas*, 1999, 6(11), 4137–4145.

[17] Ausloos M, Ivanova K. Multifractal nature of stock exchange prices. *Computer Physics Communications*, 2002, *147*, 1–2, 582–585.

[18] Frisch U. From global scaling, a la Kolmogorov, to local multifractal scaling in fully developed turbulence. *Proceedings of the Royal Society of London A: Mathematical, Physical and Engineering Sciences The Royal Society*, 1991, 434 (1890), 89–99.

[19] Kraichnan RH, Montgomery D. Two-dimensional turbulence. *Reports on Progress in Physics*, 1980, 43(5), 547.

[20] Meneveau C, Sreenivasan KR. Simple multifractal cascade model for fully developed turbulence. *Physical Review Letters*, 1987, 59(13), 1424.

[21] Biferale L, Boffetta G, Celani A, Devenish BJ, Lanotte A, Toschi F. Multifractal statistics of Lagrangian velocity and acceleration in turbulence. *Physical Review Letters*, 2004, 93 (6), 064502.

[22] Burlaga LF. Multifractal structure of the interplanetary magnetic field: Voyager 2 observations near 25 AU, 1987–1988. *Geophysical Research Letters*, 1991, 18, 1, 69–72.

[23] Macek WM, Wawrzaszek A. Evolution of asymmetric multifractal scaling of solar wind turbulence in the outer heliosphere. *Journal of Geophysical Research: Space Physics*, 2009, 114 (A3).

[24] Burlaga LF. Lognormal and multifractal distributions of the heliospheric magnetic field. *Journal of Geophysical Research: Space Physics* 2001, 106(A8), 15917–15927.

[25] Burlaga LF, McDonald FB, Ness NF. Cosmic ray modulation and the distant heliospheric magnetic field: Voyager 1 and 2 observations from 1986 to 1989. *Journal of Geophysical Research: Space Physics*, 1993, 98 (A1), 1–11.

[26] Macek WM. Multifractality and chaos in the solar wind. *AIP Conference Proceedings*, 2002, 622 (1), 74–79.

[27] Macek WM. Multifractal intermittent turbulence in space plasmas. *Geophysical Research Letters*, 2008, 35, L02108.

[28] Benzi R, Paladin G, Parisi G, Vulpiani A. On the multifractal nature of fully developed turbulence and chaotic systems. *Journal of Physics A: Mathematical and General*, 1984, 17 (18), 3521.

[29] Horbury T, Balogh A, Forsyth RJ, Smith EJ.). Ulysses magnetic field observations of fluctuations within polar coronal flows. *Annales Geophysicae*, January 1995, 13, 105–107).
[30] Sporns O. Small-world connectivity, motif composition, and complexity of fractal neuronal connections. *Biosystems*, 2006, 85 (1), 55–64.
[31] Dasso S, Milano LJ, Matthaeus WH, Smith CW. Anisotropy in fast and slow solar wind fluctuations. *The Astrophysical Journal Letters*, 2005, 635(2), L181.
[32] Shiffman D, Fry S, Marsh Z. The nature of code. USA, 2012, 323–330.
[33] Qian H, Raymond GM, Bassingthwaighte JB. On two-dimensional fractional Brownian motion and fractional Brownian random field. *Journal of Physics A: Mathematical and General*, 1998, 31 (28), L527.
[34] Thao TH, Nguyen TT. Fractal Langevin equation. Vietnam Journal of Mathematics, 2003, 30(1), 89–96.
[35] Metzler R, Klafter J. The random walk's guide to anomalous diffusion: A fractional dynamics approach. *Physics Reports*, 2000, 339(1), 1–77.
[36] Serrano RC, Ulysses J, Ribeiro S, Lima RCF, Conci A. Using Hurst coefficient and lacunarity to diagnosis early breast diseases. In *Proceedings of IWSSIP 17th International Conference on Systems, Signals and Image Processing, Rio de Janeiro*, Brazil, 2010, 550–553.
[37] Magdziarz M, Weron A, Burnecki K, Klafter J. Fractional Brownian motion versus the continuous-time random walk: A simple test for subdiffusive dynamics. *Physical Review Letters*, 2009, 103, 18, 180602.
[38] Qian H, Raymond GM, Bassingthwaighte JB. On two-dimensional fractional Brownian motion and fractional Brownian random field. *Journal of Physics A: Mathematical and General*, 1998, 31 (28), L527.
[39] Chechkin AV, Gonchar VY, Szydl/owski, M. Fractional kinetics for relaxation and superdiffusion in a magnetic field. *Physics of Plasmas*, 2002, 9 (1), 78–88.
[40] Jeon JH, Chechkin AV, Metzler R. First passage behaviour of fractional Brownian motion in two-dimensional wedge domains. *EPL (Europhysics Letters)*, 2011, 94(2), 2008.
[41] Rocco A, West BJ. Fractional calculus and the evolution of fractal phenomena. *Physica A: Statistical Mechanics and its Applications*, 1999, 265(3), 535–546.
[42] Coffey WT, Kalmykov YP. The Langevin equation: With applications to stochastic problems in physics, chemistry and electrical engineering, Singapore: World Scientific Publishing Co. Pte. Ltd., 2004.
[43] Jaffard S. Functions with prescribed Hölder exponent. *Applied and Computational Harmonic Analysis*, 1995, 2(4), 400–401.
[44] Ayache A, Véhel JL. On the identification of the pointwise Hölder exponent of the generalized multifractional Brownian motion. *Stochastic Processes and their Applications*, 2004, 111(1), 119–156.
[45] Cruz-Uribe D, Firorenza A, Neugebauer CJ.. The maximal function on variable spaces. In *Annales Academiae Scientiarum Fennicae. Mathematica* Finnish Academy of Science and Letters, Helsinki; Finnish Society of Sciences and Letters, 2003, 28(1), 223–238.
[46] Mandelbrot BB, Van Ness JW. Fractional Brownian motions, fractional noises and applications. *SIAM Review*, 1968, 10(4), 422–437.
[47] Flandrin P. Wavelet analysis and synthesis of fractional Brownian motion. *IEEE Transactions on Information Theory*, 1992, 38(2), 910–917.
[48] Taqqu MS. Weak convergence to fractional Brownian motion and to the Rosenblatt process. *Probability Theory and Related Fields*, 1975, 31(4), 287–302.
[49] Trujillo L, Legrand P, Véhel JL. The estimation of Hölderian regularity using genetic programming. In Proceedings of the 12th Annual Conference on Genetic and Evolutionary Computation, Portland, Oregon, USA, 2010, 861–868.

[50] Karaca Y, Cattani C. Clustering multiple sclerosis subgroups with multifractal methods and self-organizing map algorithm. *Fractals*, 2017, 1740001.

[51] Véhel JL, Legrand P. Signal and Image processing with FracLab. FRACTAL04, Complexity and Fractals in Nature. In 8th International Multidisciplinary Conference World Scientific, Singapore, 2004, 4–7.

[52] Duncan TE, Hu Y, Pasik-Duncan B. Stochastic calculus for fractional Brownian motion I. Theory. *SIAM Journal on Control and Optimization*, 2000, 38(2),582–612.

[53] Jannedy S. Rens, Jennifer H (2003). Probabilistic linguistics. Cambridge, MA: MIT Press. ISBN 0-262-52338-8.

[54] Xiao Y. Packing measure of the sample paths of fractional Brownian motion. *Transactions of the American Mathematical Society*, 1996, 348(8),3193–3213.

[55] Baudoin F, Nualart D. Notes on the two dimensional fractional Brownian motion. *The Annals of Probability*, 2006, 34 (1), 159–180.

[56] Witten Jr TA, Sander, LM. Diffusion-limited aggregation, a kinetic critical phenomenon. Physical review letters, 1981, 47(19), 1400.

[57] Meakin P. The structure of two-dimensional Witten-Sander aggregates. *Journal of Physics A: Mathematical and General*, 1985, 18(11), L661.

[58] Bensimon D, Domany E, Aharony A. Crossover of Fractal Dimension in Diffusion-Limited Aggregates. Physical Review Letters, 1983, 51(15), 1394.

Index

https://doi.org/10.1515/9783110496369-012

www.ingramcontent.com/pod-product-compliance
Lightning Source LLC
Chambersburg PA
CBHW061155240926
51846CB00006B/1020
9783110496352